# URBAN CHANGE AND CONFLICT:
# AN INTERDISCIPLINARY READER

# URBAN CHANGE AND CONFLICT: AN INTERDISCIPLINARY READER

EDITED BY ANDREW BLOWERS, CHRISTOPHER BROOK, PATRICK DUNLEAVY
AND LINDA McDOWELL
AT THE OPEN UNIVERSITY

Harper and Row, Publishers, *in association*
*with* The Open University Press

**Harper & Row, Publishers**
London

Cambridge
Hagerstown
Philadelphia
New York

San Francisco
Mexico City
Sao Paulo
Sydney

First published 1982

Harper & Row Ltd
28 Tavistock Street
London WC2E 7PN

*British Library Cataloguing in Publication Data*

Urban change and conflict.
    1. Cities and towns – Study and teaching
    I. Blowers, Andrew
    307.7′6        HT109

    ISBN 0-06-318203-3
    ISBN 0-06-318204-1 Pbk

Typeset by Inforum Ltd, Portsmouth
Printed and bound by The Pitman Press, Bath

*The papers in this reader have been edited to reflect the needs of Open University students and our readership generally. We thank the authors concerned for their co-operation and advice in the editing process and hope that our efforts do justice to the original text. (Eds.)*

# Open University Course Team

Melanie Bayley
Andrew Blowers
Christopher Brook
Giles Clark
Allan Cochrane
Patrick Dunleavy
Mary Geffen
Chris Hamnett
Linda McDowell
Jerry Millard
Stephen Potter
Eric Reade
Enid Sheward
Paul Smith
Eleanor Thompson
Graham Turner

*Other contributors*

Mike Bateman
Bob Colenutt
Peter Daniels
Anne Jones
Camilla Lambert
John Lambert
Bill Lever
Martin Loney
Rosemary Mellor
Kenneth Newton
Chris Pickvance
John Raine
Peter Saunders
Nigel Spence
Ray Thomas
Bob Wilson

# Contents

# List of units in the Open University Course, Urban Change and Conflict

# Preface

Although urban problems have long been the focus of academic and political concern, urban studies have only recently emerged as a major field of interdisciplinary activity. Recent developments in the subject have been so rapid and wide ranging that it is appropriate to assess the field and restate its major concerns. This book, which has been inspired by the Open University course on *Urban Change and Conflict*, represents, we believe, the first attempt to provide such an assessment and to bring together in one volume the different perspectives that comprise urban studies. Together with its companion volume, *City, Economy and Society: a Comparative Reader*, it will be of interest to all those who are involved in the subject whether as academics, policy-makers or citizens interested in the social, political and economic issues that constitute the urban environment.

Urban studies have a long and respectable pedigree, and can be traced back to the nineteenth-century social surveys undertaken by reformers such as Booth and Rowntree. This empirical tradition continued and was later complimented by theoretical work with its mainsprings in the Chicago School of sociology in the 1920s. Other disciplines, notably geography, economics and political science, added new and distinctive dimensions to both theoretical and empirical study in the succeeding decades. Contemporary concern with theories of society and the State has exposed the limitations of earlier work and encouraged new directions for study, particularly related to contemporary urban problems. Interest in the relationship between social problems and the urban environment can also be traced back to the nineteenth century giving rise to the utopian vision of creative thinkers such as Howard and Unwin. This work had a profound influence on policy-making resulting in Britain in the rise of the town planning movement and throughout the world in the development of new communities designed to overcome problems of inequality, segregation and congestion in the cities. It is only now that these various traditions have begun to converge to constitute a coherent interdisciplinary field of study.

There are three reasons for this development. One is the modern tendency for social scientists to draw on the work of colleagues working in different disciplines. Interdisciplinary cooperation yields insights that cannot be achieved by a narrow disciplinary focus. A second reason stems from the widely observed association between urban development and social deprivation which has provoked various political responses aimed at alleviating the problem. The role of the state has, in consequence, become a major area of debate within urban studies and has provided a focus for a variety of viewpoints. A third reason is the revival of Marxist schools of theoretical work which have proved stimulating to those working in that tradition and which have provoked a response from those who adopt other perspectives. Social scientists have been forced to rethink their positions and have begun to appreciate the broader interdisciplinary context of their work in the process. The interdisciplinary approach has come about through both intellectual cooperation and debate.

These developments have yet to be fully assimilated and cannot be said to embrace all the work in urban studies much of which still remains restricted in focus and subservient to disciplinary or professional imperatives. There are still major problems to be tackled before the interdisciplinary approach is firmly established. One is that while there is substantial work, some of it impressive, at a theoretical level there is a dearth of good empirical accounts with which to underpin the validity of much theorizing. Some significant empirical work is reproduced in this book but there is still much to be done and many problems simply have not been adequately researched. As a result many assumptions must remain, for the time being, mere assertions. This leads to a second problem concerning the relationship between analysis and policy-making. Our understanding of the causes of urban problems may have improved but the prescriptions are either unclear or pose severe problems for implementation. Much recent work argues that nothing short of fundamental change in social organization will provide a cure. It is not, perhaps, surprising that politicians and administrators trying to deal with immediate problems are unsympathetic to the call for revolution. But it is also clear that social science has failed to penetrate the thinking of policy-makers who remain committed to limited or simplistic solutions which often serve to compound problems.

This book represents the range of work that embodies urban studies at the present time. Its organization broadly follows the approach of the course to which it relates. There are five sections and each is preceded by an introduction, which not only summarizes the contents of the section but identifies major themes, controversies and limitations in the various readings. They are intended to alert readers to those issues which are hidden as well as those which are exposed.

Section I is intended as an introduction to the whole book and consists of an overview of developments in urban studies indicating the various traditions which compose it and which together constitute this emerging interdisciplinary field. Section II provides a broadly spatial perspective focusing on the relationships between cities as well as their internal patterning. These are now established traditions in geographical studies but the readings go beyond the simple and static description of urban patterns to consider the determinants responsible for them. Predominantly market based explanations are balanced by theoretical positions which reveal the structural processes responsible for the changes occurring in urban systems. Section III introduces the sociological tradition in urban studies, its concern with community, and the processes which influence patterns of social segregation. Again, these traditions are examined in the light of recent work on social inequality and the increasing emphasis on the role of the State in providing welfare and other services.

The role of the State is the underlying theme of the last two sections of the book. In section IV theories of urban planning are considered and empirical studies of planning serve to demonstrate that planning is relatively weak where it purports to be in control in land development and regional policy. The importance of other sectors of State investment in determining outcomes is underlined here. Two of these sectors — housing and transport — are examined in section V which is concerned with the power of the State (central and local) and the constraints of this power. A number of questions about the relationship between the State and the private sector and the social consequences of that relationship are raised. At a time when State intervention is being scrutinized, the issues raised in these final sections have a topical interest.

This reader is intended to serve a wide audience. The text has been carefully edited and complex technical and statistical work has been omitted in order to make it accessible to all who are interested in the subject. Articles which reflect the various disciplinary perspectives should satisfy specialists in specific issues. The interdisciplinary approach and emphasis on current theoretical and empirical work should appeal to teachers and students who require an introduction to urban studies and who wish to comprehend the diversity of approaches to the subject. These approaches are not purely academic since many of the readings offer criticisms of urban policies and suggest implications for future policy-making. Those working in the environmental professions (planners, architects and engineers) and officials in central or local government concerned with urban questions should find the book a source of ideas, comment and criticism of immediate relevance to their work. Although the book uses mainly British and American examples much of the material is applicable to urban problems and policies throughout the advanced industrial countries. Together with the comparative reader (which includes studies of cities in different economic systems) this book provides a comprehensive and contemporary view of the state of the subject, its strengths and limitations.

The readings included here were suggested by the many people associated with the creation of *Urban Change and Conflict*. The editors' task has therefore been relatively easy and the book represents a cooperative and interdisciplinary endeavour worthy of its subject. We are greatly indebted to all who have put forward ideas for this selection and to all those who have helped in other ways in the production of this book. There are too many to mention each by name but we are especially grateful to Stephen Potter and Enid Sheward who provided the organization and discipline such work needs, to John Taylor, Giles Clark and Melanie Bayley for their expert help with publishing and to our secretaries, Maureen Adams, Eve Hussey, Pat Cooke and Michelle Kent who bore the burden of preparing the manuscript. We trust they will not be disappointed by the result.

**Andrew Blowers**      **Patrick Dunleavy**
**Christopher Brook**      **Linda McDowell**

# Section I
# Perspectives on Urban Studies
*by Patrick Dunleavy*

## Introduction

Cities and urban phenomena have formed the focus of a distinct field of social science research for only just over half a century. In the 1890s the first awakening of interest in cities occurred amongst the major theorists of German sociology, Tönnies, Simmel and Weber, but it was not until the work of the Chicago School in the 1920s and 1930s that urban studies emerged as a major area of research attention. The rapid contemporaneous development of community studies, locational economics and geographical interest in city systems extended interest in 'urban' topics into most disciplines in social science by the beginning of the 1950s. Thereafter, while extensive empirical studies in economics, geography and political science continued apace until the mid-1960s, the more theoretical and sociological aspects of the field entered a period of decline which became more serious as the prospect of interdisciplinary research or major progress in understanding urban phenomena steadily receded. By the late 1960s urban studies had fragmented into a number of connected disciplinary areas, most of which were peripheral to the major lines of theoretical development in their subject. Only in geography was urban research a high-prestige and still evolving area of knowledge.

This situation was changed essentially by the emergence of two strong new perspectives on urban issues. In Britain a new Weberian sociology developed analyses of housing markets as distinctively 'urban' influences on life chances. And, initially in France, a vigorous neo-Marxist critique of previous urban studies was transformed into an extensively researched account of the processes structuring the production of cities. While the value of these perspectives remains controversial and fiercely contested, there is little room for doubt that the *interaction* between new and old approaches has sparked a widespread re-emergence of theoretical interest in 'urban' issues. The central grounds of controversy and debate over cities and urban phenomena have been projected into sharper focus in all disciplines and very different analytic approaches.

In this paper I shall try to summarize a view of the major theoretical problems in urban studies, concentrating on the difficulties of defining the field which have largely accounted for the variations in social science interest in urban topics over time. For unless a clear definition of the scope of urban studies can be given, the field has historically languished unanalyzed or analyzed only in rather eccentric and incoherent ways. The central problem of clarifying what is to count as falling within the scope of urban studies can be simply expressed. In contemporary advanced industrial societies a great many aspects of social life takes place in cities, towns and built-up areas. If 'urban' studies is defined so that anything taking place in these kinds of settlements is included in its scope then urban studies and social science as a whole may be virtually equivalent; hence there would be no clear-cut field of 'urban' research distinguishable from other areas of social science work. Equally, it would be as hard to describe the links between 'urban' phenomena as it is to describe the links between all social phenomena. Finally, of course, many of the activities which take place in cities or built-up areas are not meaningfully describable as 'urban', even in ordinary language, still less in social science terms. For example, the making of foreign policy by governments across the world is very highly concentrated in cities, indeed in central districts of capital cities. Yet we would not describe this kind of activity as an 'urban process', because we do not (intuitively) see much linkage between the place where foreign polices are hammered out by governments and the character or development of these policies. Equally many social scientists would find it nonsensical to talk of 'urban justice', 'urban arts', 'urban culture' or 'urban poverty' in advanced industrial states on the grounds that they cannot see what is 'urban' about such social processes — apart from in the trivial sense that they take place in built-up areas.

So the key problem of urban studies is to decide just what activities of the vast number spatially located in cities or towns are to form the focus of research attention. The Chicago School concentrated on those social processes which were most *internal to cities*, looking at aspects of social life which were most locally organized. This approach remains distinctive of 'mainstream' geographical and economic studies. A development most associated with Louis Wirth in contrast tried to identify a distinctively *'urban way of life'* which could be linked to the basic characteristics of cities as physical and social entities. This focus on 'urbanism' dominated urban sociology in the 1950s and 1960s. A quite similar strategy pursued in sociological and political science 'community' studies has been to focus on *institutions* which are said to be distinctively 'urban', such as local governments or neighbourhood organizations of various kinds. Finally, recent political economy approaches have generally retreated from trying to define 'urban' processes in spatial, socio-cultural or institutional terms. Instead they tend to prefer a focus on *particular kinds of socio-economic activity*, such as collective forms of consumption or investment in the built environment.

To understand the issues involved in these different approaches, this section pursues three different strategies. In §1.1 I set out my own view of the reasons why defining urban studies has proved so problematic. This centres on the claim that the commonsense dichotomy between 'urban' and 'rural' social processes is entirely misleading when applied to advanced industrial societies. Thus I argue here that in order to reach a useful definition of the scope of urban studies, we need to completely renounce any connection between particular spatial locations (cities and towns) and urban studies. Instead we should simply define 'urban' processes as certain types of socio-economic activity which can be studied as a connected field of research wherever they are located in geographical space. If we choose the right kinds of social process to focus on then we may well find that the results of our research are especially helpful in understanding what goes on in cities or towns, but this is something which needs to be examined empirically. We cannot simply *assume* such a result in advance. Still less can we try to define urban studies around such an assumption, for in that case we risk losing sight of it altogether and in that direction lies ideology, the implicit or unreasoned mapping out of a

particular view of the world remote from empirical testing which is the antipathy of a proper social science.

In §1.2 I bring together a number of especially significant statements of alternative theoretical perspectives on urban studies. Because some areas of social science, such as geography and economics, have been characterized by paying little attention to such questions, most of the sources used are drawn from sociology. I present them in the form of short extracts, with some intervening commentary bringing out the links and differences between them.

Finally in §1.3 I introduce a diagrammatic representation of the development of urban studies from the 1890s up to the present. Using the categories suggested by this account, I then show how the papers gathered together in sections II–V of this reader can be read as illustrations of different theoretical perspectives in urban social science. Of course, most of the papers in the rest of this volume have been chosen for their value as empirical studies. With some exceptions most of them are not intended to be explicitly theoretical papers, but all of them operate within particular assumptions and procedures which form part of much broader streams of research in urban studies. §1.3 then should help to cue readers into the basic approach and assumptions involved in each of the later papers.

## 1.1 Social development during industrialization and the urban/rural dichotomy

I examine here the changes brought about by industrialization in the space-economy of societies. The essential argument is that an urban/rural dichotomy was appropriate as an analytic framework for analyzing pre-industrial societies, for at this period the organization of economic activities was heavily influenced by two different socio-cultural systems or 'ways of life'; and these different ways of life were spatially distributed so that town and country were quite distinct in character. The process of industrialization broke down the 'urban' way of life first, restructuring social arrangements in line with the requirements of a capitalist mode of production, but as industrialization proceeded, and societies entered an advanced stage, so the distinct pattern of life associated with farming production was in turn broken down and remoulded on the same lines as the rest of society. The contemporary problem in defining urban studies is not to distinguish 'urban' from 'rural' areas of the country, for no such geographical distinction is possible. Nor can this problem be reinterpreted in terms of an 'urban/rural continuum'. Instead the key problem is to separate out a small set of 'urban' concerns from the great mass of social activities which are neither 'urban' nor 'rural'. This kind of distinction has nothing to do with spatial arrangements or settlement patterns: it involves separating one kind of social process from a number of others. It is only when the 'urban/rural' contrast is seen as completely irrelevant in any form that we can set about defining urban studies with any chance of success.

In order to summarize the characteristic pattern of changes in society brought about by industrialization I need to distinguish a number of basic types of social activity. These are: *production*, the transformation of nature or existing goods to make new goods or services; *exchange*, activities distributing products or capital resources (such as buildings or machines) between industries or areas — I shall normally consider these two together; *consumption*, the final appropriation of products or goods by people; *socio-cultural life*, all those activities involving non-economic forms of organization (such as groups or associations, cultural or religious or ideological processes, social customs etc); and finally, *state intervention* and *external affairs*, concerning those activities managing a society's development and its relations with other countries.

Using these categories, the changes wrought by industrialization can be pictured in a set of simple Venn diagrams. In the pre-industrial period the relationships between economic and socio-cultural activities can be pictured as in figure 1.1. In a country such as Britain in the early eighteenth century, socio-cultural factors heavily influenced the kinds of economic activity found in different areas of the country, and the ways in which these activities were carried out. To know how production or consumption processes were organized one would first have to separate out two different situations, an urban way of life found in towns and cities, and a rural way of life found elsewhere. Taking Britain in the early seventeenth century as a point of reference the difference was broadly as follows.

In the towns production consisted mainly of artisan and craft activities in small workshops producing for sale in the market. Towns were key exchange centres gathering in farm products from their surrounding region for export to other regions or countries, and distributing manufactured goods. Town people bought their consumption goods in markets using money earned from their own work. If they could not earn money they were unlikely to be able to subsist on their own output. So in this sense the direct overlap between

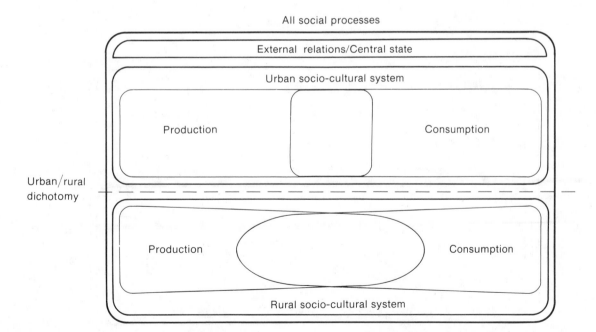

**Figure 1.1**   Socio-cultural systems and the space economy in a pre-industrial society.

production and consumption activities was small. Socio-cultural life in towns at this period was lively, with traditional and more commercial elements in a somewhat unstable combination. Municipal institutions reflected independence from the semi-feudal hierarchy still found in the countryside. Merchants' and workmens' organizations were well developed, especially in structuring and regulating markets. Some collective services were provided for town residents and significant religious, intellectual and political debate extended quite widely in the population.

In contrast rural life remained more traditional, despite changes introduced by the commercialization of agriculture. Production and consumption processes overlapped extensively since most people were still engaged in subsistence farming for their own needs. Farming for cash supported the nobility. There was a clear hierarchy with patterns of mutual obligation between masters and tenants, still quite close in some respects to the previous feudal pattern of relations between lord and peasant. People acquired their status and role in life at birth for the most part. The gentry and the clergy ran the small-scale government functions and the minimal collective services. Market patterns of relations were less prominent than in towns and political and social life was based on tradition and deference, reflecting the links between the organization of farming production and other activities.

So at this period only the growth of a centralized state and its involvements in foreign and colonial affairs would need to be analyzed outside of either the urban or the rural categories. As a basis for analyzing the domestic life of a country, the urban/rural dichotomy was still perfectly viable, and indeed essential.

Industrialization in its early stages, accomplished in Britain by the mid-nineteenth century, broke down this dichotomy by destroying the pre-existing way of life of towns and cities (figure 1.2). The massive migration of people from the countryside into the towns and cities with new industries swamped the previous population, so that the socio-cultural system of pre-industrial towns disappeared in the space of a few decades. This change was both a pre-condition for, and was powerfully accelerated by, the introduction of the factory system. The increased regulation and reduced living standards of working people this brought about went along with the growth of unregulated markets in housing organized along the same capitalist lines as industry. Workers' and merchants' organizations were destroyed or declined in the new towns; collective services were non-existent; wage labour under appalling conditions and fully monetarized consumption processes dominated city life. Only slow and piecemeal changes rebuilt a distinctively capitalist form of collective social organization as the first wave of industrialization passed. But this rebuilt socio-cultural system

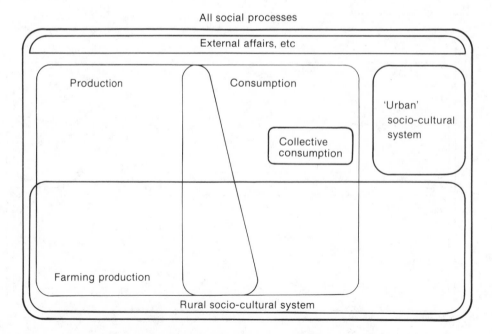

**Figure 1.2**  Socio-cultural systems and the space economy in an early industrializing society.

overwhelmingly expressed a response by different social groups to industrialization and the new modes of organizing economic activity. Thus the owners of industrial capital pushed for new local political institutions, for franchise extensions and political representations, for environmental regulation 'in the public interest', for the stabilizing of law and order and for new means of mass integration into the social system — such as basic education and evangelistic religion. Working people in turn coined their own responses to their situation, such as trade union organization, the growth of extended family networks providing mechanisms of mutual support, and a wide spectrum of voluntary associations. What emerged here, however, was not a specific product of 'urban' conditions. Rather the general outlines of a social system appropriate to industrial capitalism were defined at this period.

In the countryside the early period of industrialization had relatively small and delayed effects. Those working people who did not migrate to the towns were, of course, still socialized into a rural 'way of life', and social relations between different social groups carried on in much the same way as before. Capitalist modes of farming spread with new technologies, and more consumption goods became available — so that the overlap between production and consumption processes was in this sense progressively reduced. Thus the countryside began to be economically integrated into the wider industrialized society, while still remaining distinct in socio-cultural terms.

An industrializing society cannot be effectively analyzed using an 'urban/rural' dichotomy. For although a rural way of life partially influencing economic activity in the countryside can still be identified, there is no 'urban' counterpart. The 'way of life' of people in towns has already been remade in line with the logic of industrial capitalism and not in response to people's spatial location.

In an advanced industrial society, such as Britain in the post-war period, both economic activities and socio-cultural life are clearly established on exactly the same basis in cities or built-up areas as in the countryside (figure 1.3). Production in both areas is organized along almost exactly the same lines. Subsistence farming has declined in significance and consumption patterns in both areas are completely based on the purchase of marketed goods. The number of people still engaged in agricultural production has dwindled to a very small proportion of the population (under 3% in Britain), a level which is too small to support any distinctive socio-cultural system. Thus people in the countryside are fully integrated into a single, national 'way of life' shaped by a business culture, the mass media, advertising, state education and other quite general influences. Finally the evolution of transport and work patterns has meant that the countryside is

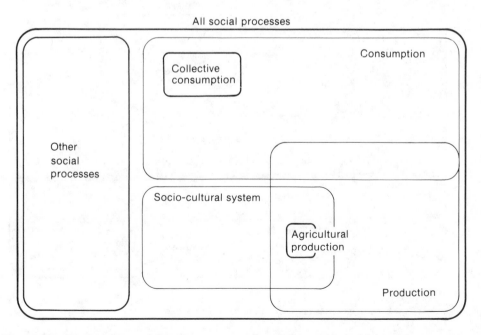

**Figure 1.3**  The socio-cultural system and the space economy in an advanced industrial society.

increasingly dominated by residents who have little connection with agricultural production.

What meaning can be given to the 'urban' and 'rural' labels in advanced industrial societies? If there are no significant differences in the organization of production, consumption or socio-cultural activities across built-up areas and the countryside, how can either of these terms define any connected set of social processes, still less form the basis of worthwhile fields of social science research? The only answer which seems defensible to me is to reject any spatial reference for the 'urban' or 'rural' labels, and instead to try to redefine these terms to apply to *particular kinds of social activity*. These may be found everywhere in some form or they may be spatially concentrated. For example, we could apply the 'rural' label to agricultural production and to the few distinctive socio-cultural processes still associated with it, and we could interpret the 'urban' label to refer to collective forms of consumption (as Castells proposes) or to investment in the built environment (as Harvey proposes) or to the 'management of scarce resources' by local authorities (as the neo-Weberians suggest). Figure 1.3 shows that if we interpret 'urban' social processes on these lines as those involved in collective consumption, and 'rural' processes as those involved in farming, then there can be no 'urban/rural' dichotomy — rather our attention is directed to widely separated sub-sets of consumption and production activities.

Of course, the suggestion that we should unlink the academic definition of 'urban' (and 'rural') studies from the spatial connotations fundamental to our commonsense or everyday use of these terms remains a highly controversial one. Critics respond by asking why we should retain these terms at all in social science, or alternatively complain that the avenues for reaching a sound definition in line with commonsense usage have still not been fully explored. Without going into such issues in more detail, I would like to extract one general point from the discussion of advanced industrial societies, a point summarized in figure 1.4. Put simply, this is that *any* contemporary definition of urban studies must be able to separate out a realistic sub-set of social activities from the vast mass of social processes which are not 'urban'. At the same time, to know how to define 'rural' processes will not help to make the distinction between what is 'urban' and what is not. For on any conceivable definition 'rural' processes will be a negligibly small part of those activities which are not 'urban'. Most aspects of economic activity and social life in an advanced industrial society on any defensible definitions of the 'urban' and 'rural' labels will be both non-urban and non-rural. So however we may define urban studies there can be no role

All social processes

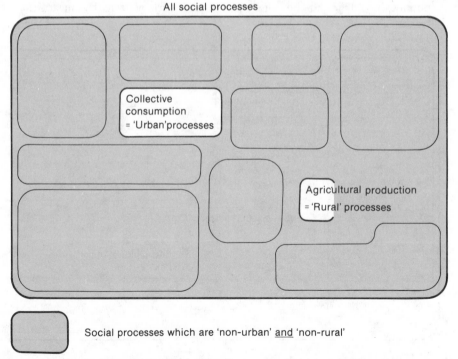

Collective consumption = 'Urban' processes

Agricultural production = 'Rural' processes

Social processes which are 'non-urban' <u>and</u> 'non-rural'

**Figure 1.4**   The non-existence of an 'urban/rural' dichotomy in advanced industrial societies.

for an 'urban/rural' dichotomy in contemporary social science. No such dichotomy exists in an advanced industrial society, nor does the attempt to postulate a finely graded continuum of spatial variations from 'urban' to 'rural' locations have any relevance to the problem of defining urban studies. We need a *social* definition of urban studies, not one phrased in terms of different geographical areas.

## 1.2  Defining the scope of urban studies — liberal sociology and Marxist approaches

In § 1.1, I argued in general terms about the problems which have marked theoretical attempts to develop a basis for urban studies in social science. In this part I turn to a more detailed and substantive consideration of the positions which have shaped analysis in the past or which have been influential in the recent period. This review takes the form of some short extracts drawn from five writers: R E Park, Louis Wirth and John Rex in liberal approaches to urban sociology, and Manuel Castells and David Harvey who are prominent writers in the newer Marxist stream.

Robert Park was in many ways the central figure in the Chicago School, defining the concern with cities as examples of 'human ecology' which influenced much North American work from the 1920s to the 1960s. Park was fascinated by two aspects of city life: the impact which he thought that living in cities had on people's behaviour; and the ways in which human adaptation to the physical environment helped to differentiate cities into 'natural areas' appropriate for different kinds of people. This extract, drawn from a seminal article (Park 1915), sums up his orientation well.

*R E Park: The city: suggestions for the investigation of human behaviour in the urban environment*†

The city, from the point of view of this paper, is something more than a congeries of individual men and of social conveniences — streets, buildings, electric lights, tramways, and telephones, etc; something more, also, than a mere constellation of institutions and administrative devices — courts, hospitals, schools, police and civil functionaries of various sorts. The city is, rather, a state of mind, a body of customs and traditions, and of the organized attitudes and sentiments that are inherent in these customs and are transmitted with this tradition. The city is not, in other words, merely a physical mechan-

† Reprinted with permission of Macmillan Publishing Company Inc., from *Human Communities* by R E Park (1915). Copyright 1952 by The Free Press, a Corporation.

ism and an artificial construction. It is involved in the vital processes of the people who compose it; it is a product of nature, and particularly of human nature.

The city has been studied, in recent times, from the point of view of its geography, and still more recently from the point of view of its ecology. There are forces at work within the limits of the urban community — within the limits of any natural area of human habitation, in fact — which tend to bring about an orderly and typical grouping of its population and institutions. The science which seeks to isolate these factors and to describe the typical constellations of persons and institutions which the cooperation of these forces produce, is what we call human, as distinguished from plant and animal, ecology.

The city, particularly the modern American city, strikes one at first blush as so little a product of the artless processes of nature and growth, that it is difficult to recognize it as a living entity. The fact is, however, that the city is rooted in the habits and customs of the people who inhabit it. The consequence is that the city possesses a moral as well as a physical organization, and these two mutually interact in characteristic ways to mold and modify one another. It is the structure of the city which first impresses us by its visible vastness and complexity. But this structure has its basis, nevertheless, in human nature, of which it is an expression. On the other hand, this vast organization which has arisen in response to the needs of its inhabitants, once formed, imposes itself upon them as a crude external fact and forms them, in turn, in accordance with the design and interests which it incorporates. Structure and tradition are but different aspects of a single cultural complex which determines what is characteristic and peculiar to city life, as distinguished from village life, and the life of the open fields. The old adage which describes the city as the natural environment of the free man still holds so far as the individual man finds in the chances, the diversity of interests and tasks, and in the vast unconscious co-operation of city life the opportunity to choose his own vocation and develop his peculiar individual talents. The city offers a market for the special talents of individual men. Personal competition tends to select for each special task the individual who is best suited to perform it.

The approach embodied in Park's work was open to criticism on a number of grounds, chiefly the eclectic nature of the 'urban' phenomena he studied, which seemed to encompass virtually all aspects of modern life, and the rather deterministic view which Park took of human behaviour — producing the typical passages above where Park seems to ascribe social agency (the ability to act) to cities. Critics objected that these tendencies lead to abstract concepts (like 'the city') being thought of as real things, or even as 'things with human qualities'. In response a demand arose for a more genuinely sociological or social definition of urban studies, providing a framework within which human behaviour could be studied in a methodical way free of deterministic assumptions. A key contribution here was made by Louis Wirth, himself a prominent figure in the Chicago School, but one remote from Park's ecological approach. His paper summarized in the extract below (Wirth 1938) was an especially influential statement of his approach, focusing on 'urbanism'.

## Louis Wirth: Urbanism as a way of life†

As long as we identify urbanism with the physical entity of the city, viewing it merely as rigidly delimited in space, and proceed as if urban attributes abruptly ceased to be manifested beyond an arbitrary boundary line, we are not likely to arrive at any adequate conception of urbanism as a mode of life.

For sociological purposes a city may be defined as a *relatively large, dense, and permanent settlement of socially heterogeneous individuals*. On the basis of the postulates which this minimal definition suggests, a theory of urbanism may be formulated in the light of existing knowledge concerning social groups.

The central problem of the sociologist of the city is to discover the forms of social action and organization that typically emerge in relatively permanent, compact settlements of large numbers of heterogeneous individuals. We must also infer that urbanism will assume its most characteristic and extreme form in the measure in which the conditions with which it is congruent are present. Thus the larger, the more densely populated, and the more heterogeneous a community, the more accentuated the characteristics associated with urbanism will be.

A number of sociological propositions concerning the relationship between (i) numbers of population, (ii) density of settlement and (iii) heterogeneity of inhabitants and group life can be formulated on the basis of observation and research.

*Size of the population aggregate* Ever since Aristotle it has been recognized that increasing the number of inhabitants in a settlement beyond a certain limit will affect the relationships between them and the character of the city. Increase in the number of inhabitants of a community beyond a few hundred is bound to limit the possibility of each member of the community knowing all the others personally. Max Weber, in recognizing the social significance of this fact, explained that from a sociological point of view large numbers of inhabitants and density of settlement mean a lack of that mutual acquaintanceship which is ordinarily inherent between the inhabitants in a neighbourhood. The increase in numbers thus involves a changed character of the social relationships. As Georg Simmel points out: 'If the unceasing external contact of numbers of persons in the city should be met by the same number of inner reactions as in the small town, in which one knows almost every person he meets and to each of whom he has a positive relationship, one would be completely atomized internally and would fall into an unthinkable mental condition.' This is not to say that the urban inhabitants have fewer acquaintances than rural inhabitants, for the reverse may actually be true; it means rather that in relation to the number of people whom they see and with whom they rub elbows in the course of daily life, they know a smaller proportion, and of these they have less intensive knowledge.

The superficiality, the anonymity and the transitory character of urban social relations make intelligible, also, the sophistication and the rationality generally ascribed to city-dwellers. Our acquaintances tend to stand in a relationship of utility to us in the sense that the role which each one plays in our life is overwhelmingly regarded as a means for the achievement of our own ends.

*Density* As in the case of numbers, so in the case of concentration in a limited space certain consequences of relevance to sociological analysis of the city emerge. Of these only a few can be indicated.

On the subjective side, as Simmel has suggested, the close physical contact of numerous individuals necessarily produces a shift in the media through which we orient ourselves to the urban milieu, especially to our fellow-men. Typically, our physical contacts are close but our social contacts are distant. The urban world puts a premium on visual recognition. We see the uniform which denotes the role of the functionaries and are oblivious to the personal eccentricities hidden behind the uniform. We tend to acquire and develop a sensitivity to a world of artifacts and become progressively farther removed from the world of nature.

Diverse population elements inhabiting a compact settlement become segregated from one another to the degree in which their requirements and modes of life are incompatible and to the measure in which they are antagonistic. Similarly, persons of homogeneous status and needs unwittingly drift into, consciously select, or are forced by circumstances into the same area. The different parts of the city acquire specialized functions, and the city consequently comes to resemble a mosaic of social worlds in which the transition from one to the other is abrupt.

*Heterogeneity* The social interaction among such a variety of personality types in the urban milieu tends to break down the rigidity of caste lines and to complicate the class structure; it thus induces a more ramified and differentiated framework of social stratification than is found in more integrated societies. The heightened mobility of the individual, which brings him within the range of stimulation by a great number of diverse individuals and subjects him to fluctuating status in the differentiated social groups that compose the social structure of the city, brings him toward the acceptance of instability and insecurity in the world at large as a norm.

On the basis of the three variables, number, density of settlement and degree of heterogenity of the urban population, it appears possible to explain the characteristics of urban life and to account for the differences between cities of various sizes and types.

Wirth still went on to discuss a wide range of modern social activities as 'urban' processes forming part of this 'city life' which he claimed to have distinguished. For example, he included extensive discussions of political parties, newspapers and the mass media, and even a money-based economy as integral elements of 'urbanism'. His paper had explicitly cautioned:

It is particularly important to call attention to the danger of confusing urbanism with industrialism and modern capitalism. The rise of cities in the modern world is undoubtedly not independent of the emergence of the modern power-driven machine technology, mass production and capitalistic enterprize; but different as the cities of earlier epochs may have been by virtue of their development in a pre-industrial and pre-capitalistic order from the great cities of today, they were also cities.

† Reprinted from Urbanism as a way of life by L Wirth, by permission of the University of Chicago Press. *American Journal of Sociology* **44** (1938).

Yet within a few pages of this remark Wirth can be found apparently attributing changes in the class structures of industrial societies not to capitalism or industrialism but to the pressures and 'logic' of city life. For example he argued:

While the city has broken down the rigid caste lines of pre-industrial society, it has sharpened and differentiated income and status groups. Generally, a larger proportion of the adult urban population is gainfully employed than is the case with the adult rural population. The white-collar class, comprising those employed in trade, in clerical and in professional work, are proportionately more numerous in large cities and in metropolitan centres and in smaller towns than in the country.

Even sympathetic writers have found it hard to defend Wirth's empirical definition of 'urban' phenomena. Writing thirty years after Wirth, Gans (1968, pp 96–7) described Wirth's paper as 'a classic in urban sociology' but commented:

Despite its title and intent, *Wirth's paper deals with urban industrial society, rather than with the city.* This is evident from his approach. Like other urban sociologists, Wirth based his analysis on a comparison of settlement types, but unlike his colleagues, who pursued urban–rural comparisons, Wirth contrasted the city to the folk society. Thus, he compared settlement types of pre-industrial and industrial society. This allowed him to include in his theory of urbanism the entire range of modern institutions which are not found in the folk society, even though many such groups (e.g. voluntary associations) are by no means exclusively urban. Moreover, Wirth's conception of the city dweller as depersonalized, atomized and susceptible to mass movements suggests that his paper is based on, and contributes to, the theory of the mass society.

Many of Wirth's conclusions may be relevant to the understanding of ways of life in modern society. However, since the theory argues that all of society is now urban, *his analysis does not distinguish ways of life in the city from those in other settlements within modern society*. In Wirth's time, the comparison of urban and pre-urban settlement types was still fruitful, but today the primary task for urban (or community) sociology seems to me to be the analysis of the similarities and differences between contemporary settlement types.

Gans went on to compare Wirth's hypothesized 'urban' lifestyle, which was formulated with what would now be regarded as inner-city areas in mind, with his findings from research in US suburbs. Not surprisingly he found few continuities in some of Wirth's key areas, while other phenomena supposedly linked to area characteristics were much the same in high-density, large-population, and socially heterogeneous inner cities as in low-density, low-population and very homogeneous suburbs. Although Gans himself continued to try to specify reasons for different lifestyles in terms of settlement characteristics, most subsequent studies have not uncovered any clear spatial mechanisms independent of social factors. (A good example of later studies in this vein is the comparison by Quinton of social life in Inner London and the Isle of Wight presented in Section III of this reader.)

Partly in reaction to the disappointing results produced by studies of the alleged behavioural consequences of cities, another group of British sociologists switched their attention to the origins of social conflicts in cities which seemed remote from any direct influence by capitalism or industrialism. Rex and Moore in particular located 'distinctively urban' issues in the housing field. John Rex set out his approach as follows.

*John Rex: Housing classes and the 'zone of transition'*†

Max Weber, it will be remembered, relativized Marx's view of the nature of social classes by suggesting that any market situation and not only the labour market led to the emergence of groups with a common market position and common market interests which could be called classes. We need only qualify this slightly to include groups differentially placed with regard to a system of bureaucratic allocation to arrive at a notion of 'housing classes' which is extremely useful in analyzing urban structure and processes.

Some Marxists may argue that such housing classes do nothing more than reflect the class struggle in industry and clearly they are partly right in that there is some correlation between the two. But it is also the case that among those who share the same relation to the means of production there may be considerable differences in ease of access to housing. This is a part of the 'superstructure' which manifestly takes on a life of its own. A class struggle between groups differentially placed with regard to the means of housing develops, which may at local level be as acute as the class struggle in industry. Moreover, the independence of this process is emphasized the more home and industry become separated.

Rex identified these 'housing classes' in the main with the basic forms of tenure in Britain, namely private home ownership, local authority (public housing) rental and rental from a private landlord — although he also defined small landlords themselves as an additional housing class. He went on:

In the class struggle over housing, qualification either for a mortgage or a council tenancy are crucial. They are, of course, awarded on the basis of different criteria. In the first case size and security of income are vital. In the second 'housing need', length of residence and degree of affiliation to politically powerful groups are the crucial criteria. But neither mortgages nor council tenancies are available to all so that either position is a privileged one compared with that of the disqualified. It is likely, moreover, that those who have council houses or may get them soon will seek to defend the system of allocation which secures their privileges against all categories of potential competitors. Thus local politics usually involves a conflict between two kinds of vested interest and between those who have these interests and

---

† Reprinted from The sociology of a zone of transition in *Readings in Urban Sociology* ed. R E Pahl (Oxford: Pergamon) pp 214–5, 230–1.

outsiders. As with classes generated by industrial conflict, however, there is always some possibility of an individual moving from one class to another.

These observations were then used to analyze the social organization of part of Birmingham where there was a high concentration of coloured people, mostly immigrants from Britain's former colonial territories. Rex argued that an understanding of coloured people's housing situation was crucial to understanding the social organization of this area, which he characterized as a 'zone of transition' serving to accommodate new arrivals in the city. Without the capital necessary for home ownership or the 'steady' jobs necessary to acquire mortgages, and excluded from public housing by rules requiring a long period of local residence to qualify, coloured immigrants were concentrated in those areas of Birmingham where rented housing was still available, some of them making the transition to small landlord in order to be able to afford house purchase. Thus, Rex concluded, the housing class notion allowed a distinct analysis to be made of the social organization of the 'zone of transition'. He summarized his findings as a series of general propositions:

(1) urban development in advanced industrial societies divides men into classes differentially placed with regard to housing;

(2) the zone of transition is that area of the city where the least privileged housing classes live, especially the landlords and tenants of lodging-houses;

(3) the community life of the zone of transition is shaped by conflicts between these housing classes;

(4) since many of the under-privileged are newcomers to the city their organizations will also perform functions in the re-orientation of men from foreign, traditional and rural societies to urban life;

(5) the actual associations such as churches, political parties and clubs will perform important functions in regard to both (3) and (4) above;

(6) they will perform these functions relatively imperfectly both because of their functional diversity and because of their responsiveness to outside pressures from parent organizations;

(7) community-wide organizations may arise which have the function for the local community of peaceful conflict resolution and for the city as a whole of mediating and facilitating mobility between its problem areas and the larger urban society;

(8) the situation in the zone of transition is a highly unstable one and in any sudden crisis ethnic and class conflicts which are temporarily contained may crystallize and be pursued by more violent means.

These are propositions which relate the study of the urban zone of transition dynamically to a wider concept of the city as a social system.

Marxist writers have mounted a much more thorough-going and drastic attack on previous urban sociology than that implicit in Rex's analysis. They argue that writers such as Park and Wirth confused aspects of the social organization of capitalism with the consequences or influence of particular social locations. The effect of this confusion, they claim, is to represent *social* problems or arrangements as if they were partly natural, spatial or physical phenomena — caused by particular locations or environments. This kind of effect they claim to find in everyday uses of the 'urban/rural' dichotomy, with many adverse societal features such as crime, poverty or bad housing being presented as the product of 'urban' conditions or locations — instead of being accurately seen as contradictions inherent to capitalism. This view has been particularly expressed by French writers since 1968, amongst whom Manuel Castells has been prominent, as illustrated by the following extract (Castells 1977).

### Manuel Castells: Defining urban studies: spatial factors versus economic processes†

Let us begin, then, with *space*. This is something material enough, an indispensable element in all human activity. And yet this very obviousness deprives it of any specificity and prevents it from being used directly as a category in the analysis of social relations. In fact, *space*, like *time*, is a physical quantity that tells us nothing, in itself, about the social relation expressed or as to its role in the determination of the mediation of social practice. A 'sociology of space' can only be an analysis of social practices given in a certain space, and therefore in a historical conjuncture. Just as in speaking of the nineteenth century (itself, one might remark, a questionable expression), one is not referring to a chronological segmentation, but to a certain state of social formations, so in speaking of France, the Auvergne, the quarter of Ménilmontant, the Matto Grosso or the Watts district, one is referring to a certain social situation, to a certain *conjuncture*. Of course, there is the 'site', the 'geographical' conditions, but they concern analysis only as the support of a certain web of social relations, the spatial characteristics producing extremely divergent social effects depending on the historical situation. From the social point of view, therefore, there is no space (a physical quantity, yet an abstract entity *qua* practice), but an historically defined *space-time*, a space constructed, worked, practised by social relations. Does it not, in turn, have an effect on the said social relations? Is there not a spatial determination of the social? Yes. But not *qua* 'space' — rather as a certain efficacity of the social activity expressed in a certain spatial form. A 'mountain' space does not determine a way of life: the discomforts of the physical milieu are mediatized, worked, transformed by social conditions. In fact, there is nothing to choose between the 'natural' and the 'cultural' in social determination, for the two terms are indissolubly unified in the single material reality of the social point of view: *historical practice*. Indeed, all the 'theories of space' that have been produced are theories of society or specifications of these theories.

---

† From M Castells 1977 *The Urban Question* (London: Edward Arnold) pp 442–6.

Castells proceeds to discuss whether particular types of socio-economic activity can be considered as especially important in structuring the social or spatial organization of cities. He begins by considering production processes:

In fact, I think that the means of production are not organized on the spatial plane at the company level in an economy so complex as that of advanced capitalism. The milieu of technological interdependencies, common resources, 'external economies', as the marginalists say, are realized on a much broader scale. On the scale of an urban area, then? Not always. For although certain urban areas possess a specificity at the level of the organization of the production apparatus (within, of course, a generalized interdependence), other residential units (urban areas) are no more than an entirely heteronomous cog in the process of production and distribution. The organization of space into specific, articulated units, according to the arrangements and rhythms of the means of production, seems to me to refer back to distinctions of practice in terms of region. If we consider the regional question, expressed in terms of economic imbalances within the same country, the reality immediately connoted is what the Marxist tradition treats as the effects of the unequal development of capitalism, that is to say, unequal development of the productive forces and specificity in the organization of the means of production according to a differential rhythm linked to the interests of capital. Unequal development of the economic sectors, the unequal value placed on natural resources, the concentration of the means of production in the most favourable conditions, the creation of the productive milieux or 'complex units of production' — these are the economic bases of what are called the regions and regional disparities.

Castells then turns to consider consumption processes, which he describes in Marxist terms as the 'reproduction of labour power'. The idea here is that consumption can be understood as ensuring that today's labour force will be renewed (*re*-produced) and available for productive work tomorrow. Reproduction of labour, of course, does not mean simply physiological reproduction, but the continuing process of turning out a whole labour force with the skills, attitudes, motivations and organization appropriate for a modern industrial state.

The spatial organization of the reproduction of labour power seems, on the other hand, to lead to very familiar geographico-social realities, namely, the urban areas, in the banal statistical sense of the term. What is an 'urban area'? A production unit? Not at all, in so far as the production units are placed on another scale (on a regional one, at least). An institutional unit? Certainly not, since we are aware of the almost total lack of overlap between the 'real' urban units and the administrative segmentation of space. An ideological unit, in terms of a way of life proper to a 'city' or to a spatial form? This is meaningless as soon as one rejects the culturalist hypothesis of the production of ideology by the spatial context. There is no 'Parisian bourgeoisie' except in terms of semi-folkloric details. There is international capital and a French ruling class (insofar as there is the specificity of a state apparatus); there are *regional* (not city) ideologies in terms of spatial specificity in the organization in the means of production. But there is no cultural specificity of the city as a spatial form or of a particular form of residential space.

What then is an urban area? It would be easy enough to agree that this term of social and administrative practice designates a certain residential unit, an ensemble of dwellings with corresponding 'services'. An urban unit is not a unit in terms of production. On the other hand, it possesses a certain specificity in terms of residence, in terms of 'everydayness'. It is, in short, the everyday space of a delimited fraction of the labour force. This is not very different from the definition, current among geographers and economists, of an urban area on the basis of the map of commutings. But what does this represent from the point of view of segmentation in terms of mode of production? Well, it is a question of the process of reproduction of labour power: that is the precise definition in terms of Marxist economics of what is called 'everyday life'. On condition, of course, that we understand it in the terms explicitated, namely, by articulating with it the reproduction of social relations and pacing it according to the dialectic of the class struggle.

However, we must differentiate between two broad types of process in the reproduction of labour power: collective consumption and individual consumption. Which of the two structures space? Around which are the urban areas organized? It goes without saying that both processes are articulated in practice; consequently, the one that dominates the process as a whole will structure the other. Now, the organization of a process will be all the more concentrated and centralized, and therefore structuring, as the degree of objective socialization of the process is advanced, as the concentration of the means of consumption and their interdependence is greater, as the administrative unity of the process is more developed. It is at the level of collective consumption that these features are most obvious and it is therefore around this process that the ensemble of consumption/reproduction of labour power/reproduction of social relations is structured.

We can, therefore, retranslate in terms of the collective reproduction (objectively socialized) of labour power most of the realities connoted by the term urban and analyze the urban units and processes linked with them as units of the collective reproduction of labour power in the capitalist mode of production.

Indeed, an intuitive allusion to the problems treated as 'urban' in practice is enough to observe the overlap (we have only to think of the structural meaning for the mode of production of such questions as housing, collective amenities, transportation, etc).

But is there not, then, to be any separation between 'town' and 'country'? Is it a matter of 'generalized urbanization'? In reality, this problem has no meaning (other than an ideological one) posed in the terms in which it is generally posed. For it already presupposes the distinction and even the contradiction between rural and urban, an opposition and a contradiction that have little meaning in capitalism. The spaces of production and consumption in the monopoly phase of capitalism are strongly interpenetrated, overlapped, according to the organization and unequal development of the means of production and the means of consumption and are not frozen as definite spaces in one of the poles of the social and technological division of labour.

An alternative neo-Marxist approach put forward by the geographer David Harvey defines urban studies in terms both of consumption processes and equally importantly in terms of investment in the built environment: both these, he claims, are distinctively 'urban' elements of a fundamentally capitalist system.

This extract (Harvey 1978) sets out his rationale for this approach, and briefly indicates how it can be applied to the analysis of housing processes.

## David Harvey: The urban process under capitalism†

My objective is to understand the urban process under capitalism. I confine myself to the capitalist forms of urbanization because I accept the idea that 'urban' has a specific meaning under the capitalist mode of production which cannot be carried over without a radical transformation of meaning (and of reality) into other social contexts.

Within the framework of capitalism, I hang my interpretation of the urban process on the twin themes of *accumulation* and *class struggle*. The two themes are integral to each other and have to be regarded as different sides of the same coin — different windows from which to view the totality of capitalist activity. The class character of capitalist society means the domination of labour by capital. Put more concretely, a class of capitalists is in command of the work process and organizes that process for the purposes of producing profit. The labourer, on the other hand, has command only over his or her labour power which must be sold as a commodity on the market. The domination arises because the labourer must yield the capitalist a profit (surplus value) in return for a living wage. All of this is extremely simplistic, of course, and actual class relations (and relations between factions of classes) within an actual system of production (comprizing production, services, necessary costs of circulation, distribution, exchange, etc) are highly complex. The essential Marxian insight, however, is that profit arises out of the domination of labour by capital and that the capitalists as a class must, if they are to reproduce themselves, continuously expand the basis for profit. We thus arrive at a conception of a society founded on the principle of 'accumulation for accumulation's sake, production for production's sake'. The theory of accumulation which Marx constructs in *Capital* amounts to a careful enquiry into the dynamics of accumulation and an exploration of its contradictory character. This may sound rather 'economistic' as a framework for analysis, but we have to recall that accumulation is the means whereby the capitalist class reproduces both itself and its domination over labour. Accumulation cannot, therefore, be isolated from class struggle.

*Accumulation and the urban process* The understanding I have to offer of the urban process under capitalism comes from seeing it in relation to the theory of accumulation. We must first establish the general points of contact between what seem, at first sight, two rather different ways of looking at the world.

Whatever else it may entail, the urban process implies the creation of a material physical infrastructure for production, circulation, exchange and consumption. The first point of contact, then, is to consider the manner in which this built environment is produced and the way it serves as a resource system — a complex of use values — for the production of value and surplus value. We have, secondly, to consider the consumption aspect. Here we can usefully distinguish between the consumption of revenues by the bourgeoisie and the need to reproduce labour power. The former has a considerable

† From D Harvey (1978) The urban process under capitalism *International Journal of Urban and Regional Research*, **2**, 101–2, 114–5, 126–7.

impact upon the urban process, but I shall exclude it from the analysis because consideration of it would lead us into a lengthy discourse on the question of bourgeois culture and its complex significations without revealing very much directly about the specifically capitalist form of the urban process. Bourgeois consumption is, as it were, the icing on top of a cake which has as its prime ingredients capital and labour in dynamic relation to each other. The reproduction of labour power is essential and requires certain kinds of social expenditures and the creation of a consumption fund. The flows we have sketched, insofar as they portray capital movements into the built environment (for both production and consumption) and the laying out of social expenditures for the reproduction of labour power, provide us, then, with the structural links we need to understand the urban process under capitalism.

It may be objected, quite correctly, that these points of integration ignore the 'rural–urban dialectic' and that the reduction of the 'urban process' as we usually conceive of it to questions of built-environment formation and reproduction of labour power is misleading if not down-right erroneous. I would defend the reduction on a number of counts. First, as a practical matter, the mass of the capital flowing into the built environment and a large proportion of certain kinds of social expenditures are absorbed in areas which we usually classify as 'urban'. From this standpoint the reduction is a useful approximation. Second, I can discuss most of the questions which normally arise in urban research in terms of the categories of the built environment and social expenditures related to the reproduction of labour power with the added advantage that the links with the theory of accumulation can be clearly seen. Third, there are serious grounds for challenging the adequacy of the urban–rural dichotomy even when expressed as a dialectical unity, as a primary form of contradiction within the capitalist mode of production. In other words, and put quite bluntly, if the usual conception of the urban process appears to be violated by the reduction I am here proposing then it is the usual conception of the urban process which is at fault.

Harvey then goes on to consider a number of examples of how his approach might be applied. This is how he sketches in an analysis of 'the housing question':

The demand for adequate shelter is clearly high on the list of priorities from the standpoint of the working class. Capital is also interested in commodity production for the consumption fund provided this presents sufficient opportunities for accumulation. The broad lines of class struggle around the 'housing question' have had a major impact upon the urban process. We can trace some of the links back to the workplace directly. The agglomeration and concentration of production posed an immediate quantitative problem for housing workers in the right locations — a problem which the capitalist initially sought to resolve by the production of company housing but which thereafter was left to the market system. The cost of shelter is an important item in the cost of labour power. The more workers have the capacity to press home wage demands, the more capital becomes concerned about the cost of shelter. But housing is more than just shelter. To begin with, the whole structure of consumption in general relates to the form which housing provision takes. The dilemmas of potential overaccumulation which faced the United States in 1945 were in part resolved by the creation of a whole new life-style through the rapid proliferation of the suburbanization process. Furthermore, the social unrest of the 1930s in that country pushed the bourgeoisie to adopt a policy of individual home ownership for the more affluent workers as a means to ensure social stability. This solution had the added advantage of opening up the housing sector as a means for rapid accumulation through commodity production. So successful was this solution that the housing sector became a Keynesian 'contra-cyclical' regulator for the accumulation process as a whole, at least until the *débâcle* of 1973. The lines of class struggle in France were markedly different. With a peasant sector to ensure social stability in the form of small-scale private property ownership, the housing problem was seen politically mainly in terms of costs. The rent control of the inter-war years reduced housing costs but curtailed housing as a field for commodity production with all kinds of subsequent effects on the scarcity and quality of housing provision. Only after 1958 did the housing sector open up as a field for investment and accumulation and this under government stimulus. Much of what has happened in the housing field and the shape of the 'urban' area that has resulted can be explained only in terms of these various forms of class struggle.

A comparison between this passage from Harvey and the earlier quotation from Rex should suffice to indicate one of the main lines which divide contemporary Marxist and Weberian writers. While Harvey assimilates housing conflicts fairly directly into what he sees as the fundamental conflict between owners of capital and working people, Rex sees multiple housing classes overlaying and cross-cutting production cleavages.

Before leaving these extracts for a broader characterization of the theoretical content of other papers in this reader, it is worth mentioning one final area of disagreement between Marxist and liberal writers of all kinds. As Harvey's article makes clear, Marxists see contemporary urban problems in the West as shaped not just by industrialization, but by specific features of *capitalist* industrialization. Weberian and other writers fundamentally deny this claim, pointing to the occurrence of quite similar kinds of urban problems in non-capitalist countries such as in eastern Europe. They also deny that all aspects of social life can be attributed to the *mode* of production, arguing that technological change and 'human nature' (for example) can exert far-reaching effects on urban development. The neo-Weberian writer Ray Pahl (1977) remarks sceptically, for example, on the influence of the automobile on socio-economic life:

I am not convinced that the dependence of the advanced industrial societies on fossil fuels is directly related to *mode* of production. I know of no evidence from existing societies nor of any political manifesto where a reduction in the level of material production and a reversion to simpler technologies is seriously advocated. Whilst small groups may rely on wind or water power or experiment with the use of solar energy, the world's population is too large and too concentrated to move back from the level of production which demands considerable supplies of coal and oil. Whilst there may be ideological preferences by some for the bicycle and the public 'bus, I know of nowhere in the world where more lavish private transport is not

preferred by those who can obtain access to it. There is a further argument that if technology were used for social purposes and not simply as a tool in the process of the accumulation of capital, then other inventions would have been developed and, so it is hoped, the noxious, noisy, polluting internal combustion engine would not have been developed and instead some pure, silent engine based on something like stored solar energy or the force of the tides would have emerged. I am extremely sceptical about such arguments yet it seems to me that it is on such lines that those who believe in the dominance of mode of production must argue.

No-one doubts that a motor car is simply a motor car: it is not an object specifically related to any one mode of production. Similarly, its propulsion is not fundamentally linked to the necessity for capital to accumulate and reproduce itself.

For Weberian writers then, an adequate explanation of urban change and urban conflicts is likely to appeal to a general logic of industrial, technological and social development. And, they argue, this logic of development is likely to be substantially the same across very different modes of production. These questions, about whether urban problems are products of specifically capitalist development or are general corollaries of industrialization, are discussed in detail in the companion reader to this volume (Cochrane *et al* 1982).

## 1.3 Theory into practice: the wider scope of urban social science

The differences between liberal sociological and Marxist approaches to urban studies documented above, of course in no way exhaust the variations in perspective across urban social science as a whole. Beyond sociological and political economy accounts, there is a large spectrum of divergent analyses, each of which views urban studies in a more or less distinctive way. Thus geographers have operated predominantly within a spatial approach to urban studies, focusing on the internal arrangements of city areas, but also looking at the growth of networks or systems of cities and their arrangement in space; liberal or neo-classical economics has concentrated largely on the influence of land markets in structuring city areas and distributing people and functions to particular locations; political science approaches have focused on local governments to a large extent, following an institutional strategy for defining urban studies; environmental psychologists have carried on the previous sociological concern with examining the patterning of human behaviour by the physical environment of towns and cities; and the list could be still further extended. Each of these approaches has a long history and most have evolved from long and complex combinations and developments of earlier approaches.

To describe this evolution in narrative terms would be a lengthy and difficult job, one which I have already attempted elsewhere (Dunleavy 1982). However, it is worth presenting a summary of such an account in diagrammatic form, see figure 1.5. In this figure I assume that most readers are likely to have some acquaintance with a number of broad approaches to urban studies, or will have the opportunity of forming such an acquaintance by reading the papers given in later sections of this volume. Figure 1.5 then presents a (subjective) picture of the main currents in urban social science over the last ninety years. The horizontal axis is a time dimension, and the main currents of analysis

appear on the figure as white areas, with nodes indicating where different influences converge. The vertical width of each current indicates in a rough manner my estimate of the volume of work and the significance of the work being done in any particular current of analysis at each point in time — so an expanding current suggests that the strand of research in question was of growing importance, while a narrowing vertical width suggests that it had entered a period of decline. The majority of the most influential schools of urban research have been included in figure 1.5, and I have added the names of some representative authors at points which may help readers to fix more clearly the character of the different currents.

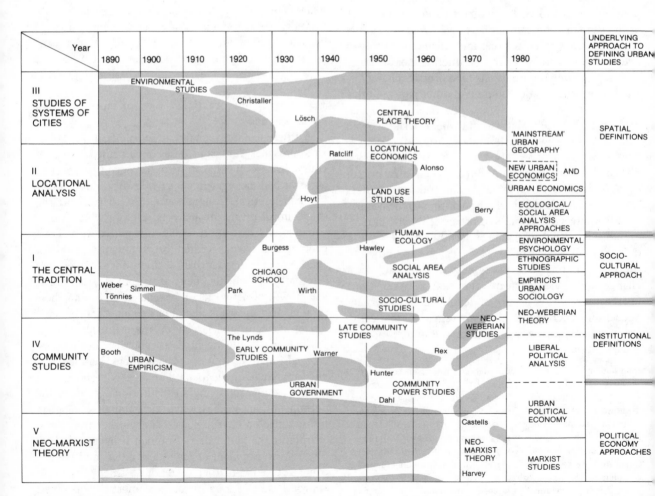

**Figure 1.5**  The development of urban studies, 1890–1980.

**Table 1.1**   The theoretical approaches of papers in this reader

| Type of analysis | Explicitly theoretical papers | Applied papers |
|---|---|---|
| 1. STUDIES OF SYSTEMS OF CITIES | SIMMONS: a 'mainstream' geographical analysis examining conceptual models of city systems' development. | COATES *et al*: focuses on the comparative analysis of variations in social welfare across different urban areas. |
| | PRED: a mixed geographical/marginalist economics paper examining the spatial interactions creating systems of metropolitan inter-dependence. | DAVIES: an analysis of retailing development in Britain from a mixed geographical/marginalist economics perspective. |
| | ALONSO. 4: a mixed geographical/marginalist economics paper examining the reasons for the cessation of metropolitan growth in North America. | KEEBLE: a mixed geographical/economic paper operating within liberal assumptions and putting forward a conventional reformist view of regional planning policies. |
| 2. LOCATIONAL ANALYSIS | ALONSO. 6: a classic application of marginalist economics to the analysis of city housing markets. | HALL *et al*: an evaluation of the British urban planning system from within a mixed geographical/marginalist economics perspective. |
| 3. CENTRAL SOCIOLOGICAL TRADITION | QUINTON: a modern empirical treatment of the 'urbanism as a way of life' perspective. | RABAN: a modern ethnographic or narrative sociological account, presenting an overtly subjective view of city life. |
| | THORNS: a critical analysis of myths of a distinctive suburban 'lifestyle'. | |
| 4. COMMUNITY STUDIES AND LIBERAL POLITICAL SCIENCE | ABRAMS: a Weberian critique of the concept of 'community'. | BLOWERS: an elite theory or managerialist view of urban planning expressed in a local case study.<br><br>MEEHAN: ⎫<br>HEADEY: ⎬ liberal political science accounts of<br>PAINTER: ⎭ urban policy making at a national level, mostly critical of pluralist theory. |
| | DUNLEAVY: a comparison of pluralist and Marxist views of urban protest movements, with an empirical case study. | HUDSON: a radical critique of KEEBLE (above) and of the efficacy of conventional regional planning policies. |
| | KIRK: a review of liberal pluralist and elite theory interpretations of urban planning contrasting with Marxist approaches. | |
| 5. NEO-MARXIST THEORY | HARVEY: a critique of marginalist urban economics.<br><br>MASSEY: a reinterpretation of processes of uneven industrial development as produced not by firms' spatial behaviour but by a-spatial processes of capital restructuring.<br><br>COCKBURN: a Marxist reinterpretation of 'community conflicts' in terms of divergent interests of social classes. | DICKENS: a case study of the introduction of rent controls in Britain in 1915, undertaken from a Marxist view of the State. |

For readers already well versed in the history of urban studies, figure 1.5 may hold few surprises — except perhaps where they may disagree with my judgements about the patterns of influences involved. For readers coming fresh to the subject, figure 1.5 may at least help to give a feel for the different influences at work which have produced the complex mapping of contemporary developments in urban studies. But since this is also an inter-disciplinary sampler of the current situation, figure 1.5 also suggests some hints which may help to bring out the underlying theoretical thrust of each of the papers dealing with empirical subjects in sections II–V of this volume. The right-hand vertical side of figure 1.5 groups the different strands of urban studies into five broad categories: studies of systems of cities; locational analysis; a central, sociological tradition; community studies and other institutional approaches; and neo-Marxist analysis. In table 1.1 these same categories are used to order the articles in this reader. Of course, not all the papers chosen to describe important aspects of contemporary urban development have an explicit theoretical content — some indeed are very applied pieces — but the vast majority do involve a particular set of assumptions or way of looking at urban phenomena, which I have tried to capture in a brief description or labelling in table 1.1.

I hope the discussion in this section will have stimulated readers to think about some of the most neglected and at the same time fundamental questions in urban studies — such as, how can we define a coherent focus of study? What are the implications for the substantive analysis of 'urban' issues of different definitions of our field of study? Why do approaches vary so markedly and seem sometimes to conflict? What are the underlying presuppositions and assumptions of different empirical analyses? The discussion here is intended to encourage readers to use table 1.1 as a map of the intellectual as well as empirical disagreements which form the fabric of contemporary urban studies. I believe that table 1.1 demonstrates the broad intellectual balance of the readings brought together here, and I hope that readers will be encouraged to try to define

their own inter-disciplinary synthesis of the themes presented, taking account of the problems raised earlier about the applicability of 'urban/rural' contrasts to western societies and of divergent sociological theorizations of the proper focus of urban research. This may seem a difficult task, but it is hard to deny its importance. The penalty of not trying at all is intellectual confusion which can be fatal for research progress. As Pahl remarked (in 1969): 'in an urbanized society, "urban" is everywhere and nowhere'.

## References

Castells M 1977 *The Urban Question* (London: Edward Arnold) pp 422–6

Cochrane A *et al* (1982) *City, Economy and Society: a Comparative Reader* (London: Harper and Row)

Dunleavy P 1982 The scope of urban studies in social science *Urban Change and Conflict* D202 Units 3/4 (Milton Keynes: The Open University Press)

Gans H 1968 Urbanism and suburbanism as ways of life *Readings in Urban Sociology* ed. R E Pahl (Oxford: Pergamon)

Harvey D 1978 The urban process under capitalism *International Journal of Urban and Regional Research* **2** 101–31

Pahl R E 1977 Stratification, the relations between States, and urban and regional development *International Journal of Urban and Regional Research* **1** 13

Park R E 1915 The city: suggestions for the investigation of human behaviour in the urban environment, reprinted in Park R E 1929 *Human Communities* (New York: Free Press)

Rex J 1968 The sociology of a zone of transition *Readings in Urban Sociology* ed. R E Pahl (Oxford: Pergamon) pp 214–31, 230–1

Wirth L 1938 Urbanism as a way of life *American Journal of Sociology* **44**

# Section II
# Patterns and Processes of Urban Change

## Introduction

Over the last two decades urban society in advanced economies has become almost synonymous with society itself. We have seen the way of life of the major towns and cities invading the countryside to such an extent that they can hardly be defined or contained physically any more. Furthermore, some pressing national issues have been associated increasingly with urban areas, particularly the inner city. As a result the traditional separation of urban research between studies focusing on the internal structure of particular cities and those dealing with wider regional issues has appeared more and more anachronistic, and there is increasing overlap and interdependence between geographical scales of analysis as reflected, for example, in the recent interest in city-systems analysis. At the same time the ever increasing complexity and indeed variety of contemporary urban society has created further difficulties for any description or explanation, demanding different and often more sophisticated interpretive frameworks.

These changes are reflected in the range of papers selected for this section, which examine the spatial and economic processes at work in contemporary urban societies and some of the patterns that these processes appear to give rise to. Despite the substantial differences that exist between and within towns and cities most places exhibit, superficially at least, a certain amount of order. For example, there are the traditional links between size and spacing of settlements expounded in central-place models, with smaller centres located closer together and fulfilling more modest needs than larger centres with their more wide-ranging functions. Within urban areas we also find recurring patterns, for example with most large office blocks and major retail outlets still being located close to the city centre, surrounded in turn by older industrial and residential areas often in need of renovation and in many instances replaced by council estates, and then by inter-war followed by post-war suburbs interspersed with industrial estates and suburban shopping centres.

Although such traditional urban models still have a general validity, many have required substantial refinement as a result of dramatic changes in the scope and nature of the forces moulding urban development, particularly those changes brought about by technological development and political action. Perhaps the most profound and physically apparent change to have taken place in modern times is the emergence of a complex system of interdependent settlements, where the repercussions of economic, social and political change at one place is felt throughout the country and where extraneous international forces often have more impact than essentially domestic factors. Here improved transport and communications facilities have been crucial, since they have facilitated greater information flow and interaction generally.

The papers included in this section draw mainly on North American and British experience and are primarily concerned with the role of market processes in urban development, though clearly it is impossible to differentiate exactly between market and governmental forces in the reality of contemporary 'mixed' economies. However, not all such studies approach issues in the same way and can vary markedly in their interpretation and viewpoint. Some, such as Alonso's extract about the urban land market, focus more on the analysis of individual competition and the operation of the free market. Others, Pred's paper for instance, place more stress on the role of major institutions and organizations in determining contemporary urban changes. Yet others, such as the paper by Massey, emphasize the links between urban change and the operation of the capitalist economic system as a whole. The structure of this section reflects these different perspectives as well as the division between spatial and sectoral studies. Specifically, the readings are organized to distinguish between: (i) broadly based and essentially spatial studies (based on systems and locational analyses) which attempt to set out some of the general processes responsible for the changing relationships between cities (Simmons, Pred, Alonso 4) and the locational patterns within cities (Alonso 6); (ii) those studies which focus on more structural factors that underlie patterns and processes in capitalist economies (Massey, Harvey); and (iii) those studies which concentrate on individual sectors of the urban economy (Davies, Headey). The readings have also been selected to include a balance of theoretical and empirical work throughout the section.

The first paper by *Simmons*, intended very much to set the scene, examines some of the conceptual models which have been used to explain general city-systems development, though focusing on North American experience. Simmons divides these theories into four broad categories arguing that each is more appropriate to particular development phases and locations. Thus the so-called 'frontier–mercantile' model has been applied to newly developing areas such as colonial America of the eighteenth and nineteenth centuries, while ideas linking 'industrial specialization' to urban growth are more appropriate to industrial heartlands during the industrial revolution era, and 'social change' models are closely linked to contemporary urban development in post-industrial societies. The 'staple export' model, though not tied exclusively to particular development phases, is closely linked to (primary) resource-rich areas and as such is more appropriate to the early development of peripheral regions. It should be stressed that Simmons' paper gives only a general overview of theories on city-systems development and this is particularly apparent in the case of his 'social change' model which really constitutes an umbrella for miscellaneous social factors whose impacts and associations are often too complex to tie down. Also, no explicit consideration is given to the more structural neo-Marxist interpretations of urban development in capitalist economies, nor to the relationship between city-systems development, the production of profits and the power of business organizations, financial institutions and State agencies, though it must be added that research in this area is still underdeveloped.

The increasingly complex processes, economic, social and political, at work in all post-industrial economies are reflected in the ever increasing scale and varying nature of contacts between towns and cities. These complex linkages, and particularly the dramatic influence of multilocational firms on contemporary urban geography, are the subject of the second paper by *Pred*, which illustrates how multinational organizations produce a complex network of interactions which have increased dramatically metropolitan interdependence, supplementing the traditional essentially hierarchical linkages of central-place systems. Empirical evidence is drawn mainly from the United States, but it is argued that these trends are applicable to all advanced economies. In the final section some consideration, albeit brief, is given to the regional planning implications of multilocational organizations.

Another major characteristic of many advanced market societies is the demographic decline in metropolitan areas (the so-called population 'about turn'). *Alonso's* paper examines this recent phenomenon in relation to national social and economic trends, again focusing on the United States. In particular, he argues that the decline in the population of metropolitan areas is not a random fluctuation but a more general trend, which must be viewed against the major social and economic changes of recent decades (the decline in birth rate, changes in age structure and the number of retired persons, the increasing number of footloose industries and so on). Of course, major out-migrations and urban decline have occurred before but they have tended to be more localized, and have normally been associated with relatively short-lived economic downturns, rather than linked to a general decline in birth rate. Alonso also considers the definitional problems encountered when attempting to compare the population of towns and cities, and stresses that any interpretation of such statistics should be viewed with caution, since these data collecting units vary widely in geographical extent, contain variable amounts of built-up area, and exclude areas which though superficially rural are now essentially urbanized as reflected for example in their occupational structure.

The next paper by *Massey* considers (albeit tangentially) some of the points made by Pred and Alonso in assessing the locational implications of contemporary changes in industrial structure and its employment consequences, using the electrical and electronics industries in the United Kingdom during the 1960s and

1970s as case studies. Massey examines the nature of the restructuring processes at work in these industries and contends that these processes have worked against the hitherto more prosperous regions, which have been hit hardest by employment decline during this period. But she is also at pains to point out that this trend does not necessarily signal the end of a spatial division of labour. For example, it was found that control over investment was being concentrated increasingly in southeast England (a finding in line with Pred's observation that more and more head offices are being located in the information-rich environments in and around metropolitan areas), and there was some evidence of a similar spatial polarization of research and development and more skilled activities generally. Furthermore it is revealed that the traditional broad social and economic division between the southeast and Midlands regions on the one hand and the rest of the United Kingdom on the other is both oversimplistic and indeed out-dated. In particular, the pattern of territorial inequality is often more marked between the largely stagnating conurbations, containing areas of acute social deprivation particularly in the inner city, and the less populated and quasi-rural areas of the rest of the United Kingdom, than between broad regional divisions (a development clearly linked to the general counter-urbanization trends considered in the preceding article by Alonso). Massey found that in the case of the electrical and electronic industries the conurbations of London, Birmingham, Manchester and Liverpool each suffered a much higher than average employment loss over the survey period, both in terms of complete industrial shutdowns and of plant relocations, with greater militancy, unionization and labour costs suggested as major disincentives to investment in these areas.

Massey's paper, like those by Simmons, Pred and Alonso, takes the view that urban patterns and processes are better explained by examining changes in the national, and indeed international, economy rather than looking for explanations within particular cities. But in doing so such studies tend also to focus on regional as opposed to local issues, though national explanations and local issues are by no means mutually exclusive. In contrast the next two papers focus more on patterns and processes within cities and give as much consideration to local as to national features.

The second paper by *Alonso* and the extract by *Harvey* illustrate the changing ideas about land-use patterns and processes. Alonso's article, first published in 1960, draws on the economic logic of von Thünen's model of agricultural land use to produce a theory for the urban land market. Alonso also follows the von Thünen model in assuming that all decision-making is perfectly rational and based on complete information, and that the area under consideration is perfectly uniform with equal accessibility in all directions. The key to this theory is the concept of bid-rents. Using this concept Alonso was able to demonstrate that urban land use in an equilibrium situation will consist of a series of concentric zones focusing on the city centre with businesses at the centre and residential areas further out. He argues that similar reasoning can be applied within land-use groupings and illustrates this by showing that even the seemingly paradoxical situation of the urban poor inhabiting more expensive inner-city land fits logically into this model. He also contends that this theory can be developed to allow for multiple centres, restricted transport patterns and other real world factors, but fails to show how this can be effected.

Like all micro-economic theories, Alonso's paper can be criticized for its simplistic, and even false, assumptions, and this is the main theme of Harvey's extract which critically appraises the qualities and validity of traditional land-use theories, as well as touching on the more descriptive models formulated by geographers and sociologists. Harvey identifies two major inadequacies in such formulations. Firstly, they fail to take account of the varying behaviour, attitudes and goals of the diversity of actors and interest groups influencing land-use decisions, with the assumption of utility-maximizing behaviour having little apparent validity. Secondly, their dismissal of historical factors (because of their static equilibrium framework) inevitably leads to an inability to deal adequately with the inherent monopolistic characteristic of land ownership and the way this is institutionalized in capitalist economies. Harvey acknowledges however that it would be extremely difficult to combine these factors to form a single coherent land-use theory, and also remarks that micro-economic theories have provided surprisingly accurate predictions of real world land-use patterns although seemingly misunderstanding the processes involved.

The general pattern of urban land uses and values are, of course, both the product and determinant of a number of more specific land uses and activities. The final two articles examine trends in two such sectors, retailing and residential housing, focusing on their locational patterns and institutional structure respectively, but also giving consideration to policy implications. The paper by *Davies* examines contemporary patterns of retailing in and around urban areas of Britain, concentrating on two major issues: the need to decentralize retail facilities to reflect demographic changes and the need to redevelop the depressed parts of the inner city. It is argued that though both are vital, the existing planning system has been almost exclusively concerned with the central city, while a heavily restrictive attitude has been imposed on retailing in suburban areas. It is stressed that the losers are the increasing numbers of suburban consumers, who are provided with only limited local convenience facilities, while many specialist shops are inaccessibly located in city centres. The paper also argues for a more rational and balanced approach to future retail planning policy.

One of the most politically sensitive urban issues is housing provision and access. The final paper in this section, an extract from a book on UK housing policy by *Headey*, surveys the domestic housing situation between World War I and the mid-1970s focusing on three different tenure groups: private tenants, local authority tenants and owner-occupiers. Here, changes in a number of seemingly pertinent housing market characteristics, ranging from general building rates to housing age and condition and from the level of housing choice to housing cost-to-income ratios, are examined systematically, and the implications for different sectors of society noted where appropriate. An underlying theme of this extract is that policy-makers should no longer consider working- and middle-class housing as separate issues, particularly when they emphasize the assistance and security given to tenants while neglecting to recognize the subsidies and privileges given to owner-occupiers. Headey argues that this is of particular concern considering that owner-occupiers appear to have been in a consistently more favourable position.

The extract by Headey is very much in the institutional tradition, looking as it does at three tenure groups, which are subject to different institutional structures and to different interplay between bureaucratic institutions and individuals. However, the urban housing market, whether in the United Kingdom or elsewhere, can be perceived, approached and analyzed in other ways; in some cases the approach is more market orientated assuming competition between consumers in a free market, in other cases it is more Marxist orientated stressing the links between housing problems and trends and the capitalist economic system itself. The fact is that no single approach has all the answers, or so it appears, because whether we like it or not reality is not straightforward and pigeon-holed for perfect comprehension, but must first be interpreted and analyzed. One could even argue that the various approaches are as much complementary as competing, shedding different light on the subject in question, whether it is the housing market or other areas of urban research, or indeed social science in general.

**Christopher Brook**

# 2    The Organization of the Urban System

## by J W Simmons

## Introduction

In its narrowest and most traditional sense, the urban system refers to the set of cities in a region or nation and its attributes. The system is simply the aggregate of cities; no attempt is made to identify relationships among them. But when the concept is developed more fully, the urban system can embrace the totality of activities in a nation, account for the observed relationships among regions and provide a model for the analysis of spatial variations of growth and change in the system.

The urban system, in this broader sense, is still based on urban nodes, that is on spatial concentrations of people and activities within the region or nation, but it also includes the relationships of the nodes to their surrounding areas and particularly the linkages among nodes. What are the patterns of connections and flows among cities and how do they cumulate into growth impulses to be transmitted through the system? In this essay, then, we are concerned with the description of the organization of a nation's territory and how it evolves over time.

The focus throughout this discussion is on the urban system as a whole. How can the urban system as a complex entity be described? How does it work as a unit? What parameters best describe it in its entirety? The approach here is holistic and operates on the premise that the system is more than an aggregate of cities and that individual elements can only be examined in relation to other elements. The *distribution* of attributes of cities in the system — their average values and variability — are of paramount interest, as is the covariation among different attributes. When examining a particular city those properties held in common with other cities are the object of principal concern.

Many different principles (or forces) combine to determine the spatial organization of an urban system. The military, the church and the process of public administration each creates its own pattern of structure, flow and growth. In Canada, for example, each primary industrial activity has its own distinct pattern of economic relations, its own geography. As other industrial activities are added, they transform the inputs (coming from primary products) and outputs (going to markets) to alter the interdependency matrix. The expansion of public-sector activities further modifies the growth relationships by translating part of the growth of the national economy into growth at certain specific locations, such as a military base or a new prison. The urban system helps to integrate all or at least most of these diverse processes.

When the process of spatial organization (or urbanization) is examined over the long run, it becomes evident that the context provided by one organizing principle (e.g. the fur trade) inevitably affects the distribution of elements of those that follow (the timber trade and the railway, for example). This effect is cumulative over a succession of organizing principles until what was initially an economic sector-specific spatial system becomes a complex, multipurpose urban system, coordinating flows among many different economic sectors and affecting social relationships as well. As the range in size of urban regions increases and the limiting effects of distance and the local resource base decline, the urban system becomes more and more the dominant factor in the location and growth of any new economic or social activity — a principle of spatial organization in its own right.

Source: *Systems and Cities: Readings on Structure, Growth and Policy* edited by L.S. Bourne and J.W. Simmons. Copyright © 1978 by Oxford University Press Inc. Reprinted by permission.

Another area of concern is the relationship between the urban system and the rest of the world. That is, how open or closed is the national urban system? At what points do the external contacts occur, in what sectors and with what impacts? The future of an urban system, particularly in countries with small populations such as Australia, Sweden or Canada, depends very much on these exogenously derived inputs. The future is essentially as unpredictable as these inputs are. The variety of external influences is endless. Inflation, trading patterns, immigration, and preferences for environmental amenities or attitudes against pollution are just a few examples of such influences. When the system is defined on a world scale, the cycles of growth or decline become explicable; when treated from the viewpoint of one nation, they often seem random.

**Models of system organization**

Numerous commentators have described and discussed the development process of national urban systems. Depending on their discipline and the scale in time and space on which they focus, these researchers have drawn on different interpretive themes. These themes contain implicit models of system organization, such as the size, spacing, interaction and particularly the growth focus in the system. The latter is the major taxonomic criterion used below. For present purposes we shall focus on the North American example where at least four models can be recognized:

(1) *the frontier* — the growth of cities is initiated externally by investment decisions from a previously developed urban subsystem;

**Table 2.1**  Models of urban-system organization

| | Frontier–mercantile | Staple export | Industrial specialization | Social change |
|---|---|---|---|---|
| *Location* | Edge of growing periphery | Periphery | Heartland | Universal |
| *Period of greatest relevance in North America* | 1740–1910 | 1700 to the present | 1840–1920 | 1880 to the present day |
| *Source of growth* | Growth of core region | Resource base. External markets | Agglomeration advantages | Technology. Labour supply |
| *Degree of openness* | High. Depends on capital and labour transfer | High. Depends on external demand | Low | High. Responds to information and technological change |
| *City size distribution* | Primate (single, very large city) | Rank size rule | Not specified | May change rapidly |
| *Economic specialization* | Little, except by scale. Cities are service centres | All cities share the same staple specialization | High | Specialization evolves over time |
| *Interaction patterns* | Links to core region dominate | Heirarchical, modified by staple | Highly linked | Varies with mode |
| *Settlement pattern* | A long frontier arc | Dispersed, but depends on staple | Clusters. Agglomerarions. Transport sensitive | Corridors or nodes specified by leading edge of economy |
| *System characteristics* | Investments by older centres send stimuli down to frontier area | Growth stimuli move up the heirarchy. (The central-place system) | Complex market linkages diffuse growth throughout | Growth responds to the removal of technical/physical constraints |
| *Examples* | Colonial America. Prairie settlement | Southern US, British Columbia | Manufacturing belt until World War II | US at the present |
| *Authors* | Taaffe, *et al* (1963), Lukermann (1966), Pred (1973) | North (1961), Christaller (1933), Innis (1957) | Lampard (1968), Thompson (1965), Pred (1967) | Borchert (1967), Berry and Neils (1969), Dunn (1971) |

(2) *staple export* — the growth of cities depends on the growth of primary activities in the areas they serve;

(3) *industrial specialization* — the growth of cities depends on the growth of the national economy and the relative strength of the cities' sector of economic specialization;

(4) *social change* — the growth of cities is largely unpredictable in the face of rapid social and technological change.

Most discussions of national urban systems as complex as those in North America contain some elements of each model, but for the most part they focus on one or two. A summary of each model is given in table 2.1.

Each model has its own era and location of particular relevance, hence the tendency of those writers who are concerned with the rise of cities in Britain or the US manufacturing belt in the nineteenth century to stress factors of industrial advantage and to draw from their studies an emphasis on system stability and stages of urban growth within an essentially closed system. More recent commentators are impressed with the volatility of the system and its increasing response to consumption preferences rather than to production characteristics.

*The frontier—mercantile model*
The frontier has long been a fundamental theme in American history (Turner 1894, Mikesell 1960, Innis, as discussed by Neill, 1972), but its explicit relationship to the urban system has, until recently, received less attention (Gras 1922, Careless 1954, Wade 1959, Vance 1970, Pred 1973). At the continental level urban growth in the US has occurred in a regular spatial pattern, beginning in the northeast and moving inland in regular rings of change over time. Lukermann (1966) points out that the 'urban frontier' must be defined carefully. It is only partly identified by growth rates since older cities supplying higher-order services to the new frontier may grow rapidly as well. It is measured better by the generation of new cities (those crossing the urban size threshold) and particularly by the geographic pattern of increases (or decreases) in the rank ordering of cities during any time period.

The growth wave of cities follows the settlement frontier and ties each new regional urban subsystem into the existing national system, with the nature of this linkage having important implications for the older cities. To what eastern US cities will the Midwest be linked, for instance? Over a long period of time the extension of the frontier creates a continual tension between the older, high-order urban places and the newer subsystems. The former attempt to maintain their economic control and to extract income for the supply of services; the latter strive for economic, political and social independence. This power struggle has been described by later authors as the 'core–periphery relationship' (Friedmann 1966), or the 'heartland–hinterland pattern' (Ray 1972). It implies economic differentiation between the frontier and older areas, with the former producing staples and the latter largely industrial goods and services.

The frontier depends on growth, since it is characterized by relatively high levels of capital investment and in-migration. Its rate of expansion is determined by a complex of forces. The rate of growth in the national or international economy is obviously a driving factor in the expansion of the frontier as a whole, determining such conditions in the source areas as employment, wage levels, rates of natural population increase, out-migration, etc. Yet the operation of the international market in staple products, the development of new modes of transportation and the conflicts among nations are also important at various times for various frontiers.

When frontier expansion takes place on a broad front, as it did in North America, the spatial sequence of expansion, hence of initial advantage in urban development, requires an intricate model of the development process (for example, Meinig 1972). Relative differences in current states of information or knowledge (Pred 1973) regarding accessibility to source areas, political conditions and the resource base (e.g. gold) often determined the growth response at any one time. Wade's (1959) comparison of the early growth of cities in the US Midwest portrays the moment-to-moment evolution of a city's changing growth prospects. The timing of growth or specific entrepreneurial decisions become as important as the spatial situation.

Out of this kind of micro-analysis, however conducted, emerges an urban subsystem which has considerable stability, even as the entire region evolves in its role, and which has a cumulative impact on the urban system as a whole through its pattern of linkages to the earlier system and its degree of interdependence (i.e. whether that subsystem is economically specialized and in what way). A less transient version of the frontier model but one using the same set of forces to explain the location of growth is the mercantile model described by Vance (1970) and Pred (1973). Urban growth in this model is determined by accessibility to the sources of merchandise, information and new technology, and it specifically responds to capital-investment decisions and transportation improvements.

*The staple export model*

The staple export model describes the pattern of regional growth and organization in certain peripheral regions of the continent — the South, the Pacific Northwest and, in particular, Canada — where expansion is determined by external demands for primary products or staples. The urban system supplies goods and services in response to the growth and prosperity of the region. There is some evidence (see Neill 1972) that the staple theory is essentially a reinterpretation of the frontier model where external market demands and technological conditions governed the pattern of exploitation of the wilderness, but the staple theory is better viewed as a framework that describes the mechanism driving a continuing and stable economy, growing perhaps, but not necessarily expanding in space. It can be conceived of as a static equilibrium model, and may be valid for large portions of the non-western world.

The difference between the frontier model and the staple model then rests largely in the interpretation of the dynamics of the process involved. The former requires expansion of a core area and the process is governed by investment decisions and expectations; the latter is governed by year-to-year fluctuations in productivity. Compare, for example, the California gold rush with the growth of the Cornbelt. Also, a staple economy is driven by external markets while the frontier economy depends on development decisions made in the source region.

Watkins (1963) makes the basic assumption of the staple export theory: 'Staple exports are the leading sector of the economy and set the pace for economic growth.' The spatial distribution and production function of a resource in turn determine the location of its impact; the character of the staple production determines the distribution of the income it generates and the rewards to factors of production, and its bulk and perishability account for the linkages it shows with the rest of the economy.

Several types of linkages can be defined. Backward linkages are inputs from other industrial sectors required for production of the staple, such as the purchase of mining machinery. Forward linkages are the outputs from production of the staple when those outputs are used by other firms, such as the conversion of iron ore to pig iron and steel. Final-demand linkages trace the impacts on other sectors of the staple income, creating a tertiary economy in the form of a local system of central places.

North (1961) employs the same resource-based ('staple' is the Canadian term) argument to explain the growth of the United States economy from 1790 to 1860. During the first thirty years the significant resource was New England shipping. The merchant marine provided the main source of external revenue, but its prosperity was initially controlled by foreign legislation and by European and domestic wars. After 1820 peace brought about an enormous surge in world trade, with the key contribution from the US being cotton. Very rapidly a three-region system emerged: the South became specialized in cotton production, having backward linkages to the Midwest in order to provide food for the increasingly labour-intensive (slave) production, while the cities of the Northeast provided final-demand linkages to both regions. The latter linkages were very complex, including high-order goods and services for consumption as well as business services, such as capital, shipping, insurance and the organization of trade. Finally, forward linkages developed between the South and the North as the textile mills of New England expanded with the capital and entrepreneurship of the New England merchants. During the period prior to the Civil War, North shows how closely national growth and prosperity were linked to the balance between external demand and the internal production of cotton. Approximately twenty-year growth cycles developed, based on surges of expansion of the Cottonbelt.

Tangentially, North (1961) also mentions the role of the urban system in this expanding economy. The major urban functions of finance, trade and service, transportation and manufacturing went to the Northeast. Despite the great surge of growth in the South virtually no urbanization (except for New Orleans) took place. The production of cotton was intensely rural, even hostile to the growth of cities (see Wade 1964). The contrast with the West was notable. There, a broad agricultural base rapidly developed a complete urban subsystem, with a full range of service linkages. By 1870 there were four cities west of the Alleghenies with populations greater than 100 000.

Lampard (1960) carries the theme of the regional economic base through to the present, although his interpretation is not that of a staple theory in the sense used by Watkins (1963). No regional economy is assumed to be entirely dependent on a single staple and its production function, but the significance of primary products (one or more, in sequence or simultaneously) for regional growth is evident. Repeated examples show how a local resource development transforms and reshapes the economic and social basis of a region. One such example is the Pacific Northwest, which has undergone the staple sequence of furs, fish and timber.

Because most US primary production is marketed within the nation a certain stability emerges. Market demands depend primarily on the national growth rate and each region competes with other regions in the country for a share of that growth. Market shares can be protected by legislation or by transportation preferences. Some analysts treat primary resource specialization at this level as another form of industrial specialization.

The major differences between the Canadian and American discussions of staple theory arise from the implications of specialization in a staple product for growth and prosperity. North (1961) suggests that dependence on an export base can only bring a temporary prosperity and that the boom must eventually collapse as the staple market declines or the resource is over-exploited. He notes that (p 3) 'Regions which remain tied to a single export staple almost inevitably fail to achieve sustained expansion.' In his view a more complex economic base is required to sustain growth. The logic is similar to discussions of the dangers of extreme economic specialization among cities. The Canadian literature, however, is more optimistic. A history of sustained national growth over 150 years stimulated largely by a succession of staples suggests that a viable economy and a stable urban society is possible (though possibly not preferable) under these conditions.

*The industrial specialization model*
The industrial specialization of the nineteenth century undoubtedly dominates most discussions of urbanization and the evolution of urban systems, overwhelming most students of the US urban experience (e.g. Lampard 1968, Pred 1967, Thompson 1965). They tend to equate the urbanization process with industrialization and to associate urban growth with industrial growth.

Certainly this period and these processes did create the basic outline of the North American urban system. First, New England and the growth of the great coastal ports occurred and then the giant industrial cities of the Midwest — Pittsburgh, Cleveland, Detroit, Chicago and St Louis — emerged. In each case a city's position as a regional leader was converted into national metropolitan status by the development of a powerful and integrated industrial economy. At the same time scores of smaller places suddenly grew several times in size as a direct result of the location of factories within their borders. The ability of industry to impose a whole new urban pattern on top of the existing structure was indeed impressive, although the specific location of manufacturing activities within any given region is not readily explained.

Studies of urban growth based on the theme of industrialization, however, tend to be aspatial. There are occasional references to the regional resource base which differentiates the industrial structure of cities, such as St Louis in leather goods, Minneapolis–St Paul in grain milling, and Pittsburgh and its coal-based economy. Harris (1954), for example, has demonstrated the close relationship between the extent of the manufacturing belt and access to the national market. For the most part, however, analysis is focused on a particular urban place. It is usually assumed that growth in industrial production (serving a national market of infinite capacity) is primarily due to characteristics peculiar to that city. The role of 'initial advantage' and 'circular and cumulative growth' (Pred 1967) are posited, which can be tied to other concepts such as economies of scale, agglomeration and urbanization, the growth pole and the industrial complex. Entrepreneurial competence, the structure of capital investment and the ability to innovate and to adopt innovations have also been suggested as local factors explaining urban growth. In this American framework every individual, every firm, every city has the potential for success, if only the will is there.

The industrial model of the organization and evolution of urban systems stresses size, internal industrial structure, and recent growth trends (momentum) rather than regional linkages, hierarchic level and relative location within the national system. However determined, the resultant linkages among cities are complex, cutting across size hierarchies, and they emphasize scale but not distance. This model has its own form of evolution as well. Thompson (1965) has described the process by which the oldest (hence highest-order) cities lose their older and less technically innovative activities to smaller and more peripheral places as a 'trickle-down' process. The movement of textile firms from New England to the South is an obvious example. The old firms are replaced in New England by new growth industries requiring high skills, risk capital and research capacity, as Perloff *et al* (1960) have documented.

In fact the urban system emerging from the nineteenth century was quite complex. Duncan and Lieberson (1970) stress the dual structure of closely linked manufacturing cities with weak hierarchic relationships such as Pittsburgh and Cleveland, coexisting with a commercial–financial urban system operating very much on central-place principles of the range and threshold of a good or service. In their words, 'Large-scale manufacturing became an alternative to commerce as a city-building activity. The graded order of commerical centres was no longer adequate to account for the role and sizes of cities in the national economy.'

system is comprehensible only *ex post*.

This last view of the American urban system has clearly been reinforced by the events of the 1950s and 1960s. Examples are numerous: the stimulation of urban growth in selected areas by amenity (climate, recreation) factors (Ullman 1954); the impact of the massive federal investment in space and defence programmes on peripheral areas with no observable location logic (Berry 1973); the tendency of further investments in transportation and communication facilities increasingly to channel social contacts rather than to diffuse them; and the largely unanticipated decay of the central cores of many large urban areas. Urbanologists monitoring such current phenomena turn for explanations to the past and observe the impact of new technologies such as steamboats, railroads and canals. They note the inability of statistical models of any kind consistently to predict variations in urban growth rates and they conclude that the evolution of the urban system is virtually incomprehensible.

## The social change model

Those authors whose analyses of the city are firmly rooted in mid-twentieth century phenomena (e.g. Borchert 1967, 1972, Berry and Neils 1969, Berry 1973, Meier 1962) see the growth of an urban system primarily as a reflection of a complex and evolving society in which an urban place plays many roles simultaneously and operates under a wide variety of interwoven forces. In their view one must add to the evolving economic forces the characteristics of an evolving society with increasingly higher incomes and levels of education, new preferences for places to live and new life-styles. Both the national economy and these cultural variables operate on and through the urban system, often through intermediaries that are themselves subject to change, such as the technology of communication, transportation and production processes, and the institutional infrastructure (political systems and legislation, educational and research procedures, service-delivery systems and the like). Social change becomes a determinant rather than a consequence of other changes.

The advocates of this approach treat the urban system as open in the fullest sense. For them, it is open not only to commodity and financial input and output from abroad, but also to forces of change and evolution whose effects are largely unpredictable at any point in time. The temporal (and spatial) path of the urban

## References

Berry B J L 1973 Contemporary urbanization processes *Geographical Perspectives on Urban Problems* (Washington: National Academy of Sciences) pp 94–107

Berry B J L and Neils E 1969 The urban environment writ large *The Quality of the Urban Environment* ed. H S Perloff (Baltimore: Johns Hopkins)

Borchert J 1967 American metropolitan evolution *Geographical Review* 57 301–322

——1972 America's changing metropolitan regions *Annals of the Association of American Geographers* 62 352–73

Careless J M S 1954 Frontierism, metropolitanism and Canadian history *Canadian Historical Review* 35 1–21

Christaller W 1933 *Central Places in Southern Germany* (Englewood Cliffs, NJ: Prentice-Hall)

Duncan B and Lieberson S 1970 *Metropolis and Region Revisited* (Los Angeles: Sage Publications)

Dunn E L 1971 *Economic and Social Development: A Process of Social Learning* (Baltimore: Johns Hopkins)

Friedman J 1966 *Regional Development Policy: A Case Study of Venezuela* (Cambridge, MA: MIT Press)

Gras N S B 1922 *An Introduction to Economic History* (New York: Harper)

Harris C D 1954 The market as a factor in the localization of industry in the United States *Annals of the Association of American Geographers* 44 315–48

Innis H 1957 *Essays in Canadian Economic History* (Toronto: University of Toronto Press)

Lampard E E 1960 Regional economic development, 1870–1950, in *Regions, Resources and Economic Growth* ed. H S Perloff *et al* (Baltimore: Johns Hopkins) pp 109–292

——1968 The evolving system of cities in the United States *Issues in Urban Economics* ed. H S Perloff and L Wingo (Baltimore: Johns Hopkins) pp 81–140

Lukermann F 1966 Empirical expressions of nodality and hierarchy in a circulation manifold *East Lakes Geographer* **2** 17–44

Meier R L 1962 *A Communications Theory of Urban Growth* (Cambridge, MA: MIT Press)

Meinig D W 1972 American Wests: preface to a geographical introduction *Annals of the Association of American Geographers* **62** 159–84

Mikesell M W 1960 Comparative studies in frontier history *Annals of the Association of American Geographers* **50** 62–73

Neill R 1972 *A New Theory of Value: The Canadian Economics of Harold Innis* (Toronto: University of Toronto Press)

North D C 1961 *The Economic Growth of the United States, 1790–1860* (Englewood Cliffs, NJ: Prentice-Hall)

Perloff H S *et al* (eds) 1960 *Regions, Resources and Economic Growth* (Baltimore: Johns Hopkins)

Pred A 1967 *The Spatial Dynamics of US Urban and Industrial Growth* (Cambridge, MA: MIT Press)

——1973 *Urban Growth and the Circulation of Information: The United States System of Cities, 1790–1840* (Cambridge, MA: Harvard University Press)

Ray D M 1972 The Economy *Studies in Canadian Geography: Ontario* Ch 3 ed. L Gentilcore (Toronto: University of Toronto Press) pp 45–63

Taafe E J, Morrill R L and Gould P R 1963 Transport expansion in underdeveloped countries: a comparative analysis *Geographical Review* **53** 503–29

Thompson W R 1965 *A Preface to Urban Economics* (Baltimore: Johns Hopkins)

Turner F J 1894 *The Frontier in American History* (New York: Henry Holt)

Ullman E L 1954 Amenities as a factor in regional growth *Geographical Review* **44** 119–32

Vance J E Jr 1970 *The Merchant's World: The Geography of Wholesaling* (Englewood Cliffs, NJ: Prentice-Hall)

Wade R C 1959 *The Urban Frontier* (Chicago: University of Chicago Press)

——1964 *Slavery in the Cities* (New York: Oxford University Press)

Watkins M H 1963 A staple theory of economic growth *Canadian Journal of Economic and Political Science* **24** 141–58

# 3

# On the Spatial Structure of Organizations and the Complexity of Metropolitan Interdependence

## by A Pred

In the United States and other highly industrialized or post-industrial countries the economy is dominated by large private-sector and government organizations that are normally composed of a number of functionally differentiated and spatially separated units. A variety of statistics indicates that the relative and absolute economic power of such organizations has been rapidly expanding in recent decades (Pred 1974). Insofar as these organizations dominate the economy they are the most important generators of flows of goods, services, information and capital. In other words, in an economically advanced system of cities large multilocational organizations are the major source of inter-metropolitan and inter-urban interdependencies. Despite this fact, relatively little is known of the spatial characteristics of the city-system interdependencies created by the intra-organizational and inter-organizational relationships of major business corporations and government activities.

The aims of this article are threefold:

(1) to conceptually outline the characteristics of metropolitan interdependence arising from the spatial structure, or *intra*-organizational linkages, of major job-providing organizations;

(2) to examine empirical material[1] on intra-organizational and inter-organizational interdependencies and to ascertain whether or not these interdependencies are consistent with those assumed in the author's model of the process of city-system growth and development in advanced economies; and

(3) to consider very briefly the regional planning implications of the spatial structure of large private and public organizations.

Before these objectives can be dealt with directly, it is necessary to sketch the aforementioned model and to say something more precise about the growing role of large multilocational organizations in shaping the overall pattern of metropolitan interdependence.

### Synopsis of the model

A probabilistic model describing the means by which the inter-urban circulation of specialized information and the spatial structure of organizations supposedly feed back upon one another to influence the process of city-system growth and development in advanced economies has been tentatively presented at length elsewhere (Pred 1973a, 1974, 1975). In addition to the subprocess of locally propagated growth in each metropolitan complex and city, the model incorporates three inter-related subprocesses which function in a circular and cumulative manner (figure 3.1). These are the intra-organizational and inter-organizational generation of nonlocal employment multipliers after each birth or large-scale expansion of an organizational unit, the diffusion of 'growth-inducing' innovations from one metropolitan complex to another[2] and the accumulation of 'operational' decisions, which usually have an implicit rather than an explicit locational dimension.[3] When observing the national city system as a whole — as opposed to the situation in figure 3.1 where only a few larger metropolitan complexes are considered — the entire process can be summarized in the following

Source: Pred A 1975 *Papers, Regional Science Association* 35 115–42. Reprinted by permission of the Regional Science Association.

admittedly oversimplified terms (figure 3.2).

The totality of intra-organizational and inter-organizational linkages existing at a given date $(t_1)$ creates a goods-and-service interaction matrix between the metropolitan complexes and lesser cities of the city system in question. Each of the component linkages has a dual over which some form of specialized information flows (cf Dunn 1970). The totality of these specialized linkages can be visualized as a probability matrix that influences both the places to which new growth-inducing innovations diffuse and the places identified for the implementation of new operational decisions.[4] (In other words, because of limited search behaviour, uncertainty-reducing behaviour and other factors, organizations tend to be greatly influenced by their existing network of market and other specialized information contacts when making explicit and implicit location decisions.) Each innovation adoption and operation decision either creates new intra-

organizational and inter-organizational linkages or fortifies such existing linkages. Thus, by a later date $(t_1 + \Delta t)$, a new slightly modified goods-and-services interaction matrix has emerged. The new counterpart specialized information flow probability matrix influences the loci of subsequent explicit and implicit location decisions until a later date $(t_1 + 2\Delta t)$, when yet another goods-and-service interaction matrix has evolved, and so on.

In developing the model, various pieces of very general and coarse-grained evidence suggested that:

(a) the most important nonlocal intra-organizational and inter-organizational linkage within a city system are those between large metropolitan centres; and

(b) intra-organizational and inter-organizational city-system interdependencies are complex, since metropolitan centres of a given size may provide goods,

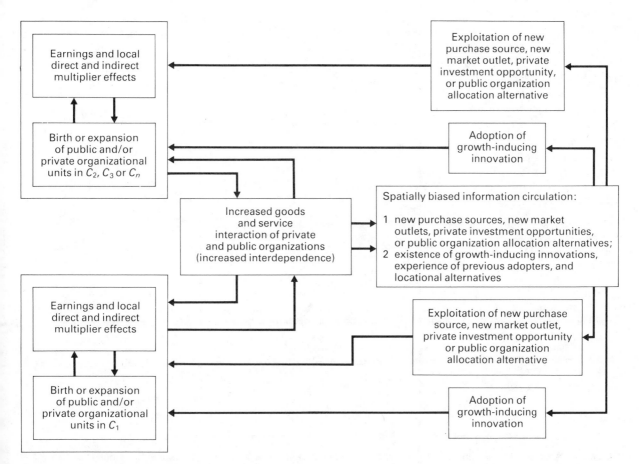

**Figure 3.1** The circular and cumulative feedback process of urban-size growth for large metropolitan complexes $C_1, C_2, C_3, \ldots, C_n$ in advanced economies.

services and specialized information to nearby and distant larger metropolitan complexes, and since there exists an extensive criss-crossing of linkages between large metropolitan complexes on the one hand and medium- or small-sized metropolitan centres on the other.

Thus, the assumed linkages and resulting probabilities are such that over time the rank stability of the largest metropolitan complexes is either maintained or only slightly altered. This, of course, is consistent both with the considerable body of empirical evidence concerning long-term rank stability among the largest units of national and regional systems of cities (for example Pred 1973b) and with the fact that in the United States, France, Great Britain, Canada, Sweden and other advanced economies recent population and employment growth has been largely concentrated in previously large metropolitan complexes and their expanding peripheral areas or urban fields. Also, in accordance with reality, the probabilistic qualities of the model are such that medium- or small-sized cities can occasionally make relatively rapid progress through the size ranks of a system of cities. This would occur when important explicit or implicit location decisions are implemented at organizational units in medium- or low-probability places. Progress would be accounted for by sizeable and temporally concentrated local multipliers, increased interaction with other cities and a higher probability of being associated with future explicit or implicit locational decisions.

## The growing role of private-sector organizations in metropolitan interdependence

It is widely documented that since the end of World War II large multifunctional and multilocational organizations have been gaining an ever larger share of the economy in the United States and other post-industrial

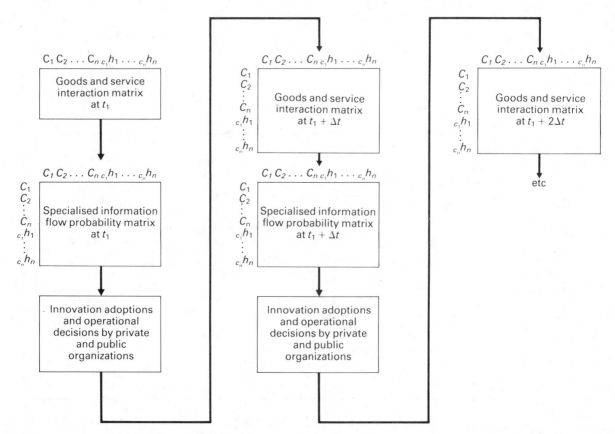

**Figure 3.2**  A simplified description of the growth and development process for an entire system of cities in an advanced economy. $C_1, C_2, \ldots, C_n$ refer to large metropolitan complexes, $c_1 h_1, c_2 h_2, \ldots, c_n h_n$ refer to all medium- and small-sized cities within the regional 'hinterlands' of large metropolitan complexes.

countries due to mergers, acquisitions, the expansion of existing units and capital investment in new facilities (Blair 1972, Reid 1960). The almost frenzied pace of merger and acquisition activity, particularly during the late 1960s, was motivated primarily by a desire to spread risks in an era of high technological and economic instability and uncertainty, by the identification of personal management goals with organizational size and growth and by 'synergy,' or the belief that the effectiveness of an enlarged corporation somehow becomes greater than that of the sum of its previously separately operating parts (Lorsch and Allen 1973, Reid 1960). Only during the early 1970s did government decisions, tight money and lower stock prices really cut into the number of corporate acquisitions and mergers. (Each acquisition or merger, by definition, affects the spatial structure and intra-organizationally generated inter-metropolitan linkages of the growing

organization.) Nevertheless, there is ample evidence that large organizations are continuing to expand their economic dominance. For example, in 1969 the 500 largest industrial corporations in the United States[5] accounted for 74% of the profits of all such corporations, while in 1973 the profit share of the 'top 500' had risen to 79% of the total, and between 1972 and 1973 alone, the number of people employed by the 500 leading US manufacturing corporations increased by almost 900 000 to over 15.5 million (*Fortune* 1974).

Lacking any locational specificity, general statistics of the type just presented only provide suggestions as to the growing role of large private-sector organizations in metropolitan interdependence. More specific statistics referring to the growth and *location* of organizational activities are not readily accessible in the literature except in those headquarters-location studies based either on the uncomprehensive annual directory of

**Table 3.1**  Number of jobs controlled by multilocational private-sector organizations in selected metropolitan complexes of the western United States (after Dun and Bradstreet 1974, US Bureau of the Census 1971, and from a direct survey of Seattle – Tacoma organizations)

| Metropolitan complex | Estimated number of employees controlled by locally headquartered multilocational organizations[a] | | | Population | | |
|---|---|---|---|---|---|---|
| | 1959[b] | 1973 | Percentage increase 1959–73 | 1960 | 1970 | Percentage increase 1960–70 |
| Los Angeles[c] | 539 341 | 1 252 478 | 132.2 | 6 742 696 | 8 452 461 | 25.4 |
| San Francisco– Oakland – San Jose[d] | 572 101 | 892 451 | 56.0 | 3 291 077 | 4 174 235 | 26.8 |
| Seattle–Tacoma[e] | 165 210 | 204 411[f] | 23.7[g] | 1 428 803 | 1 832 896 | 28.3 |
| Portland SMSA | 55 732 | 146 234 | 162.4 | 821 897 | 1 009 129 | 22.8 |
| Phoenix SMSA | 18 625 | 123 182 | 561.4 | 663 510 | 967 522 | 45.8 |
| Honolulu SMSA | 18 885 | 93 492 | 395.1 | 500 409 | 629 176 | 25.7 |
| San Diego SMSA | 39 682 | 73 561 | 85.4 | 1 033 011 | 1 357 854 | 31.4 |
| Boise City SMSA | 12 702 | 66 891 | 426.5 | 93 460 | 112 230 | 20.1 |

[a] Employment totals are based on units completely owned by corporations and firms with local headquarters. They do not include employment partially steered by locally present corporate divisional head offices. Both 1959 and 1973 totals are restricted to organizations with multilocational characteristics that employed at least 400 people as of 1973. Some no longer existing firms are included in the 1959 totals.

[b] Some of the organizations included in the 1959 totals did not provide Dun and Bradstreet with employment figures until subsequent years. In those cases figures from 1960 or later were substituted. Thus, the 1959 total for each of the eight metropolitan complexes is somewhat exaggerated and the percentage increases for 1959–73 are even greater than those indicated.

[c] Los Angeles–Long Beach SMSA + Anaheim–Santa Ana–Garden Grove SMSA.

[d] San Francisco–Oakland SMSA + San Jose SMSA.

[e] Seattle SMSA + Tacoma SMSA.

[f] 1974 datum.

[g] See note 7.

industrial and non-industrial corporations published in *Fortune*, or on similar listings for other countries (e.g. Armstrong 1972, Pred 1974). Thus, the data presented in table 3.1 for selected western US metropolitan units provide a picture of 1960s' and 1970s' trends that, if still imperfect, is probably clearer than those elsewhere available.

Although somewhat different time spans are involved, it would seem quite apparent from table 3.1 that, in most metropolitan complexes, the number of jobs associated with locally based multilocational business organizations has been growing at a rate of increase that far outstrips that of local population growth.[6] The only anomaly, Seattle–Tacoma, is attributable to the Boeing Company's unusual loss of almost 30 000 jobs between 1959 and 1973.[7] Even allowing that some of the indicated employment increases have occurred at organizational units in foreign countries, there would appear to be little question that, at the very least, there has been a significant burgeoning in those metropolitan and other city-system interdependencies which involve intra-organizational linkages between the headquarters offices of multilocational private-sector organizations and their subordinate domestic units. If this is true, then it follows that in many, perhaps most, metropolitan complexes and cities there has been a simultaneous absolute and relative increase in the number of local jobs controlled from elsewhere.[8] Obviously, while lacking in detail, these observations hint at a mounting complexity of metropolitan interdependence arising from the intra-organizational linkages of multilocational business organizations.

The complexity of metropolitan interdependencies shaped by increasingly dominant multilocational business organizations stems from both the division of labour within such organizations and the more widely treated market-coordinated division of labour between all corporations and firms. On the intra-organizational side, the necessity of coping with environmental diversity and instability has required that large modern business (and governmental) organizations develop increasingly intricate links of interdependence between their component units, whether spatially proximate or dispersed (Lorsch and Allen 1973). These links may involve the flow of various services and decision-making information, control information or coordination information between a headquarters unit and other organizational units.[9] They may also subsume the flow of goods, services and information between two or more units without involving an organization's head

offices. (Even when a multilocational business organization is not of the diversified conglomerate variety, there will be numerous functionally specialized units that must interact with one another. For example, large organizations whose sole or principal function is manufacturing usually are composed of some combination of main and branch plants in several product lines, management and administrative offices with different levels of authority, research and development units and marketing, warehousing and transportation facilities; cf Parsons 1972.) On the inter-organizational side, technological advances in production, transportation and telecommunications have created an ever greater array of linkages between multilocational private-sector organizations. These ties take the familiar form of physical input–output relationships, financial and commercial service linkages and other specialized backward and forward service linkages (cf Wood 1969).

## The spatial structure of multilocational organizations

### Hierarchical structures in general

Whatever their particular multifunctional attributes, private (and public) multilocational organizations are usually spatially structured along intentional or *de facto* hierarchical lines.[10] Normally, the hierarchy consists of three or more tiers, with each unit at a successively higher tier serving a successively larger area. Thus, when a nationally functioning organization purposefully assigns precisely outlined areas to its component units, or sets up geographical divisions, the hierarchy might take on the following appearance. At the peak of the hierarchy sits the controlling national headquarters. In most instances this coordinating and planning unit is situated in a metropolitan complex of national importance, but this need not invariably be so. At the next hierarchical level are the divisional offices, which are normally located either in metropolitan areas of regional significance or in larger national-level metropolitan complexes which also function as regional-level foci. If the organization includes production units designed to serve the entire market area of each geographical division, they may be found in cities of varying size. However, if the plants in question either require several thousand workers, or are dependent on any of a variety of agglomeration economies, they also are apt to be located in populous metropolitan areas.

Whether performing marketing, service, production or low-level administrative functions, the subregionally and locally oriented units at the bottom of the organizational hierarchy may have a wide spectrum of locations within the system of cities, ranging from small towns and cities to metropolitan areas of regional and national significance.

Most private-sector organizations are *not* explicitly or intentionally organized geographically. Instead, the individual units or multi-unit divisions of private-sector organizations are typically coordinated along product or other functionally specialized lines.[11] Under these circumstances, in a nationwide organization the head offices, with their great demand for specialized nonroutine information, are also most likely to be found in a metropolitan complex of national stature. Despite quite different fundamental organizing principles, the subordinate components of these organizations as a matter of practice also operate on a national, regional or local scale, and are accordingly distributed among differently sized urban centres. Here too, those units having high information demands, large market thresholds or sizeable labour-force requirements are normally placed in large or very large metropolitan complexes, while other units tend to have a greater range of locations.

Since the most important private-sector organizations in advanced economies usually are either conglomerates with many more or less completely unrelated divisions (ITT, for example, contains some 200 divisions), or multidivisional corporations composed of semi-autonomous and highly diversified vertically integrated divisions, it is important to distinguish between 'polycentric' and 'unitary' hierarchical spatial structures. In conglomerates and other large multidivisional organizations with great functional diversity, each division, or 'profit centre' is itself spatially structured along hierarchical lines with a quasi-independent divisional national headquarters unit at its peak. As a result of this, the hierarchical spatial structure of the organization as a whole is 'polycentric', with the divisional headquarters serving as the second highest level of the organization's overall internal structure. The fact that an organization is 'polycentric' does not preclude one or more of its divisional headquarters from having the same metropolitan location as that held by the organizational head office. Although the degree of divisional self-containment varies from case to case in the type of organization under discussion (Lorsch and Allen 1973), there are virtually always important link-

ages between organizational headquarters on the one hand and divisional headquarters and other subunits on the other hand. Thus, while the multidivisional organization evolved largely because only a finite span of control over routine operational activities is possible,[12] head offices in organizations with 'polycentric' spatial structures usually are still at the very least responsible for determining and coordinating strategic objectives, general long-term planning, resolving conflicts, granting approval of capital and major expense projects, and the allocation of funds and resources among competing operating divisions and subunits (Williamson 1970, Lorsch and Allen 1973). From the standpoint of city-system growth, the last two frequently overlapping functions are crucial insofar as they directly and indirectly affect where new jobs will be created. In a nationally functioning 'polycentric' organization the spatial distribution of new jobs determined by the head office over a given time period is very likely to be quite widespread, not only because of the nonlocal multiplier effects stemming from each birth or large-scale expansion of a unit, but also because units brought into an organization by merger or acquisition very often add to the scatter of the organization's existing locations. This is so since mergers and acquisitions are usually made on the basis of growth and other criteria, not locational criteria (cf Chapman 1974).

In organizations with few products, services or functions, there is generally little discretionary authority delegated to subheadquarters units — even where routine operational activities are involved. This is also true of some organizations with many products, services or functions. In both cases this centralization of authority requires that intra-organizational control be set up within a very strict hierarchical framework that is primarily based on chain-of-command principles rather than spatial designations. That framework in turn means that — regardless of the total number of subordinate organizational units — the management of all units is responsible to a *single* national level headquarters and that the linkages within the organization's intentional or *de facto* hierarchical spatial structure are very strong. Hence, such organizations may be described as having 'unitary' hierarchical spatial structures.

*Hierarchical structures at different scales*
Thus far our discussion has been confined to nationally functioning organizations. Private-sector organizations with hierarchical spatial structures exist at three other

scales. First, there are multi-unit retailing or service-providing organizations whose operations are restricted to a single metropolitan complex or urban field. Here the hierarchy is most often only two-tiered: headquarters and district or local outlets (furniture stores, speciality clothing stores, real estate offices, etc). Obviously, these types of multilocational organizations may make nonlocal purchases, but for the most part they are of relatively little interest in terms of their impact on metropolitan interdependence and the process of city-system growth and development. Secondly, there are those regional-scale business organizations whose units are found either in a single state (such as banking organizations), or within the limits of a traditionally defined major metropolitan hinterland or multistate area. In these instances the structural hierarchy typically is made up of regional, district and local units. A similar three-level spatial structure is often found for state- or provincial-level government agencies. Hierarchical spatial structures at this scale can bring intermediate- and small-sized metropolitan areas into interdependent relationships with larger metropolitan complexes. Finally, in most cases very large nationally functioning business organizations simultaneously operate at a multinational scale. This condition adds an international tier to their hierarchical spatial structure. The metropolitan interdependencies propagated by this extra tier are of rapidly increasing importance. Little is said here of these interdependencies, partly because of my preoccupation with those interdependencies which are internal to national systems of cities and their regional subsystems and partly because the topic merits extensive treatment on its own.

*Asymmetrical organizational spatial structures and metropolitan interdependence*

The spatial structure of nationally functioning multilocational organizations would be symmetrical if all organizational and national divisional headquarters were in the same metropolitan complex, and if the regional- and local-level units of each and every organization were present in the identical subset of cities. In addition, symmetry would require that any plant or other nonadministrative unit serving the entire nation also be located in the single metropolitan complex where all headquarters units were concentrated. Similarly, the spatial structure of multilocational organizations functioning in a given physically extensive region would be symmetrical if the regional, district (or subregional) and local units of each and every organization

were distributed among the identical subset of cities.

Since this is quite clearly not the case, organizational spatial structures being *asymmetrical*,[13] one would expect *intra*-organizationally based metropolitan and city-system interdependencies to be *complex*. More specifically, it is to be anticipated that the intra-organizational linkages of organizations with 'polycentric' and 'unitary' hierarchical spatial structures are synonymous with a high degree of interdependence among large metropolitan complexes. That is, it is to be expected that headquarters dominance and job-control linkages extend from large metropolitan complexes of national or regional significance to even larger metropolitan complexes, and not merely from the single largest metropolitan unit to less populous metropolitan complexes. Likewise, it is to be expected that headquarters dominance and job-control linkages run both between metropolitan areas of comparable size (regardless of population class) and even from metropolitan areas of small or intermediate size to much larger metropolitan complexes. Moreover, to the extent that multilocational organizations operate production units and various local-scale units throughout a country, one should find headquarters dominance and job-control linkages extending from metropolitan complexes of varying size to smaller towns and cities within what are normally considered to be the hinterlands of other distant metropolitan complexes.

*Interdependencies: asymmetrical organizational spatial structures versus central-place theory*

Conventionally, interdependencies within a system of cities are intentionally or unintentionally depicted in the hierarchical terminology of central-place theory, with the greatest emphasis being placed on the one-way dependence of cities of a given size upon cities of a larger size. As has been demonstrated at length elsewhere (Pred 1971, 1973a,b) a literal central-place theory interpretation of city-system relationships logically eliminates any possibility of two-way interdependencies between large metropolitan complexes. (Rather general evidence presented by Pred (1974) and others indicates that there is an intricate web of economic interdependence between large metropolitan complexes in advanced economies.) Furthermore, the other types of complex interdependencies just enumerated as being suggested by asymmetrical organizational spatial structures are not feasible under Christallerian or Löschian central-place theory. Actually, although it is infrequently stated, Löschian central-place theory does allow two-way interdependencies (interdependencies between comparably sized places) and the dependence of centres of a given size upon smaller places. However, these interdependencies *cannot* occur between the largest cities in a Löschian system and are limited to places within the same or adjacent 60° sectors of the economic landscape.[14]

Central-place theory stresses the conflux of consumers to supply points, or market-area and city–hinterland relationships, but neither job-control nor decision-making relationships, nor inter-urban input–output relationships. Hence, it should come as no surprise that the interdependencies it generates are not fully compatible with those supposedly following from the asymmetry of organizational spatial structures. In light of the evidence about to be presented,[1] and in light of the dominant role played in the economy by multilocational business organizations, it is contended that for planning and some other purposes the central-place theory image of city-system interdependencies must be displaced from its position of primacy.

## Empirical observations

Detailed examination of the Seattle–Tacoma, Phoenix and Malmö metropolitan complexes[1] tended to confirm some of the assumptions underlying the model of city-systems growth and development outlined earlier. In particular the evidence demonstrated that:

(a) the most important *nonlocal* linkages are not those between metropolis and 'hinterland' — as in central-place theory — but those between large metropolitan complexes; and

(b) the overall pattern of metropolitan interdependence springing from the asymmetrical spatial structure of organizations is complex, both because metropolitan centres of a given size frequently provide job control and other links to nearby and distant larger metropolitan complexes, and because there is an extensive criss-crossing of economic ties between large metropolitan complexes on the one hand, and medium- and small-sized metropolitan centres on the other.

The evidence also revealed a perhaps unexpected degree of complex interdependence between large metropolitan centres and smaller cities and towns in the 'hinterlands' of other metropolitan centres. In addition rather limited evidence from Malmö seems to indicate that specialized information flows and goods and service flows follow parallel paths.

Of course, more conclusive statements cannot be made regarding the impact of organizational spatial structures on metropolitan interdependence until job-control statistics and other data are amassed for other metropolitan complexes. However, it is to be recognized that, if anything, the pattern of metropolitan interdependence is probably even more complex than indicated here. For one thing, Seattle–Tacoma has an unusually high percentage of 'local' jobs associated with its locally headquartered corporations. Also, despite the tempo of foreign investments and acquisitions maintained by multinational corporations, international linkages have been ignored except for those between the United States and Canada. Moreover, no empirical consideration has been given to the interdependencies fashioned by the divisional headquarters within 'polycentric' organizational spatial structures. Interdependencies growing from intra-organizational relationships that do not involve organization-wide or divisional headquarters have also been neglected. Finally, there has been no treatment of the interdependencies emerging either from joint ventures or from relationships between units belonging to large multilocational corporations and units belonging to their *partly owned* and *not fully controlled* subsidiaries.

## Regional planning implications: some questions

In an era when the number of jobs under the direct and indirect influence of metropolitan based multilocational organizations has been rapidly expanding, it can be said that, in a very real sense the ultimate objective of planning for 'backward,' 'lagging' or 'depressed' regions in advanced economies is the creation and maintenance of new employment. In the context of both this observation and the well known increasing relative importance of contact-intensive office activities and services in advanced economies, it becomes apparent that there may well be considerable regional planning implications in what has been said here of the asymmetrical spatial structure of organizations and the complexity of intra-organizational and inter-organizational metropolitan interdependencies. A lengthy exposition on these implications would be out of place in a paper of this nature, particularly since it would require a detailed evaluation of problems pertaining to specialized information circulation. Instead, a series of questions are delineated that I believe are deserving either of extended academic discussion and research, or of direct consideration by politicians and regional planning policy-makers.

How can there be an avoidance of the great intra-organizational and inter-organizational multiplier linkages known to occur from 'backward' and 'depressed' regions consequent to the establishment of organizational production units in those areas? Can these linkages which usually terminate in administration-rich, large metropolitan complexes, be significantly reduced by the creation of intra-organizational interdependencies between the small- or intermediate-size metropolitan centres of the 'backward' and 'depressed' regions in question?

How can places whose economies are dominated by factories or other subordinate units belonging to multilocational organizations be shielded against worker lay-off decisions reached at distant head office metropolitan locations?

Since the absolute number of manufacturing production jobs is decreasing, how can the cities and lesser metropolitan centres of 'backward' and 'depressed' regions obtain organizational units other than 'trickled-down' industrial plants that often only offer a limited number of low-wage jobs and that are frequently short-lived due to demand factors or technological change?

Given that the already uneven spatial distribution of office occupations is a major source of regional economic and social inequalities, what strategy is necessary to counteract the tendency of headquarters units and other contact-intensive organizational activities to concentrate in the major metropolitan complexes of the United States, France, Sweden and other advanced economies (see Armstrong 1972, Beaujeu-Garnier 1974, Engström and Sahlberg 1973, EFTA Economic Development Committee 1974)? In other words, how can a greater regional equality of labour-market conditions be induced through altering, rearranging and manipulating the spatial structure of numerous multilocational organizations?

How can the costs of information gathering and processing be reduced in the metropolitan centres of 'backward' and 'depressed' regions so as to increase their attractiveness to private and public organizational white-collar units that are involved in nonroutine, or nonprogrammed, decision-making as well as routine decision-making and management? What costs and problems are 'decentralized' organizational office units likely to encounter in trying both to maintain established contact relationships and create new ones? (Cf Isard 1969, pp57–115.) In the face of the rapid growth

of inter-regional specialized information exchanges, is the cost attractiveness of 'backward' region metropolitan centres to organizational office units to be enhanced by improving their air connections with metropolitan complexes throughout the national system of cities (Engström and Sahlberg 1973, Törnqvist 1973)?

In a similar vein, how, if at all, can the accessibility of 'backward' region metropolitan centres to increasingly specialized business services be improved so as to magnify their attractiveness to organizational administrative units? Is it necessary to concentrate new or relocated office units from several organizations in order to make business service external economies available? (Cf Goddard 1973, and EFTA Economic Development Committee 1974.) Is office-unit dispersal alternatively possible through the internalization of business services? Or, does the answer lie in encouraging business-service providing firms to come into the metropolitan centres of 'backward' and 'depressed' regions with branch offices that have good intra-organizational communications with their elsewhere located main offices?

Does not the demonstrated ability of organizational headquarters and other high-level administrative units to function in the suburbs of major metropolitan complexes (Manners 1974) suggest that the face-to-face contact advantages traditionally attributed to major city-centre office locations may be somewhat exaggerated, either because of psychological factors, or because simple proximity encourages face-to-face meetings of a routine nature that could readily be replaced by telephone transactions (Goddard 1973)? If this is so, does it not follow that many organizational headquarters and other high-level administrative units can function in peripheral or 'backward' region metropolitan locations? If three large multifunctional organizations with nationally dispersed units and foreign operations can cluster in the Boise City SMSA (1970 population 112 230), with its remote location within the US–Canadian city system, why is not a similar clustering of important management functions possible in one or several more centrally located 'backward' region metropolitan centres?

Will communications satellites, conference-television facilities, and other new telecommunications technology help to facilitate adjustments in the spatial structure of multilocational organizations? Or, will the initial introduction of costly telecommunications innovations in already nationally dominant metropolitan complexes only serve to reinforce existing spatial structures and, consequently, existing intra-organizational and inter-organizational city-system interdependencies?

Finally, there is a single question that summarizes most of the above posed problems. How can explicit and implicit organizational location decisions be influenced or controlled in order to help bring about greater regional equality in terms of per capita income, labour market alternatives and public service accessibility? This is the most important and difficult question of all, for it is undeniable that: 'Public intervention in the locational behaviour of private and corporate enterprise has been viewed traditionally with suspicion and even distaste, in the United States' (Manners 1974, p109).

## Notes

[1] The detailed empirical studies are not reproduced here. See pp 126–36 of the original text.

[2] 'Growth-inducing' innovations fall into three often interdependent broad categories: those that involve the provision of new goods or services; those that involve new production processes; and those that affect the structural relationships, operating procedures or planning and policy-making procedures of organizations.

[3] The 'operational' decisions specified in figure 3.1, as well as any other type of organizational decision involving the allocation of funds, always may be viewed as implicitly locational when they are not explicitly locational. This is so because, whatever the motivation, fund allocations involve the implementation of action in some place(s) as opposed to others.

[4] For simple equations expressing the probabilities involved see Pred (1973a, pp 34–5, p 41).

[5] Industrial corporations include all those corporations deriving 50% or more of their sales from manufacturing and/or mining.

[6] Given the population of Boise, its 1959–73 job-control increase is remarkably high. However, without investing a considerable amount of time in compiling data for other US metropolitan complexes with populations exceeding 500 000, there is no way of precisely determining the representativeness of the remaining data contained in table 3.1. All the same, even if it is conceded that many of the job-control increases recorded for Los Angeles and Phoenix are the result of headquarters location shifts (e.g., the move of Greyhound Corporation from Chicago to Phoenix), other more crude data appearing in *Fortune* (Pred 1974) provide reason to believe that during the 1960s extremely few US metropolitan complexes had a population growth rate that surpassed the job-control growth rate of their multilocational private-sector organizations.

[7] If the Boeing Company's employment is subtracted from the 1959 and 1974 Seattle–Tacoma totals, the resulting figures are 72 332 and 141 211. Thus, exclusive of the Boeing Company, the 1959–74 increase in employees controlled by multilocational business organizations headquartered in Seattle–Tacoma was no less than 95.2%.

[8] There is ample documentation of such recent developments in Sweden (Godlund 1972, Godlund *et al* 1973, Nordström 1974).

[9] Apparently, the greater the volume of information that special-

ized organizational units acquire from the environment, the greater their need to exchange information with their head office or other sister units (Persson 1974).

[10] The position taken in this section represents an elaboration and modification of the stance taken in Pred (1973a, 1974).

[11] In some instances nationally functioning private-sector organizations have some of their units or divisions structured in geographically defined terms and other components or divisions based on functional specialization, with units that are implicitly local, regional or national in purpose.

[12] For comments on the development of multidivisional corporations see Chandler (1962), Blair (1972) and Reid (1960).

[13] See the discussion of headquarters location patterns and other evidence in Pred (1974). Maps of the component units and linkages of individual Swedish business organizations (Godlund *et al* 1973) also succinctly capture the existence of asymmetrical organizational spatial structures.

[14] For other comments on the limits of Löschian central-place theory see Parr (1973).

[15] See original text for full list of references.

## References[15]

Armstrong R B 1972 *The Office Industry: Patterns of Growth and Location* (Cambridge, MA: MIT Press)

Beaujeu-Garnier J 1974 Toward a new equilibrium in France? *Annals of the Association of American Geographers* **64** 113–25

Blair J M 1972 *Economic Concentration: Structure, Behavior and Public Policy* (New York: Harcourt, Brace Jovanovich)

Chandler A D Jr 1962 *Strategy and Structure* (Cambridge, MA: MIT Press)

Chapman K 1974 Corporate systems in the United Kingdom petrochemical industry *Annals of the Association of American Geographers* **64** 126–37

Dun and Bradstreet Inc. 1974 *Million Dollar Directory* New York: 1959–, annual

Dunn E S 1970 A flow network image of urban structures *Urban Studies* **7** 239–58

EFTA Economic Development Committee 1974 *National Settlement Strategies: A Framework for Regional Development* (Geneva: Secretariat of the European Free Trade Association)

Engström M G and Sahlberg B W 1973 *Travel Demand, Transport Systems and Regional Development: Models in Coordinated Planning*, Lund Studies in Geography, Series B, Human Geography **39**

*Fortune* 1974 **84** (5) May

Goddard J B 1973 Office linkages and location: a study of communications and spatial patterns in central London *Progress in Planning* **1** (2)

Godlund S 1972 *Näringsliv och styrcentra, produktutveckling och trygghet: Förändringar beträffande struktur, ägande och sysselsättning inom industrin belysta med exempel från Norrköping* Meddelanden från Göteborges Universitets Geografiska Institutioner, Series B, No. 25 with abridged English translation

Godlund S, Nordström L, Godlund K and Lorentzon S 1973 *Örebro kommun: Utvecklingsmöjligheter och handlingsprogram för en bättre struktur* (Göteborg: Regionkonsult AB) (mimeographed)

Isard W 1969 *General Theory: Social, Political, Economic, and Regional* (Cambridge, MA: MIT Press)

Lorsch J W and Allen S A III 1973 *Managing Diversity and Interdependence: An Organizational Study of Multidivisional Firms* (Boston: Harvard University Graduate School of Business Administration, Division of Research)

Manners G 1974 The office in metropolis: an opportunity for shaping metropolitan America *Economic Geography* **50** 93–110

Nordström L 1974 Mellanregionala beroenden — maktens regionala koncentration *Produktionskostnader och regionala produktionssystem: Bilagedel II till Orter i regional samverkan* Statens offentliga utredningar (SOU) 3 (Stockholm: Allmänna Förlaget) 345–69

Parr J B 1973 Structure and size in the urban system of Lösch *Economic Geography* **49** 185–212

Parsons G F 1972 The giant manufacturing corporations and balanced regional growth *Area* **4** 99–103

Persson C 1974 *Kontaktarbete och framtida lokaliserings förändringer: Modellstudier med tillämpning på statlig förvaltning* Meddelanden från Lunds Universitets Geografiska Institution, Avhandlingar 71

Pred A R 1971 Large-city interdependence and the preelectronic diffusion of innovations in the US *Geographical Analysis* **3** 165–81

—— 1973a The growth and development of systems of cities in advanced economies in Pred A R and Törnqvist G E *Systems of Cities and Information Flows: Two Essays, Lund Studies in Geography, Series B, Human Geography* **38** 1–82

——1973b *Urban Growth and the Circulation of Information: The United States System of Cities, 1790–1840* (Cambridge, MA: Harvard University Press)

——1974 *Major Job-providing Organizations and Systems of Cities*, (Washington, DC: Association of American Geographers, Commission on College Geography)

—— 1975 Diffusion, organizational spatial structure, and city-system development *Economic Geography* **51**

Reid S R 1960 *Mergers, Managers and the Economy* (New York: McGraw-Hill)

Törnqvist G E 1973 Contact requirements and travel facilities: contact models of Sweden and regional development alternatives in the future in Pred A R and Törnqvist G E *Systems of Cities and Information Flows: Two Essays, Lund Studies in Geography, Series B, Human Geography* **38** 83–121

US Bureau of the Census 1971 *1970 Census of Population: Number of Inhabitants, Final Report PC (1)-A1, United States Summary* (Washington, DC: US Government Printing Office)

Williamson O E 1970 *Corporate Control and Business Behavior: An Inquiry into the Effects of Organization Form on Enterprise Behavior* (Englewood Cliffs, NJ: Prentice-Hall)

Wood P A 1969 Industrial location and linkage *Area* **1** 32–9

# 4 The Current Halt in the Metropolitan Phenomenon

*by W Alonso*

The 1970s saw a number of startling societal changes. There was of course, the energy crisis and the changed awareness it symbolized of man's relation to the material environment. Within American society itself, profound and rapid changes were manifested in the role of women as people and as workers, the modes by which individuals organized themselves into families and households, the frequency with which they changed these arrangements and in a lowered rate of reproduction. In our human geography, for the first time we have seen a net migration out of metropolitan areas to more rural ones and, indeed, for the first time an absolute decline in population in a great many metropolitan areas, especially among the largest. These changes are not unrelated. This extract will focus on the changes in the human geography, but also will try to signal a few of the connections to other changes.

## Metropolitan growth and policy debate in the 1960s; changes in the 1970s

During the late 1960s, the national pattern of growth of US metropolitan areas seemed clear and well established. The major metropolitan areas were all growing in population at about the nation's pace, give or take a few per cent, carried by the momentum of their intricate economies. Some of the medium-sized metropolitan areas, especially in the West and Southwest, were growing faster and beginning to take their place as national, rather than merely regional, urban centres. The smaller metropolitan areas, having economies that

were not only smaller but also more specialized in particular activities, showed a much greater variety of growth rates; some were growing very fast indeed and some losing population according to their economic fortunes. Meanwhile, nonmetropolitan areas continued to send their young to the metropolises as they had for the two centuries since the industrial revolution. The South continued to send its youth, both white and black, to the North and increasingly to the West.

In this context, the federal government instituted some programmes in the US departments of agriculture and commerce to promote development in the lagging rural areas and in small nonmetropolitan cities. Many observers, moreover, were concerned not only with the decline of these rural areas but also with the population explosion in the nation and its crowding into urban areas. Influential individuals, organizations, and government officials from the president down worried about such overconcentration ('human anthills' was a common epithet) and called for national policies of population dispersion. The US Department of Housing and Urban Development established a programme to assist the developers of new towns. To these views, many students of urbanization replied that the existing trends were firmly rooted in demographic, economic and technological realities, and that trying to stop the trend was akin to King Canute commanding a stop to the swelling ocean tide.

Those who favoured decentralization policies replied that by the end of the century there would be an additional 100 million Americans, and asked 'Where shall they live?' They pointed to well established policies of population decentralization in the major European countries as possible models for the US. The others replied that the European experience illustrated their point, because after more than two decades of effort

Reprinted by permission of the publisher, from *The Mature Metropolis* edited by Charles L. Leven (Lexington, Mass., Lexington Books, D.C. Heath and Company, copyright 1978 D.C. Heath and Company).

neither France nor Britain had made much progress towards decentralization. Further, those politically to the right viewed the establishment of such a policy as another incursion by government into personal freedom, while those politically to the left viewed it as a form of escapism to avoid facing the cruel difficulties of the problems of the cities, most notably those of poverty and race.

The mid-1970s witnessed some startling changes in the trends in the US territorial development. What some had viewed as inevitable, and others as a dreadful fate which called for vigorous action if it were to be avoided, now seems a very doubtful outcome. These changes are the subject of this extract as they affect the patterns of growth and decline among metropolitan areas and between metropolitan and nonmetropolitan portions of the country.

The changes are many and become fascinating in their richness as one enters into details, but some of the broad outlines may be suggested. Metropolitan population has slowed its growth, and for the first time a great many US metropolitan areas, including many of the largest, are experiencing a decline in population. This applies to entire metropolitan areas, not just to their central cores. Most of the vigorous population growth is taking place in the Sunbelt, and many areas of chronic distress and population loss have turned around and are growing faster than the nation in both respects. Most startling, rural to urban migration has reversed and the balance of migration now goes from metropolitan to nonmetropolitan areas. One may even note that, whereas in the past population was shifting out of the centre of the country and concentrating within fifty miles of the ocean shores, since 1970 even this trend has reversed and the interior of the country is growing faster than its marine periphery (US Bureau of the Census 1976a).

These changes result from a variety of causes, many of which are not yet understood. Moreover, as will become clear upon examination, some of these are not changes at all but continuations of long-standing trends which were not sufficiently recognized, such as basic demographic changes, changes in the location and sectoral composition of economic activity, the rise in the importance of transfer payments and changes in lifestyles.

## The nature of metropolitan decline

Before proceeding to discuss the details of metropolitan decline, some definitional cautions are necessary. Of necessity we perceive complex societal events through the lens of statistical data gathered according to certain conventions and definitions. Therefore, we often err when we assign a commonsense interpretation to these numbers. When we say that a metropolitan area has $X$ number of people, the image leaps to the mind that all of these are urban, and that they live in the central city or in the suburbs; but this is not the case.

The principal data we will use are gathered by SMSAs (standard metropolitan statistical areas), which are essentially defined as a central city of no less than 50 000 population and its surrounding counties if they meet certain criteria of continuity of urbanized development and commuting patterns. The other principal classification of data is by 'nonmetropolitan areas' which are merely the areas which are not included in SMSAs.

While in general, we will equate SMSAs with urban areas or metropolises, and nonmetropolitan areas with rural districts and small towns, the realities encompassed in these definitions are not so simple. For instance, on the whole, only one-tenth of the land within SMSA boundaries is urbanized and the rest is rural. About one-third of the nation's rural population lives within SMSA boundaries and they produce an even larger share of agricultural production. On the other hand, within the nonmetropolitan areas about two-fifths of the population is urban and nearly one-fifth of the country's urban population lives in nonmetropolitan areas. Finally, the image of rural people as agricultural is misleading. Only 4% of the nation's labour force is engaged in agriculture, forestry or fishing; while as of 1970, 26% of the population was rural, indicating that about 85% of the rural population earned its living in pursuits other than agriculture, such as running shops and gas stations, working in factories and teaching in schools and universities.

For convenience we will often use terms such as 'metropolitan' or 'large urban area' for SMSA, and 'rural' for nonmetropolitan area, exercising due care to make clear where other meanings or definitions are involved, but the reader should keep in mind the nature of the underlying statistical bases.

## Metropolitan population decline

It had been rare in the past for metropolitan areas to decline in population. Some of the small ones sometimes have, as a result of local economic adversity, but large urban areas have tended to grow from their own momentum, based on larger, more diversified, more adaptable and more innovative economies. If one economic sector suffered a reverse, another was likely to be growing in compensation (Thompson 1965). Indeed, a common pattern for areas such as Boston and New York was to specialize in innovation in electronics, fashion apparel, professional services and the like. When these new activities matured, they tended to move to other areas, while the metropolis developed new activities. Such metropolises, while always losing industries to others, based much of their economic vitality on this seedbed function (Hoover and Vernon 1960). Hence, all of the large metropolitan areas tended to grow at rates that did not differ markedly from the nation's. Only Pittsburgh among the large urban areas lost population in the 1960s, and it is notable for being a metropolis specializing in traditional industry.

Since 1970, however, the number of metropolitan areas which are losing population has increased markedly. One-third of the nation's metropolitan population now lives in areas which experienced population decline in the 1970–4 period. Overall, one-sixth of all metropolitan areas lost population from 1970 to 1974; in the last year of that period the proportion increased to over one-fourth, so that the trend was accelerating.[1]

History has conditioned us to take metropolitan population increase for granted, but by now, in terms of the number of people exposed, metropolitan population decline is as common as growth. The consequences of this novel condition are hard to anticipate, both because it has been a rare one in the past and because it was unforeseen and so little thought has been devoted to it. Neither formal theories of academics nor practical rules for action of worldly men are at hand to deal with it. Both theorists and doers have assumed until now that the arrow of time always pointed to increasing numbers. While population growth will still be the common case in terms of numbers of areas until the turn of the century, more and more people and metropolises will find the arrow pointing the other way.

Many have associated the phenomenon of decline in the rate of metropolitan population increase with the unexpected reversal of rural to urban migratory flows which has occurred since 1970. While this reversal is both startling and important, it should not be thought of as the sole, or even the major, reason for the decline in metropolitan population growth; there are two other more important reasons. One, the more powerful, is the decline in the national rate of population growth.

The other is the continuing flow of people from one metropolitan area to another, which redistributes population among metropolitan areas to the gain of some and the loss of others.

*Decline in birthrate*

The principal reason for the decline in metropolitan population growth is simple: the rate of natural increase has plummeted. It is still positive on the whole, but a great many metropolitan areas now have yearly rates of natural increase of only one- or two-tenths of 1%. Three metropolitan areas actually had fewer births than deaths. Two of these (Sarasota and Tampa–St. Petersburg) are in Florida and are in fact growing briskly through immigration. But many of these migrants are old, and have few babies and high mortality. The third is more typical of what will become more common; it is northeastern Pennsylvania, and encompasses Scranton and Wilkes-Barre-Hazelton. Here persistent out-migration has reduced the number of fertile young people and left behind an aged population. Ironically, some of the economic changes described later have resulted in a small net immigration in the past few years, so that this area has grown slightly in population (US Bureau of the Census 1976b).

When very small rates of natural increase, to say nothing of decline, are coupled with continuing differential exchanges among metropolises, it is a mathematical certainty that many of the metropolitan areas will experience population losses, even in the absence of the reversal of rural to urban migration.

The decline in the natural increase of the US population has begun to be noticed by the public and termed the 'baby bust' by contrast to the 'baby boom' which followed World War II; but the magnitude of this decline has not sunk in. The crude birthrate now stands below fifteen yearly babies per thousand population, a historic low. By comparison, the birthrate at the top of the baby boom in the later 1950s was over twenty-five per thousand. Thus, taking account of deaths, the natural increase in the US population, which had reached 1.6% per year, stands now at less than 0.6%.

This drop in the crude birthrate is dramatic in itself, but a further look suggests that the intrinsic change is even greater. The crude birthrate is a measure of the current rate of reproduction, but it can mislead as to long-term trends, because there can be temporary increases in the numbers of potential mothers, as is the case now as a result of the maturing of the children of the baby boom.

Let us look at the total fertility rate. It is the lifetime number of children per woman which would be expected if women continued to have children at current rates for particular ages. A rate of 2.1 children per woman would replace the population exactly in the long run. The fertility rate was as high as 3.7 in the late 1950s, but now stands under 1.8, well below replacement level.

Further, it seems plausible that 1.8 may be an overestimate of the actual number of children per woman which will be borne by today's young women during the course of their lives. The manner of computing this rate attributes in the future, rates of childbearing to today's young women which are based on today's rates for older women. But younger women today are remaining unmarried in larger numbers, and the intentions of those who are married as to how many children they intend to bear in total show a strong decline. It is reasonable, therefore, to expect that the total number of births will continue to decline in the coming years since there will be fewer women of prime childbearing age and they appear to want fewer children.[2]

In short, given a slight decrease in the death rate and a sharp decrease in the birthrate, the rate of natural increase is only slightly above one-third of what it was in the late 1950s, and bids fair to continue to decline. As the benchmark of natural increase comes closer to zero, metropolitan areas as a whole will grow more slowly, and the influence of forces which differentiate among them will ensure that negative population growth will be commonplace.

In other words, the experience of continued population growth in the past was based on a vigorous rate of natural increase and on a steady stream of rural to urban migrants. By the 1960s, however, the migration rate into metropolitan areas was small, and threequarters of metropolitan population growth was based on natural increase, and only one-ninth on migration from nonmetropolitan areas, the balance resulting from immigration from abroad. Now the decline in the rate of natural increase has cut the growth rate sharply, and this has been accented by the reversal of net migration into nonmetropolitan areas. Further, now that population growth for all metropolitan areas is so small, the migratory streams along metropolitan areas add up to a zero-sum game, driving many areas into the category of population losers.

**Migration into, out of and among metropolitan areas**
Out-migration from metropolitan areas was already a common experience in the 1960s, when nearly 40% of SMSAs had more people leaving than arriving. The percentage of metropolitan areas with net out-migration rose to 44% by 1974, as did the overall rate of outmigration. But whereas the high rates of natural increase in the 1960s masked such out-migration, the current low ones reveal it as a receding tide bares the rocks along a shore.

Where do the people who leave metropolitan areas go? Many would have it that they are going back to rural America, and this is partially true. But there are rich and complex crosscurrents of people of different kinds doing different things, and the resulting pattern is a complicated one. Some of what is happening may be gleaned from figure 4.1 which shows the broad outlines of population movements in the United States in the period from March 1975 to March 1976.

Note that 2 477 000 people left metropolitan areas for nonmetropolitan destinations, and that 2 081 000 went in the other direction, for a net balance of 396 000 in the year in favour of nonmetropolitan areas. But note that total departures from metropolitan areas amounted to 6 877 000, of which 4 400 000 (64%) went to other metropolitan areas. By comparison, out of 4 619 000 departures from nonmetropolitan counties, 2 538 000 (55%) went to other nonmetropolitan areas.

Thus, taking account of the respective metropolitan and nonmetropolitan populations, in that year 1.8% of the metropolitan population moved to nonmetropolitan areas while 3% of the nonmetropolitan population moved into metropolitan areas. Simply put, the chances that a nonmetropolitan person will move to a metropolitan area are 1.7 times higher than they are for the reverse move. On the whole, then, the rate of metropolitanization of population is higher than that of demetropolitanization, but the absolute magnitudes of their bases result in a larger absolute flow out of metropolitan areas. It is thus a matter of viewpoint as to which way people are moving: in absolute terms they are moving out of metropolitan areas; on a proportional basis, the ordering is reversed.

*Migration out of metropolitan areas*
Yet the very fact that there is net out-migration from metropolitan to nonmetropolitan areas is so surprisingly contrary to two centuries of experience in the US and in Europe, that it has mobilized a great deal of

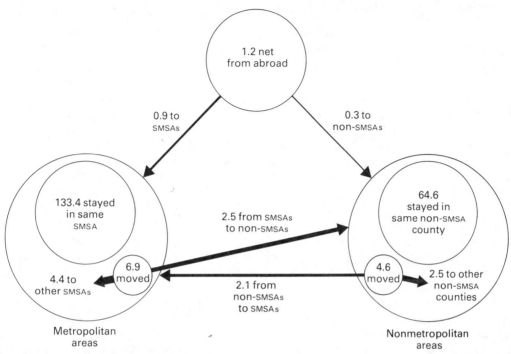

**Figure 4.1**  Population movements (in millions) in the United States, March 1975 to March 1976. Computed and adapted from the US Bureau of the Census, *Current Population Reports* Series no 305 p20 'Geographical Mobility: March 1975 to March 1976', Washington: Government Printing Office 1977.

recent research and even found its way repeatedly into the popular press. Some hold that a new millenium has arrived, a profound change in consciousness among millions of Americans who, tired of the pressures, pollution and crime of the cities, alienated by the corruption of our institutions and imbued with a new land ethic, leave the metropolis for rural areas in search of a simpler, saner, healthier life.

The explanation for the reversal seems to lie in the coincidence of several distinct effects, which include the geographic extension of the functional field of urban areas beyond the officially defined SMSA, shifts in the location of manufacturing industry, the continued expansion of the recreation and resource industries, the greater numbers of retired people and the recent recession. It is not clear, however, how much each of these contributes in diverse nonmetropolitan areas.

To some degree, the growth of nonmetropolitan areas is a continuation of the extension of the functional geographic range of urban areas. The eighteenth- and nineteenth-century cities were extremely compact, based principally on pedestrian ranges of movement. With improvements in transportation, principally the private automobile, the effective range of movement was vastly expanded, giving rise to massive suburban-

ization and to the metropolitan phenomenon. But, especially since World War II, it is not only commuters who have moved to the suburbs, but shops and industry; manufacturing led the way and now the service sectors are busily suburbanizing. In consequence, only about a fifth of suburbanites now commute to central cities, and the preponderant majority lives and works in the suburbs. The metropolitan phenomenon, then, consists not only of the diffusion of workers' homes but of the formation of strong suburban nuclei for services and employment. Thus, it is possible, and does happen, that people live beyond the suburbs of a metropolitan area (i.e. in a nonmetropolitan area) and commute to the suburbs of the nearest SMSA. Indeed, employment and services, by the same linked processes, can also move beyond these boundaries. These expanding urban fields exceed the boundaries of the SMSA, but are functionally part of the same urban system.

Current estimates indicate that over 60% of metropolitan to nonmetropolitan migrants move into counties immediately adjacent to SMSAs. The 1970 census classified nonmetropolitan counties by the percentage of their workforce which commuted to a metropolitan area. The data show a consistent pattern: counties in which over 20% of the workforce commuted to metropolitan areas in 1970 had a yearly net immigration since 1970 of 1.5%; counties with 10–20% commuters had a rate of 0.8%; those with 3–10% had a rate of 0.7%; and those with less than 3% commuters had an immigration rate of 0.5%.[3] It is clear that the expansion of urban fields, washing over earlier censual metropolitan definitions, is a large part of the apparent reversal. To a degree, what is happening is not so much that people are leaving the city for the country as that the city itself is moving to the country. It is worth recalling Wirth's (1956) aphorism that 'urbanism is a way of life'.

This expansion of urban fields, in a way, the strong tail of a movement from greater to lesser concentrations. Out-migration and population decline are more prevalent among the larger metropolitan areas, while as a group those with populations under three-quarters of a million are still growing substantially, with a net in-migration. The expansion into the nonmetropolitan periphery appears as another form of this general process of population and economic diffusion.

But this is not the entire explanation. Counties well beyond any metropolitan field are, for the first time in memory, experiencing net in-migration. This is true for every region of the country with the exception of the old Tobacco-belt and Cottonbelt extending from the North Carolina Cape to the delta of the Mississippi River.

One principal reason has been the continuing shift of manufacturing production out of metropolitan areas to small cities and rural areas. For instance, from 1969 to 1973 personal income in durable manufacturing increased by 46% in nonmetropolitan areas in comparison to a 25% increase in metropolitan areas; comparable figures for nondurable manufacturing are 33% and 24% (Bureau of Economic Analysis 1976). This follows a long-standing process whereby production manufacturing has been moving from large to small areas, down the hierarchy of sizes of metropolitan areas and into nonmetropolitan areas. Neither has this process stopped here; this flow of manufacturing production from centre to periphery has extended beyond the national borders to the northern Mexican cities, to Taiwan and Korea and to other countries of low labour costs.

Nonmetropolitan areas outgained metropolitan areas in almost every category of source of income, especially in goods-producing industries; that is, manufacturing, mining and contract construction. Also, as a consequence of their new economic and demographic

vigour, finance, insurance and real estate (FIRE) are growing very rapidly in these areas.

Recreation also has flourished in nonmetropolitan areas, often in association with an influx of retired people who not only enjoy some of these facilities but often supplement their retirement income by running them. The increased importance of transfer payments, especially to the aged, has made them more footloose. Whether on social security or on private pensions, a growing number of older people are increasingly able to move, since their income is not tied to their location; from 1970 to 1975 nearly half a million left metropolitan areas (US Bureau of the Census 1975). Some pulled up stakes and went to retirement communities in the Sunbelt, others returned to places where their roots were, so that some of this migration is, in effect, the echo four decades later of earlier rural to urban migrations.

No up-to-date information appears to be available on the expansion of non-SMSA employment linked to resource development and environmental preservation, but it is substantial. This includes coal mining and oil exploration, associated construction and services, water development, land reclamation, construction of atomic facilities and other energy-related activities, sewage treatment facilities and others. Much of this activity is distinct from other bases for the rebirth of population in rural areas, because it lasts at high levels only for the period of construction, involves to a large degree a population for whom moving from place to place is a way of life and in the end it leaves the local area. Although it may move on to another nonmetropolitan district, it results in cycles of local boom and bust, with the attendant hardships for the original and remaining population.

Another cause for rural population rebirth appears to be the current recession. If times are hard, many earlier migrants to metropolitan areas go back home, where it is easier to manage and where friends and relatives can help. There is anecdotal evidence that this is happening, and it is suggestive that the only other period when there was something like a reversal of migratory flows was during the Great Depression, but no figures are available to determine its importance. Although the economic cycle may be playing a part, it must be remembered that the reversal had begun by 1970, when economic times were good.

Finally, just to be clear about it, one thing that is *not* happening is a rediscovery of the joys of working the land and a new ethic leading the young to agriculture.

From 1970 to 1976 the farm population declined by 1.4 million or 14%, much of the drop taking place in 1975–6 (US Bureau of the Census and US Department of Agriculture 1977).

It must be remembered that national trends are manifested quite differently in different places. A particular area in the South may be seeing labour-intensive industry moving into town, absorbing some of the workers released from agriculture. In another, some whites and blacks who had moved North may be returning, some because they are now retired and find the living cheaper and more congenial where they grew up, and some because the social and economic progress of the South now makes it an acceptable alternative for them. In another area, possibly in the Great Plains, coal mining may have brought boom conditions to a tiny town, with workers moving in, living in trailers and bringing with them shopkeepers and entrepreneurs of diverse sorts. In yet other areas, the basis of growth may be a new water-based resort, a think tank, an expanded state university campus, a defence installation or an environmental project. And, of course, a fair number of nonmetropolitan counties are still declining.

The available statistics on the characteristics of the migrants support the picture outlined as to the economic bases for the revival of nonmetropolitan population growth. Those moving from the metropolises to rural areas are younger, better educated and more skilled on the average than present residents of those areas. But they are older, less educated and less occupationally skilled than those leaving the rural areas for the metropolis. Thus, though 100 people arrive in nonmetropolitan areas for every 84 who leave, it appears that the nonmetropolitan areas are still losing qualitatively, sending somewhat better educated and occupationally qualified youth to the metropolitan areas than they are receiving.

Finally, as in most American social phenomena, race must be taken into account. Although for the first time there are important return flows among blacks, on the whole their main direction is still toward metropolitan areas; for every 100 blacks who left metropolitan areas, 142 entered them.

*Migration among metropolitan areas*
As has been noted, migration *among* metropolitan areas is an important factor in their differential growth. Returning to figure 4.1, note that the number of moves among metropolitan areas was only a shade lower than the number of moves between metropolitan and non-

metropolitan areas. These flows involve a great many moves among pairs of SMSAs, with some coming and some going. The net effect on differential growth, therefore, for the year 1975–6 is far less than the total 4 400 000 intermetropolitan moves. Direct data are not available, but, based on past experience, the contribution to differential metropolitan growth would be in the order of 350 000 to 400 000 net migrations. This figure is comparable to the net loss of 396 000 migrants from metropolitan to nonmetropolitan areas in that year.

It is well known that the SMSAs that are the principal gainers in these exchanges are in the South and West. Of the thirteen SMSAs growing fastest in percentage terms from 1970 to 1974, seven were in Florida, and two each in Colorado, Arizona and Texas.

No simple generalizations can be made as to the economic bases of the growth of the so-called Sunbelt metropolises. It probably is true that many Americans want a better climate and a new way of life, but except for those who are retired or of independent means this is not enough. Jobs and businesses are needed, and, indeed, for the enjoyment of the climate so is air-conditioning, which is becoming expensive.

Some of the economic bases for the observed migration shifts are apparent. Retirement living based on transfer payments is part of the base in Florida, together with the influx and dynamism of the Cuban refugee population. Oil and gas, both as an industry and as a form of Ricardian wealth, is part of the story in Texas. Military establishments and procurements are an important component in California and in Washington. The regional centre function is important to certain metropolises of the South which are situated in regions of rapid economic growth, just as the decline of these functions accounts for the decline in growth of some of the metropolises in California's Central Valley and parts of Texas.

While no comprehensive analysis of the sectoral components of the growth of the Sunbelt is now available, it appears that the growth of its metropolises is broadly based on manufacturing services, growing in part because of competitive shifts from economic activities based in older areas and in part from specialization in economic activities which are themselves rapidly growing (Zuiches and Brown 1978).

In a very general way, part of the Sunbelt phenomenon rests on the shift to the 'service', 'post-industrial' or 'information' economy. The extraordinary advances in the transmission of information together with the implementation of standard routines for its handling, which render it subject to economies of scale and the use of semi-skilled labour, make it possible to locate an increasing number of economic activities where they are wanted, rather than where they are forced to be. It must be noted, on the other hand, that Sunbelt metropolises are more than holding their own on manufacturing employment in comparison with older urban centres.

Is it then a coincidence of factors for Florida, the South, the Mountain and Pacific States that creates the phenomenon of the Sunbelt? This seems unlikely, and it is probable that some few underlying social and economic tendencies are at the root. They may be as simple as the preference for climate and life-style, aided by technology and sectoral evolution, but there is more to it, though no grand synthesis is at hand.

In a sense, it appears that what is going on is a very long-run equilibration of the national distribution of urban centres, still trying to rectify the original mistake made by the first British settlers when they landed on the upper-right-hand corner of the US map. It is worth noting, for instance, that the growth of many of today's Sunbelt centres is not a recent phenomenon at all, but one that goes back at least to the turn of the century or earlier. It was not particularly noticed until now, because the high rates of natural increase everywhere masked this differential growth based on intermetropolitan migration.

*Immigration from abroad*
Figure 4.1 shows immigration from abroad of 1 148 000 for the year 1975–6. This must be compared to a reported net civilian immigration of 450 000 in 1975, a year when this figure was swollen by Vietnamese refugees. The difference may be accounted for by returning servicemen, businessmen and others; and some comparable number may have left our shores to take up temporary or permanent residence abroad. The census publications are singularly silent on the interpretation of this figure, yet it is not trivial since it amounts to more than the sum of the net metropolitan to nonmetropolitan migration and the net intermetropolitan exchange.

Immigration from abroad is the $X$-rated statistic in American population figures, primarily because it is terribly important and because no one knows much about it; as a statistical as well as a policy issue, it is avoided like a hot potato. It is unlikely that the number shown in figure 4.1 is accurate; the true number almost certainly is higher. It appears that there is a great deal of illegal immigration into the United States both permanent and seasonal, but it is quite unlikely that such illegal migrants would have been captured in the sample in the first place or would have replied truthfully if asked.

Various estimates place the number of illegal residents in the United States in the order of 5 to 10 million, and some go higher. Estimates in New York City alone have placed the number of illegal aliens living there at between 0.5 and 2 million. The whole matter of estimation in this case is a mare's nest of special interests, which include the politics of ethnic groups, the interests of labour and of those who hire illegal aliens at cheap wages and the vital issues of revenue sharing and other modes of distributions of money based on population estimates.

Obviously, if these estimates are anywhere near the truth, the population decline of New York is exaggerated and the population growth of the Southwest is understated. The magnitude of the numbers involved is comparable to that of the natural increase of the population. Therefore, if anything like the number of illegal migrants estimated is a reality, some important aspects of the redistribution of the country's population must be reconsidered. We cannot here resolve these uncertainties, or make up for their studious neglect. They remain a blind spot which is certain to result in errors and inconsistencies in data, in wrong diagnoses for policy and in political equivocation over figures for dollars.

## Metropolitan population decline: its sources and meaning

We have said above that there are three principal sources of metropolitan population decline. The first two affect the overall rate of population growth of metropolitan areas. The larger of these effects is the decline in the birthrate and the other is the reversal of direction in the net flow of metropolitan to nonmetropolitan migration. The third force, somewhat smaller than the others, is that of differential growth from intermetropolitan migration, which redistributes metropolitan population in a zero-sum game of losers and winners.

The importance of the decline in the birthrate is not merely statistical. It has not declined by chance, but rather as part of an interconnected set of social developments. These include the sociology and psychology of the redefined sexual roles; the increase in abour force participation of women (which raise the direct and opportunity costs of giving birth to and rearing a child); the social and personal acceptance of the mutability of family and household relationships which is manifested in more people living alone, or in various combinations of single individuals; in a soaring divorce rate and the rising number of single-parent households; and in dozens of other ways which we all experience, some subtle and some brutally forceful. These demographic changes are in turn associated with rising levels of education, with the expansion of the service sector of the economy (which employs more women), with changes in technology which range from contraception to computers and communication satellites and, in brief, with the ongoing evolution of modern society.

Once one grasps that the experience of metropolitan decline is linked mostly to the decline in the birthrate, and that this is not a matter of fashion or random fluctuation but one of structured relations to other societal changes, then the phenomenon of decline takes on another meaning, distinct from that of the 1950s and 1960s, when it only took place through major out-migrations resulting from severe economic distress (Alonso and Medrich 1972). Population stability and decline are now part and parcel of our social change; while they may not immutably be so forever, they are likely to be with us for some while.

For instance, population decline may be consistent with increasing numbers of workers, because more women are working. It is consistent with a lower dependency ratio, both because there are more workers and because there are fewer children, though there may be more older people. It may mean more money per person for the population at large and for particular people. For instance, a man who supports a wife and three children on a $25 000 a year salary results in a household with income of $5 000 per capita. On the other hand, two adults living together on salaries of $11 000 and $9000, respectively, result in a household income of $10 000 per capita. Population decline may be consistent with lowered public expenditures for schools and urban infrastructure, though it must be admitted that institutional pressures thus far have resulted in greater per capita costs for declining areas. It may mean an easier housing market, more room and more freedom, less pressure for development, and more chances to adapt; I know of no reason why less cannot be more for urban areas.

Of course, old industrial plants and established bureaucracies can result in fiscal pressures, and this seems to be what is happening to declining urban areas, so that they are today in greater fiscal difficulties than newer ones; but these may be shrinking pains. Consider, for instance, that crime, which is mostly committed by the young, is bound to go down with the proportion of those who tend to commit it. Indeed, a characteristic of metropolitan areas which lost population during the 1960s (Rust 1975) was their reduced crime rate.

Consider in particular the relation of population decline to the housing market. At first it might appear that the inevitable consequence of decline would be a softer market, and a sharp decline in construction. While it is obvious that, all other things being equal, a declining population will demand less housing than a growing one, this perspective is altered by the fact that the decline results from social changes of which the drop in the birthrate is only one manifestation. Another result is the sharp drop in average household size from 3.3 persons in 1960 to 3.1 persons in 1970 and 2.9 persons in 1976, so that a stable population would have grown by 8% in the number of households just from 1970 to 1976.

The decline of household size results in part from there being fewer children in homes, but in large part it results from more adults establishing their own households. This includes more young people leaving home, more people remaining single and more older people and divorced people maintaining their own households. All of these are manifestations of the same social transformation. In turn, smaller households result in a higher consumption of housing space per person. Even one-person and two-person households need a bathroom, a kitchen, corridor space, and so forth, and household space increases less than in proportion to the number of people in the household.

The increasing number of households and the increased space used per capita more than compensate for current rates of population decline and maintain the demand for housing. Of course, there are other effects from these changes, but this is not the place to discuss them (Alonso 1977).

To re-emphasize the general point being made, it is not as if the current halt in the growth of metropolitan population were occurring by chance in the social and economic context of the 1960s and 1970s. The decline is a manifestation of the changes in the social and economic context, and its significance and consequences can only be assayed properly when viewed in this context. Otherwise, there will be misdiagnoses from the application of sterotypes and pernicious remedies will be attempted. As mentioned at the beginning of this extract there has been precious little experience of, or thought devoted to, the phenomenon of population decline, although it is an important concern because it is already upon us.

## Nonmetropolitan growth: how permanent is the reversal?

Is the rebirth of nonmetropolitan America an episode, a temporary quirk, or is it here to stay? If the observations in this extract are accurate, it is likely to continue, if for no other reason than the continued extension of the urban fields of SMSAs beyond their defined boundaries. This diffusion of the functioning urban areas is the natural continuation of the spread from the nineteenth-century central city, fostered by improvements in transportation and communications.

Other factors present a mixed picture. The number of retired people is certain to increase and should continue to contribute to the strength of nonmetropolitan areas. Moreover, if there is a nationalization of welfare, as seems likely, this important form of transfer payment might make it still easier for its recipients to choose nonmetropolitan locations.

The exurban movement of labour-intensive manufacturing may go somewhat further, but its rate of increase may have crested. The number of production workers in manufacturing has increased only slightly since the late 1940s, and the metropolitan areas have already lost much of what they had to lose in this respect. It seems likely that, barring continued economic hardship, recreation will keep growing as an industry. It is hard to judge whether there will be increases in nonmetropolitan employment arising from mining, energy, environmental and resource improvements, and the associated construction; it seems possible, but not likely, to be large given the capital intensity of these activities. Employment in agriculture is virtually certain to continue to decline.

As we come out of the recession, some of the return migration associated with hard times will reverse again, and some more young residents of nonmetropolitan areas will try their fortune in metropolitan areas. Several of the factors noted above may be modified by the manner of our adaptation to the energy stringency. The expansion of the urban field is based in part on the cost and ease of moving people and goods; the recreation industry is in part based on the facility of travel, whether by car or by plane. And, overall, consumption of petrol for road use is very substantially higher on a per capita basis in rural areas, most probably because of the large distances involved in travel to work, shopping and services, and because many of the vehicles need the capacity to carry goods as well as people. Insofar as the energy crisis has effects such as a substantial increase in the price of petrol, it would disadvantage continued metropolitan sprawl, both through price effects which would lead to concentration in an effort to minimize distances and through an income effect because the use of petrol is so large that price increases would in fact reduce income available for other purposes in nonmetropolitan areas, and increase the effective difference in income in favour of metropolitan areas.

In the balance, it seems that for the 1980s, statistics for SMSAs and non-SMSAs will show a pattern not unlike that seen now. Whether this is interpreted as a rebirth of rural America or as the diffusion of urban areas over the countryside is a matter of degree and opinion, but the latter view appears more likely.

## Notes

[1] Computed from US Bureau of the Census (1976b).
[2] These matters are discussed in greater detail by Alonso (1977).
[3] The following section relies primarily on the data of Beale (1975, 1976); Berry and Dahman (1977); Morrison and Wheeler (1976); Zuiches and Brown (1978).

## References

Alonso W 1977 The population factor and urban structure *Harvard University, Center for Population Studies, Working Paper No.* 02

Alonso W and Medrich E 1972 Spontaneous growth centers in twentieth century American urbanization *Growth Centers in Regional Economic Development* ed. N M Hansen (New York: Free Press)

Beale C 1975 The revival of population growth in nonmetropolitan America *Economic Research Service, US Department of Agriculture Report No.* ERS 605

——1976 A further look at nonmetropolitan population growth since 1970 *American Journal of Agriculture Economics* December

Berry B and Dahman D 1977 *Population Redistribution in the United States in the 1970s* (Washington, DC: National Academy of Sciences)

Bureau of Economic Analysis, US Department of Commerce 1976 Metropolitan and nonmetropolitan non-farm personal income: growth patterns, 1969–73 *Survey of Current Business* December

Hoover E M and Vernon R 1960 *Anatomy of Metropolis* (Cambridge, MA: Harvard University Press)

Morrison P and Wheeler J 1976 Rural renaissance in America: the revival of population growth in remote areas *Population Bulletin* 31

(3) (Washington, DC: Population Reference Bureau, Inc.)

Rust E 1975 *No Growth: Impact on Metropolitan Areas* (Lexington, MA: Lexington Books)

Thompson W 1965 *A Preface to Urban Economics* (Baltimore: Johns Hopkins)

US Bureau of the Census 1975 Mobility of the population of the United States, March 1970 to March 1975 *Ser. P-20 No. 285* (Washington, DC: Government Printing Office)

—— 1976a *Statistical Abstract* (Washington, DC: Government Printing Office)

—— 1976b Estimates of the population of metropolitan areas, 1973 and 1974, and components of change since 1970 *Ser. P-25 No. 618* (Washington, DC: Government Printing Office)

—— 1977 Geographical mobility: March 1975 to March 1976 *Current Population Reports Ser. P-20 No. 305* (Washington, DC: Government Printing Office)

US Bureau of the Census and US Department of Agriculture 1977 and *New York Times* April 15

Wirth L 1956 *Community Life and Social Policy* (Chicago: Chicago University Press)

Zuiches J and Brown D (1978) The changing character of the nonmetropolitan population, 1950–75 *Rural Society in the United States — Current Trends* ed. T Ford (Ames, Iowa: Iowa State University Press)

# 5 Capital and Locational Change: the UK Electrical Engineering and Electronics Industries

*by D Massey*

## Introduction

British capital was hit early and severely by the world economic crisis. By the late 1950s the rate of profit was already beginning to decline steeply. In the face of this, UK capital had to attempt to raise the rate of exploitation, in order to raise its rate of profit and to stave off realization problems by increasing its share of declining world markets. Since the mid-1960s, the combined effects of the problem and of its attempted resolution have produced a considerable reorganization of British capital, a dramatic upward shift in the national level of unemployment, and a fall in the number of people employed in manufacturing. However, and not surprisingly (since most capitalist countries are pursuing the same course) neither of these twin and related aims, of increasing the rate of profit and increasing the degree of international competitiveness, has been achieved. The processes discussed in this paper are thus likely to continue for some time yet.

One response to this situation, and in many cases a necessary condition for reorganizing productive capital, has been restructuring. By 'restructuring' is meant the reorganization of the ownership of capital, primarily through processes of centralization. Such processes, as Marx pointed out, are typically reinforced during periods of crisis.

A significant feature of the crisis has been the important role played by the State. The Industrial Reorganization Corporation (IRC) which lasted from 1966 to 1970 was one arm of this general intervention, with the explicit aim of restructuring crucial sectors of the economy. Some aspects of its interpretation of the term 'crucial' will emerge in the discussion in this paper. Certainly, potential for internationalization and for increasing the rate of exploitation were central.[1] The cases of restructuring examined here were all carried out in a general sense under this arm of Government policy. They included mergers of whole companies, the reorganization of branches of production through the amalgamation into new companies of overlapping product ranges from a number of individual capitals, and the provision by the IRC of additional money capital for expansion of production. The IRC itself selected major branches of industry on which to concentrate, and the research presented here has focused on one of its early-established priority areas — electrical engineering and electronics.[2] In 1966, there were 1 911 000 people employed in the sector, representing about 14% of all the workforce in all manufacturing industries (data from the Department of Employment *Gazette*, October 1975). According to the Census of Production, the sector accounted in 1968 for 10% of net manufacturing output.

The decade of crisis in the UK has also been marked by significant changes in the geographical distribution of economic activity, changes which are paralleled to differing degrees in a number of countries of western Europe and in Canada and the US. Figure 5.1 indicates the basic economic geography of the UK which is relevant to these changes. Firstly, the major conurbations marked have all been losing manufacturing employment at a rapid rate throughout the period. Secondly, there are differences between development areas and non-development areas. Development areas

Source: Massey D 1978 *Review of Radical Political Economies*, Fall, pp 39–543.

Glasgow

Newcastle

Manchester
Liverpool
Birmingham

London

Development areas

**Figure 5.1** Development areas in the United Kingdom (1966).

are those which were assisted under the regional policies of the period.[3] In them, grants were available to manufacturing industry both on new capital investment and on all employment. It should be stressed that, in contrast perhaps to the Sunbelt states of the US or the predominantly agricultural periphery of France and Italy, these assisted areas are not on the whole non-industrialized. Indeed, as the map indicates, they include some of the major industrial conurbations. The development areas are defined primarily in terms of their unemployment rates, though this criterion is used in a fairly broad way. Since the war, the difference in unemployment rate between the present development areas and the rest of the country has fluctuated, but it has remained. Between 1966 and 1976 a shift in this pattern became apparent, with the manufacturing employment shares of assisted areas noticeably increasing, and those of the non-assisted areas declining. Kee-

ble (1977) gives the following figures for those regions for which data is available:

*Change in Manufacturing Employment Share, 1965–1975, in % Points*

| | | |
|---|---|---|
| South East | non-assisted | −2.37 |
| West Midlands | areas | −0.21 |
| Scotland | Development | +0.04 |
| Wales | areas | +0.60 |
| North | | +0.70 |

This 'convergence' is a new phenomenon, and specifically it is one which runs counter to the common assumption that in periods when the national economy is 'in difficulty' regional differentials widen.

The concern of this paper is to follow up this crucial element of retrenchment, to analyze the forms taken by the restructuring process at the level of individual capitals and, the central theme, to assess its impact in terms of the spatial distribution of employment. The results presented in this paper indicate some of the mechanisms through which the restructuring of capital in the face of international crisis conditions is contributing to both these aspects of geographical change.[4] The analysis is also presented to make a number of more theoretical points. Firstly, it serves to emphasize the link between national economic developments and intranational spatial employment patterns — to emphasize that regional employment changes must be understood in terms of an analysis of the economy as a whole. Secondly, the results indicate that a number of well recognized spatial changes are themselves closely bound up with the effects of the present industrial retrenchment.

The approach taken is to examine individual firms' responses to economic problems over the period 1968–72. This derives from one aim of the original research, which was to explain the locational behaviour of individual firms in relation to the development of the economy. In the established approach of industrial location theory, based on neoclassical economics, this relationship is rarely treated. At most, the structure within which the firms are operating is included merely as some vaguely conceptualized 'content', and certainly not as an integral part of the explanatory framework. But if neoclassical location theory has difficulty in moving from the individual firm to the structure, Marxist work, though correctly starting from the level of the structure as a whole, and from the overall process of accumulation, sometimes appears to neglect examination of the form taken by those structural movements at a more disaggregated level.

The research presented here has attempted firstly to look at the very different ways in which the crisis may be articulated in different sectors of the economy, and secondly to analyze, in relation to this, the response of individual capitals.[5] This process of moving beyond the level of industrial capital as a whole is essential to any political understanding of the present situation. Finally, examining spatial changes in this manner enables one to go beyond the frequent rather evocative references to 'uneven development' (which, while recognizing the fact of regional differentiation, remain at the level of description) to analyze the operation of some mechanisms which produce that differentiation.

Data were collected both from published sources and from a long series of detailed interviews with the firms concerned.[6] The firms in the survey (which included all those in the above sectors which operated in any way under the aegis of the IRC) accounted for about 20% of the sectoral employment at national level. This coverage was unevenly distributed, however, many specific product groups within these sectors having played little part in the processes we are examining. In those which are included, therefore, coverage is well above this figure, and in many product groups is 100%.

The research was structured around a series of questions, the sequence of which reflected the main directions of causality postulated and which incorporated, in a way traditional theory does not (indeed cannot), both the link between individual firm behaviour and the wider economic structure, and the centrality of the process of production. After the characterization of the general economic situation, the next stage analyzed the specific way in which the crisis was articulated in each case. This gave us, therefore, the particular reasons for the restructuring which did take place. The third stage of the analysis examined the forms of reorganization of the production and labour processes which were consequent upon/enabled by the restructuring. Finally, the last stage was concerned with the spatial impact of this reorganization primarily in terms of broad patterns of employment and unemployment.[7]

The analysis revealed the existence of three distinct groups amongst the cases of restructuring examined; groups which were distinct in terms of each of the stages of analysis: their relation to the overall crisis, their forms of production reorganization, and the spatial impact of reorganization.[8] This classification does not simply follow a sectoral pattern (partly because of differences within sectors in the effects of the crisis and partly because of the different situations of specific individual capitals); however, the breakdown is roughly on the following lines:[9]

Group 1: heavy electrical engineering, supertension cables, aerospace equipment;
Group 2: electronic capital goods;
Group 3: individual capitals from all sectors, but involved in specific conditions in relation to the international market.

In this paper, the next three sections deal with the reasons for restructuring and the form it took, in terms of the sector of product group which was dominant (because of the nature of the effects of the crisis on it) in stimulating that restructuring. These sections concentrate on establishing the particular form of the effects of the crisis in each case. It is this discussion which makes the link between the crisis, as an 'aspatial' phenomenon, and its effects on the geographical patterns of economic activity. At the end of each section, the spatial implications specific to each form of restructuring are very briefly indicated. The final two sections attempt to put all these together to assess their combined spatial implications.

## Centralization in the face of surplus productive capital

The problem of surplus productive capital dominated the first group of cases. There was simply too much capacity for it all to be employed at an adequate rate of profit. This situation typified a considerable range of the industries examined in the study. It clearly was the major factor behind restructuring in supertension cables and aerospace, and most of the mergers in the heavy electrical machinery sector. The context of the problem was the slackening in the rate of growth of demand specifically for products of the electrical sector, and the exacerbation of this situation by the turndown in worldwide industrial activity. In all cases previously secure domestic markets were both declining in size and being increasingly invaded by foreign capital, itself under similar pressures. In terms of potential exports, the problems were just as great. The Commonwealth was no longer such a protected, nor such a lucrative, market. Export outlets had to be found elsewhere, in a situation where international competition was becoming increasingly severe.

In most parts of these sectors the problem was expressed on the ground by the existence of actually unused fixed and variable capital. Some capacity was simply standing idle; orders were spread amongst the rest leaving even this capacity under-used. The problem for both the individual capitals and the sector as a whole was therefore to produce existing and projected output from a smaller amount of constant and variable capital, from which a higher rate of profit would consequently be appropriated. Without financial restructuring the situation was a competitive one of which individual capitals would disappear and which would remain. The process of centralization of capital which took place allowed the solution to this problem by enabling a coordinated (instead of a competitive) reduction in capacity whereby no individual capital went bankrupt. Instead, these capitals merely withdrew from the sector involved, allowing the money capital representing its previous interests to be reabsorbed into more profitable areas of production.

In all of the product groups it was necessary to cut productive capital (and thus costs of production) in relation to the (almost unchanged) amount of commodities being produced, thereby increasing both the rate of surplus value and the rate of profit. The IRC encouraged an orchestrated process of further centralization which allowed major amounts of unused and under-used production to be taken out of production in the sector without any individual firm going out of business. Indeed, it allowed these firms to re-allocate money capital into more profitable lines of production, and to lessen their involvement in these declining industries. The capacity cuts enabled by these mergers went as high as 50% in a number of product groups. But those firms which remained needed also to increase their share of the world market. Pressure was therefore directed towards further increasing productivity. A number of measures were adopted, with different emphases in different sectors. The measures can be broadly divided into two groups — those more or less directly associated with the reductions in capacity, and those relating to the reorganization of the labour process.

Profitability was of course increased simply by the abandonment of unused capacity, the rate of surplus value was increased by the organization of that abandonment. Firstly, the State was clearly not supporting all individual capitals; those which survived the restructuring process were those with the greatest potential for internationalization (see also e.g. Fine and Harris 1975). Secondly, it was the production processes with the highest labour productivity and technical composition of capital which were retained. These measures dominated supertension cables and were important in the heavy electrical sectors.

The second group of strategies designed to increase the rate of surplus value involved changes in the labour process. The simplest of these was the process ideologically known as 'the reduction of overmanning.' This involved increasing the intensity of labour by reducing the number of workers required to produce a given output, with no other changes in the production process necessitated. Frequently, for instance, it involved sacking indirect workers (mates, etc). More complex reorganizations of the labour process took place as a result of the introduction of new techniques. These, too, took a number of forms. In the heavy end of the power-generation product groups, the large size, small number, and individual specification of many of the products makes mass production impossible. Here productivity was increased by standardization of as many components and characteristics as possible (e.g. in switchgear), and by the introduction of conveyor production line systems and automation to those processes amenable to such forms of organization.[10] In aerospace, the small-batch, custom-built nature of much of the production meant again that automatic transfer and mass production were not possible. On the other hand, it was possible in such cases to introduce numerically controlled machine tools which allowed both the number of workers and their average wages (because of the associated deskilling) to be lowered. Finally, at the lighter end of the electrical machinery sector, where the number of products produced is larger, and their size smaller (e.g. distribution transformers), full standardization and automation has been possible. All of these changes are, at least in part, 'competitive' in nature at the level of the individual firm, and enable a relative increase in the rate of surplus value. They all also increase the technical composition of capital.

What was the locational impact of all this? First of all, two basic facts should be established: that all of these industries are located primarily in the major industrial conurbations of England, and mainly outside the development areas (the major exceptions to this being Merseyside and the area of Newcastle-upon-Tyne); and secondly that all were sectors which employed relatively high proportions of skilled craft labour.

The changes introduced into the production and

labour processes can be characterized as: (1) capacity cutting/closures; (2) reduction of labour force without closure; (3) partial standardization and automation; (4) introduction of NC (numerically controlled) machine tools; and (5) full standardization and automation. In terms of labour requirements, all are characterized by cuts in amount. The last three, moreover, are also characterized by changes in the *type* of labour required. Partial standardization, simply by reducing the individually specified nature of the product, requires a less adaptable (to different specifications — i.e. less skilled) workforce (interview). Full standardization and automation reduces most of the individual labour processes involved to assembly work. The skill requirements are again lowered and cheaper labour power, including that of women, can be introduced. The introduction of NC has a similar impact, both reducing the number of workers employed and deskilling the requirements of the remainder (see, for instance, Palloix 1976).

Moreover, because of the nature of the changes, measures (1) to (4) take place within the existing geographical confines of the industry concerned. Thus the conurbations of London, Manchester, Birmingham and Liverpool (three of which are outside the development areas) have seen a massive decline in both the absolute level of employment in these industries and, further, in the level of skill required of the (previously highly skilled) remaining labour force. Where actual closures have taken place, these have frequently been in the inner parts of major cities, where plants are older, individual labour productivity therefore lower and unionization frequently stronger.

In only one of the types of change in this group, type (5), is any major relocation of fixed capital normally produced. Cases where full standardization and automation process can be introduced more frequently merit investment in brand new plant (see the section on inter-regional relations). In such cases, a locational change may be required (see later) and the new lack of ties to skilled labour, combined with regional controls and incentives, mean that the new location is most often outside the major conurbations but within a development area. The original location will therefore lose jobs, and a development area may gain — though fewer jobs, and of lower skill requirements than the ones lost.

## Problems of technological change

The second group of industries was involved in finan-

cial restructuring as a result of pressures very different from those affecting the first group. This is reflected in the forms of reorganization of the production and labour processes, and in their spatial repercussions. All the cases in this group fall within 'electronics', the specific industries being manufacturing of computers, numerically controlled machine tools and industrial instruments. Two very different factors appear to have conditioned the restructuring process and State involvement in that process in these sectors. Firstly, at the level of the economy as a whole, there was a need both to increase and to cheapen the output of these industries, since this output constituted an important means of increasing productivity in the rest of the economy. Secondly, and located more at the level of the individual capitals involved, there was continued pressure to keep up with international competition.

In contrast to the electrical industry, the market in electronics has from its inception been an internationally competitive one. Thus, while the import share of the UK market may be higher for some products than it is in the electrical sector, this does not represent a new departure: nor does it take place against a background of decline in that home market overall. The share of the UK domestic market taken by non-UK capital is also considerably higher than indicated by the share of imports, since significant shares of home production are accounted for by foreign-owned firms. Although there are obviously considerable variations between product markets, sales by foreign-owned enterprises as a percentage of total sales are on the whole higher in electronics than in the electrical industry. This distinction is particularly true of the specific products on which attention is focused in this study.[11]

These different conditions in the electronics sectors mean that the economic crisis gives rise to different forms of problems there. Although particular periods of recession may now lead to closures, the main problems is one of competing on an international market in a highly technological and fast-changing industry. The resultant characteristics of the industry may make increasing firm size advantageous within the framework of capitalist competition particularly with regard to research and development (R and D).

The combination of all these pressures produced the following effects on the production and labour processes:

(1) consolidation of R and D into a smaller number of larger groupings;

(2) the reduction of labour content in the products, as part of the process of cheapening the output of this sector. (This, of course, is an ongoing change, and also does not apply equally to all products, but each generation of computers requires only one-tenth of the labour amount of the previous generation; and the labour content of the present (6th) generation of NC is down to 5% of costs (interviews).);

(3) reduction in the level of skill required by the production labour force, as a result of the process noted under (2). Production in many electronics industries is now largely a matter of semi-skilled assembly and simple testing. In electronics, the resultant dichotomization of the labour force in terms of skill, between production of R and D and control, epitomizes that developing within the economy as a whole;

(4) overall sectoral growth (in both output and employment).

This in turn produces the following clearly distinguishable spatial effects:

(1) the growth and concentration of industrial research and development activity primarily within southeast England. This entails also, of course, the concentration in that region of the highly qualified labour force necessary for these activities;

(2) the establishment of locational hierarchies of production. Such hierarchies have been discussed at an international level (e.g. by Hymer 1972), and at an intranational level (e.g. by Buswell and Lewis 1970). Our own work has confirmed the existence of this tendency under conditions such as those typified by the sectors in this group;[12]

(3) the long-term growth nature of most of these industries meant that new capital investment dominated closures (i.e. the opposite of the situation in Group 1). Most of the new production facilities were established in development areas, as a result either of regional policy or of relaxed labour requirements (or, more usually as a mutually reinforcing combination of the two). In these areas they employed semi-skilled labour;

(4) on-going increases in technical composition in the process of production of some (not all) of these industries led to reduced demand for labour, and a downgrading of its required skill, at existing plants. These existing plants are in both assisted and non-assisted areas, but differently distributed from similar losses in Group 1 in being far less concentrated in 'older industrial areas', and specifically in their relatively rare occurrence in inner-city areas.

## Market power

The financial restructuring which took place in Group 3 was stimulated overwhelmingly by the need to strengthen international market position and power by additions to the sheer size of the acquiring company. That is, in this group the advantages derived from financial restructuring are largely those of muscle and market share. They do not require, in order that advantage be taken of them, that the financial reorganization be accompanied by any major reorganization of production. There is consequently little effect on the labour process. Although the IRC tended to refer to potential economies of scale, the reality of which the relevant companies in interview subsequently denied, it was also quite clear that sheer size could be an advantage in its own right:

International economic competition today involves a great deal of horse trading and arm twisting: bargaining situations and influence processes in which total size of resources deployable, rather than unit costs or rate of return on capital, may be the critical factor (McClelland 1972, p 25).

In spatial–locational terms, this group may be quickly dealt with. As already outlined, the financial restructuring was primarily aimed at increasing market standing through sheer size and/or combination with potential competition. This was not the 'rationalization route' to greater international competitiveness. There were, therefore, virtually no effects on the spatial organization of production, although reorganization of marketing and overseas supply functions was significant.

## Inter-regional relations

This section will draw together some threads from the previous separate discussions of restructuring processes, and will attempt to assess their combined spatial impact.

Aggregate job changes at the national level are presented in table 5.1 in terms of the three groupings. The figures include not only specific employment results of the mergers themselves, but also effects of ongoing changes (discussed in the previous section) which are integral to overall reorganization of the production process. The table also includes employment changes in industries considered in this paper as 'secondary' sectors. So far the discussion has concerned itself entirely with those 'primary' sectors which predominated as stimuli for the financial restructuring. In the cases of large, multidivisional companies, however, such restructuring will also involve other sectors.

**Table 5.1**   Major national level employment changes as a result of restructuring

|  | Absolute loss | Absolute gain | Total gain or loss | Locational shift (transferred jobs) | Jobs lost 'in transit'[†] |
|---|---|---|---|---|---|
| Group 1 | −24 113 | +100 | −24 013 | 966 | 5 505 |
| Group 2 | −4 006 | +1 750 | −2 256 | 703 | 641 |
| Group 3 | −210 | +120 | −90 | 150 | 190 |
| Secondary sectors | −9 657 | +0 | −9 657 | 2 676 | 4 909 |
| Total | −37 986 | +1 970 | −36 016 | 4 495 | 11 245 |

† Also included in 'Absolute loss' column.

Indeed, the merger once having taken place, these secondary sectors may be subject even to drastic rationalization. These effects are listed here for completeness, since they were an integral part of the employment consequences of financial restructuring. The sectors involved were telecommunications, traction and a variety of electronic product groups.

In table 5.1 employment changes which represented a gain or loss 'to the country as a whole' are 'absolute changes'. Those employment changes which, in contrast, result from geographical movement within the country are listed under 'locational shift'. The figures in the 'locational shift' column represent the employment which was created in a new location after movement from a previous plant. This number is far smaller than the loss recorded at the original factories.

Not surprisingly the results were dominated by employment loss. However, the three different forms of financial restructuring contributed differentially to this loss, with Group 1 dominating the Group 3 producing relatively little job-loss. As would be expected, Group 1 produced least in the way of newly generated employment. These results confirm the importance of national and international economic developments as a dominant determinant of locational change.

Job movement has frequently been either part of a process of overall cutbacks or has been the occasion for cutback. First, overall cuts in capacity often entail concentrating the work of smaller factories on a reduced number of larger ones. Such moves are frequently announced as transfers, and indeed some production may well be moved. They do not, however, represent a transfer of all the jobs at the previous location. Second, locational shift may be the occasion for major changes in production technology, again leading to a reduced workforce in the recipient region. The locational shift may be brought about because the nature of the technological change demands either new

**Table 5.2**   Aggregate results for assisted and non-assisted areas

|  | Non-assisted areas | + | Development areas | = | Total |
|---|---|---|---|---|---|
| Locational gain[†] | 270 |  | 2 452 |  |  |
| Absolute gain | 210 | + | 1 760 | = | 1 970 |
| Locational loss[†] | (2 452) |  | (270) |  |  |
| Absolute loss | (27 748) | + | (10 238) | = | (37 986) |
| Net total | (29 720) | + | (6 296) | = | (36 016) |

† Excluding all moves *within* a regional category.

fixed capital or a new workforce. In the first case it may be necessary, in the second prudent, to move thus avoiding conflict with the unions. The job losses associated with transferred jobs are given in table 5.1.

Turning to the *regional level*, the first question to be addressed is whether or not there was a differential impact, in terms of employment numbers, on development areas and non-development areas. Table 5.2 gives the aggregate results by category of employment changes for development areas and non-development areas. It is immediately clear that in these simple numerical terms, the development areas have fared far less badly from the processes analyzed than have the non-development areas. This is true both of the net total change of employment (where development areas lost less than one quarter of the number of jobs lost in the rest of the country), and of each gross component of this change. In all categories the development areas record a better result. They have (1) more locational gain; (2) more absolute gain; (3) less locational loss; and (4) less absolute loss.

Table 5.3 confirms this bias in a comparison of the percentage of initial total employment in each of the two types of regions with their respective percentages of the components of change. The percentage distribu-

**Table 5.3** The major regions as a percentage of the national results, by component of employment change

| Initial employment: | Development areas | Non-development areas |
|---|---|---|
| Manufacturing | 25 | 75 |
| Total employment | 27 | 73 |
| In sample[a] | 26 | 74 |
| *Results/employment change* | | |
| Absolute loss | 27 | 73 |
| Absolute gain | 89 | 11 |
| Locational gain[b] | 90 | 10 |
| *Total net change (loss)* | 17 | 83 |

[a] This percentage distribution is for the sample as a whole. Individual industries in the sample had initial distributions which differed from this.

[b] Calculated as the percentage of jobs gained out of all those which shifted location within the UK.

tions of absolute gain, locational gain and total net change (loss) confirm unambiguously a relative improvement in the position of the development areas. Certainly at this superficial level, the processes of restructuring analyzed here do seem to be contributing to the inter-regional convergence of employment availability. It is necessary, however, to examine the results in more detail.

What is the relationship of the process of restructuring to this regionally distributed impact? Four major components of change have been analyzed by which the development areas may be gaining in employment relative to the non-development areas.[13] These are:

(1) They are losing less drastically from the capacity cuts resulting from the restructuring in Group 1. As the tables show, the losses in non-assisted areas were far greater than those in development areas, in absolute terms. This is predominantly a structural phenomenon. That is, as representative of an overall sectoral decline it is dependent on the initial geographic distribution of those sectors.[14]

(2) They are gaining new plants from production relocated after major technological change. An example here is the electrical machinery industry (MLH 361), where not only the absolute loss but also the percentage of loss (of initial total employment in this industry) is lower in development areas. This appears to be the case not so much because older plants in the development

areas are more competitive, but because to the extent that full automation is possible within this group, new plants are being established.[15] With the labour requirements of the new techniques being limited in terms of skill, a non-conurbation development area location can now suffice where previously the craft skills of an old industrial area were necessary.

(3) They are receiving the bulk of new investment in the production end of the growth in Group 2. In this group development areas gained 1650 jobs as against 100 in non-development areas. More significantly, however, and confirming the identification of the process at work, *all* the gains registered in this group took place in new plants. One must be careful not to exaggerate the generalizability of this conclusion, since all these new plants, save one, were in a single development area — Scotland — renowned for its presence of externally controlled electronics facilities. Nonetheless, the evidence is indicative.

(4) They are losing less from the labour saving technical change within electronics sectors in Group 2. The labour shedding resulting from increasing capital intensity in this group was concentrated in the non-development areas. Again there was a mixture of structural and differential effects. Thus considerable losses were associated with the running down of facilities at the older end of the electronics spectrum, mostly located in non-development areas.[16]

## Some implications

The more detailed results fully confirm the fact that the restructuring processes had differential effects on the depressed areas of the country on the one hand and on the previously relatively prosperous regions on the other, and that this differential worked in favour of the depressed areas. Such results are of immediate importance for two reasons. First, they are clear warning against easy (and apparently 'radical') assumptions that the aspatial centralization of capital is always mirrored in increased locational centralization of production. They also warn against a-historical assumptions that regional policies of employment dispersal are necessarily contradictory to the process of accumulation at the national level.

Such tendencies, however, do not mean the end of a spatial division of labour, a homogenization of the whole country. They imply, rather a different form of integration and new forms of differentiation. The most likely development suggested by the present investigation is related to the increasing dichotomization of labour skills, between a small and highly qualified personnel and a semi-skilled/unskilled majority (a process which is in itself important in reducing the costs of reproduction of the mass of the labour force, and therefore potentially further increasing the rate of surplus value over the economy as a whole). There are a number of aspects to this.

The first point is that the centralization of capital involved in the cases studied did contribute to the further spatial centralization of control functions. Control over investment and production were increasingly withdrawn from most areas of the country (save perhaps the West Midlands), and concentrated in southeast England. This process occurred at the level of divisional control as well as that of the whole firm. This is unsurprising, and corroborates other evidence, such as that of Parsons (1972) and Westaway (1974). It has not been dealt with in detail in this research because its aggregate effects on employment are not large. Nonetheless, it is an important aspect of inter-regional relations, and should therefore be noted.

The second aspect of changing inter-regional relations which emerged was the evidence of the hierarchical spatial separation of research/development and mass production. This is an intra-product hierarchy closely resembling the various formulations based on the concept of the product cycle. In the cases studied in the present research, this spatial separation was primarily between southeast England, together with some parts of central England, and the rest of Great Britain. This hierarchy mirrored, and therefore reinforced, that stemming from the spatial centralization of control functions already mentioned. It entailed not merely the build up of skilled scientific and managerial workers in the southeast, but also the loss of some employment of this type in the rest of the country.[17] In this it was part of a cumulative process, the firms' response to location factors (locating research facilities in the southeast because of the existing concentration there of suitable labour) being precisely to reinforce the differential nature of their distribution.

Lastly, there was a territorial division of labour with production at the *inter*-commodity level. Thus within a number of broad product groups, the production of commodities requiring more standardized and less skilled processes was concentrated in development areas, with the production of commodities requiring more skilled labour more frequently found outside the development areas.[18] Evidence of the formation of this pattern was identified in a number of MLHs spreading right across the sectors investigated. It was usually associated with new investment in the non-conurbation parts of the development areas; this new investment was itself the result of two distinct processes. On the one hand some such investment (and consequently the establishment of such a hierarchy) resulted from major technical change in an existing production process. It was therefore accompanied by closure in the area of initial distribution. On the other hand, some of the new investment resulted simply from the extensive growth of a particular sector, the point here being that the production process in the expanding sectors under study tended to require semi-skilled assemblers rather than skilled craft workers.

This last 'hierarchy' (that between commodities and within production) was, however, unlike the other two, clearly not increasing in intensity and distinctness. On the contrary, in this case the evidence was for a weakening of spatial employment differentiation. For at the same time as the new investment in production was taking place, changes were also occurring in the nature of labour demand at the already existing production sites — which were more dominantly located outside the development areas, and certainly outside the non-conurbation parts of the development areas. Evidence has already been presented of, on the one hand, the run-down of the skilled craft basis of electrical engineering in many of the older industrial areas of England, and, on the other hand, the declining aggregate labour demand and increasing dichotomization of skills (predominantly a process of de-skilling) in the initial areas of distribution of the production of many commodities within electronics. Thus, outside southeast England and a small number of favoured locations in central England, not only unemployment levels, but also the demand for labour skills on any scale, may be tending to converge. The ongoing changes in production (as opposed to the absolute decline of craft-based sectors such as power-engineering), themselves a result of the need to increase the rate of exploitation, are thus releasing certain branches of industries from their former locational dependence on reserves of skilled labour. Such changes are making feasible locations outside the major industrial conurbations but within the depressed areas of the country.

The locations from which these industries are being released are also those in which the major absolute declines in skilled employment have taken place. These areas are of course the major industrial conurbations;

and any consideration of the evolving spatial divison of labour must take account of the major changes taking place in the cities. The results of the present research were therefore disaggregated to examine the effects on four such conurbations: Manchester, London, Liverpool and Birmingham. The cities suffered a far higher than proportionate absolute and net loss, and the results were also consistent across the individual conurbations. The lowest percentage loss (in Birmingham) was still far higher than for any other spatial division. The performance of the cities also differed from that of the other spatial categories in that their percentage locational loss was rather higher and their share of total absolute gain was zero.[19]

There is a vast reservoir of unemployed unskilled labour[20] in these same cities that are being deserted. Why, then, the moves to areas outside conurbations? A number of indications (and they are no more than that) began to emerge from the present study. In the first place, if it were a case of major technical change and a complete switch of workforce (and consequently probably also of union) a locational shift might be used to avoid conflict with the unions. Secondly, the reserve of labour in the cities was seen as being more expensive, more highly unionized and more militant. It is labour already well integrated into the labour force. So, instead of using this labour, capital is beginning to spread out. Small towns are now favoured locations. From being centres of skill the cities are becoming a location for the continually reconstituted reserve army of the unemployed.

What conclusions, then, can be drawn from these results? Firstly, it is perhaps worth stressing again the importance of establishing the relation between locational change, and therefore the differential economic fortunes of different areas, and the overall process of accumulation. Present attempts in the UK to divert and disconcert working-class action over unemployment focus particularly on fostering the inter-area competitive approach to industrial location. The most obvious case of this is in the debate over inner cities where protest has to some extent been played off against the problems of development areas. Such tactics produce a tendency to see the causes of economic decline within the area itself. The Home Office's attitude to the Community Development Projects it had sponsored in cities grew increasingly hostile as the latter linked the problems of their areas to the problems of the national economy rather than to the skills and psychological propensities of inner-city dwellers (see,

for instance, Community Development Project 1977). The problems of particular areas result from the particular form of use by capital, at any given period, of spatial differentiation. It has been the major concern of this paper to firmly establish the link between locational behaviour, changing patterns of employment, and developments at the level of the process of accumulation.

Secondly, in terms of the empirical results concerning locational behaviour, the present study indicates that one effect of restructuring during the period has been the emergence of a change in the form of territorial inequality. This change in form has been in terms of both its geographical pattern and its nature. The general disparities between development areas and non-development areas may be becoming less marked, in terms of employment numbers, while that between conurbations and non-conurbation areas seems likely to become more so. While the pattern of unemployment differentials may be evening out between assisted and non-assisted areas, the dichotomization of skill levels within manufacturing between the metropolis and the rest of the country may be being reinforced.

Such changing forms of inequality give a hint, also, of the reasons behind these locational shifts. As numerous examples in the previous pages have indicated, certain forms of technical change are increasingly freeing parts of capital from its previous ties to the skilled labour of the cities. In this context, capital is freer to search for labour which is less unionized, less militant. Such changes go hand in hand with regional policy, and the process of restructuring and the policy of regional location appear to have been mutually reinforcing. This exodus has been reinforced in its effect on the cities by the disproportionate declines there, in these sectors, of industries suffering problems of overcapacity.

Thirdly, and finally, what of the implications of these shifts for the strength and strategy of labour? In certain ways, the inter-regional evening out of unemployment rates may make the management of high national-level unemployment easier for the State to handle. The problem is spatially diffused. On the other hand, this very diffusion at the same time should make clear the nature of the real issue — declining *national* employment opportunities, not the simple gain of one region at the expense of others. Inevitably, in a period of such high national unemployment, some areas are particularly hard hit. But the spatial pattern is changing — the 'black spots' now vary from Skelmersdale to

Coventry. Most particularly, of course, the effect is being felt in the cities, where grass-roots activity is becoming increasingly important — through Trades Councils, in action groups for particular areas, such as the Docklands in London, and through the work of the Community Development Projects. The old industrial cities have long been the bases of some of the strongest union power. The present severe decline of industries within them therefore undermines that strength. At the same time, for those industries which are not simply declining but which are moving out of the cities, decreasing demand for skilled labour means that considerations such as differential labour militancy and organization may increase in importance as location factors.

## Notes

[1] The nature of the IRC and of its intervention are analyzed more fully in Massey and Meegan (1978) which also contains more analysis of restructuring and statistical information etc on the sectors analyzed. In the present paper, the IRC's importance is as a means of identifying sectors and firms in which the restructuring process was significant during the late 1960s.

[2] The industries covered are the following (the numbers refer to the Minimum List Heading (MLH) of the British Standard Industrial Classification):

354 scientific and industrial instruments and systems;
361 electrical machinery;
362 insulated wires and cables;
363 telegraph and telephone apparatus and equipment;
364 radio and electronic components;
365 broadcast, receiving and sound reproducing equipment;
366 electronic computers;
367 radio, radar and electronic capital goods;
368 electrical appliances primarily for domestic use;
369 other electrical machinery;
383 aerospace equipment.

[3] In fact the pattern of regional policy assistance is rather more complicated than this dichotomy, but it was this division which was dominant during the period.

[4] UK regional policy was implemented particularly strongly through this period, and the 'accepted' explanation of the convergence is that it was simply the result of State policies for geographical redistribution. This issue is not addressed directly here (for a detailed argument see Massey 1978). The aim of the present paper is simply to point out how restructuring such as that analyzed here could also be a significant component of this convergence.

[5] It is the first of these which is given more attention in the present paper. The second is dealt with more fully in Massey and Meegan (1978).

[6] Points of discussion or information in the text which are based specifically on material derived from those interviews are indicated '(interview)'. We should like to thank the interviewees for this very helpful contribution, though responsibility for the conclusions drawn is entirely our own.

[7] In the rather simplified version of the framework presented here,

it appears that 'space' is merely a passive surface. In fact, of course, it is itself a social product and has its own effects.

[8] It should be stressed that although the locational changes introduced by the firms in the groups were, on aggregate, significantly different from each other, the point of the grouping is not to introduce some new form of simple correlation (restructuring of type $x$ implies locational change of type $y$). The objective is to emphasize (a) that the overall crisis hits sectors in different ways, and (b) that locational change can only be understood as part and parcel of an economic analysis which takes account of such structural determinants.

[9] This classification does not cover all the industries initially listed; telecommunications, for instance, is missing. The reason for this is that most of the firms in these sectors span a wide range of product groups. The identification of the dominant sectors was a significant part of the research and has been confirmed in all cases by published data and interviews. However, mergers of major conglomerates may have repercussions on all their divisions and not just on the sectors which provided the rationale for the restructuring. The aggregate results, which are presented in the section on inter-regional relations, therefore include all effects, in both 'dominant' sectors and in 'secondary' sectors.

[10] Standardization was a process related directly to internationalization, as well as through increasing productivity, since it involved also the alignment of British and international specifications.

[11] It should be pointed out here that, although the major part of IRC activity was directly to aid British-owned firms, and although the strength of productive capacity of foreign-owned firms within the UK may considerably have increased the degree of competition, and therefore the problems, of those British-owned firms, it does not necessarily follow that any State action will be directed *against* the foreign capital operating on UK soil (see e.g. National Economic Development Office 1973). It is not possible simply to equate 'national economy' with 'national capital'.

[12] By such a locational hierarchy is meant a spatial division of labour between, for instance, research, development and initial production, and fully finalized mass production. In this way individual firms take advantage of regional uneven development. Financial restructuring aided in a number of ways the full establishment of such hierarchies. Firstly, size in itself is necessary in order to maintain a viable development programme at all; secondly, greater size increases the feasibility of separable locations. The process again relates to the dichotomization of skills in these industries, with the dichotomization increasingly reflected in locational patterns. In a number of interviews it was stressed that research 'must' be carried on in southeast England (occasionally it was allowed that Edinburgh was also a feasible location). But the 'next stage' can rarely be full mass production, a period of developmental production often being necessary to iron out snags. It was argued in interview, therefore, that this stage should be located in close proximity to the research laboratories. Indeed this was felt to be sufficiently important to warrant making out a lengthy case for avoiding the locational requirements of regional policy. It was stressed a number of times that under present conditions 'one can train testers in development areas, but not designers'.

[13] This analysis of employment decline is based on a simple statistical disaggregation of employment change in a region within a nation into (a) that component of change which would be expected on the basis of regional *structure* (e.g. a regional employment decline representing the presence of a nationally declining industry), and (b) a

*differential* (sometimes called a 'locational') component which indicates the differential performance within a region of any given industry (in the case of decline, the fact that an industry is declining faster in the region than it is nationally).

[14] In broad historical terms, the equal representation of cuts in a declining sector constitutes a worsening of the relative situation of such areas. If such decline is becoming a more important phenomenon in non-development areas this is significant. The existing assisted areas owe a large measure of their historical employment problems to such structural changes. MLH 362 here represents supertension cables, where both the initial distribution and the losses were entirely confined to southeast England. This was a simple case of plant closures. The other two sectors are more complicated. The losses in MLH 383 in development areas were entirely in Liverpool, and again in this sector, as explained in the section on the problems of technological change. There was no relocation, for instance, as a result of technical change; all losses took place on existing sites. This conclusion is greatly heightened when it is noted that of the total development area loss, 4910 jobs were from the Merseyside conurbation.

[15] This does not necessarily mean, however, as is frequently interpreted, that the differential gainer is in some sense the more 'competitive' region — that is has a locational advantage for any specified production process.

[16] This applied particularly in instruments (MLH 354) and computers (MLH 366). In the latter, much of the large (3550) absolute loss in non-development areas was due to the closure of electromechanical plants. Electro-mechanical equipment is an early halfway stage to full electronics, and is considerably more labour intensive. Moreover, as part of the initial development of the industry, it is more concentrated in southeast England, which region consequently suffered the bulk of the losses of this type. Within the fully electronic parts of this group, job loss due to increasing capital intensity was more dispersed but still overwhelmingly within the non-development areas.

[17] It should be stressed that the spatial hierarchies under discussion were identified as general tendencies, and not as iron laws. Some of the firms surveyed are still entirely located within development areas.

[18] The old industrial conurbations of England which are within development areas — Liverpool and Newcastle-upon-Tyne — also retain much of this production. The more detailed spatial pattern is discussed later in this section.

[19] The explanation for this very specific pattern of results is once again found in looking at the individual processes of restructuring. As in the overall results, Group 1 accounted for the bulk of the absolute loss, and of the net change. Indeed it was in the cities that the proportional effect of this form of restructuring was at a maximum. In the first place, the high level of initial presence of Group 1 plants in the cities meant that these areas were vulnerable to the employment cut-backs taking place as a result of both capacity cutting *per se*, and of the various forms of technical change outlined in the section on centralization. Secondly, the fact that the inner cities accounted for a higher than average proportion of older more labour-intensive plant, meant that they were more liable to both forms of employment reduction and change. Thirdly, none of the small amount of new investment and of associated absolute gain of employment, which was generated in Group 1, was located in the cities. It should be noted that this applied — as indeed did all these conclusions — as much to the development area conurbation (Liverpool) as to those outside the

development areas. Finally, the locational losses in Group 1 were the result of two forces. On the one hand, the product line reorganization within power generation occasioned some 500 job losses within the cities. This was the result firstly of an overall programme of concentration on a few major differentiated sites, of which only one was in a conurbation. Secondly, where, within this overall programme, explicit choices were available about the location of lay-offs or closures, it was often thought easier for the company if the latter were in conurbations, where their individual effect would be more muted by the context of a large and varied job pattern, than in towns where the firm concerned dominated the labour market. Such a strategy was mentioned in a number of interviews.

The cities also suffered particularly severely from the changes in Group 2. This may seem strange since their initial presence in the industries constituting this group was not dominant. It results from a combination of causes. Firstly, the intensification of the process of technical change in this group produced labour loss. Much of this took place in production processes within electronics and located primarily outside the cities. There was also, however, a more dramatic technological change in progress, such as that from electromechanics to electronics in MLH 366, where the older production processes had been located in the cities. It was of course also these processes which were both larger employers of labour and which were closed rather than undergoing technical change on-site. The resultant losses to the cities were, therefore, again both absolutely and proportionally high. Secondly, the cities did not gain in employment terms from the consolidation of R and D activities in the group. There is a presence of such activities, and the possibilities of gain should not be ruled out, but the numbers involved would be small, and the jobs would be located outside the inner areas of the cities. Thirdly and finally, the more considerable new capital investment in this group was located entirely outside the cities, whether or not these latter were in development areas. This is evidently the other aspect of the process already referred to by which a change in labour requirements may increase capital's locational flexibility.

[20] Unskilled as unemployed labour seeking new jobs. This does not preclude the possibility that their previous employment was in fact highly skilled, but in crafts no longer demanded by capital.

# References

Buswell R J and Lewis E W 1970 The geographical distribution of industrial research activity in the United Kingdom *Regional Studies* **4** 297–306

Community Development Project 1977 *The Costs of Industrial Change*

Fine B and Harris L 1975 The British economy since March 1974 *Bulletin of the Conference of Socialist Economists* **4** 3(12) October

Hymer S 1972 The multinational corporation and the law of uneven development *Economics and World Order from the 1970's to the 1990's* ed. J Bhagwati (London: Collier-Macmillan) pp 113–40

Keeble D 1977 *Industrial Location and Planning in the United Kingdom* (London: Methuen)

Massey D B 1978 In what sense a regional problem? *Centre for Environmental Studies Working Note No.* 479

Massey D B and Meegan R A 1978 Industrial restructuring versus the cities *Urban Studies* **15** 273–88

McClelland W G 1972 The IRC 1966–71: an experimental prod *Three Banks Review* June

National Economic Development Office 1973 *Industrial Review to 1977* (London: HMSO)

Palloix C 1976 The labour process: from Fordism to neoFordism *The Labour Process and Class Struggle, Conference of Socialist Economists Pamphlet 1 Stage 1*

Parsons G F 1972 The giant manufacturing corporations and balanced regional growth in Britain *Area* **4** (2) 99–103

Westaway J 1974 The spatial hierarchy of business organizations and its implications for the British urban systems *Regional Studies* **8** 145–55

# 6 A Theory of the Urban Land Market

*by W Alonso*

The early theory of rent and location concerned itself primarily with agricultural land. This was quite natural, for Ricardo and Malthus lived in an agricultural society. The foundations of the formal spatial analysis of agricultural rent and location are found in the work of von Thünen (1826) who, without going into detail, said that the urban land market operated under the same principles. As cities grew in importance, relatively little attention was paid to the theory of urban rents. Even the great Marshall provided interesting but only random insights, and no explicit theory of the urban land market and urban locations was developed.

Since the beginning of the twentieth century there has been considerable interest in the urban land market in America. Hurd (1903) and Haig (1926, 1927) tried to create a theory of urban land by following von Thünen. However, their approach copied the form rather than the logic of agricultural theory, and the resulting theory can be shown to be insufficient on its own premises. In particular, the theory failed to consider residences, which constitute the preponderant land use in urban areas.

Yet there are interesting problems that a theory of urban land must consider. There is, for instance, a paradox in American cities: the poor live near the centre, on expensive land, and the rich on the periphery, on cheap land. On the logical side, there are also aspects of great interest, but which increase the difficulty of the analysis. When a purchaser acquires land, he acquires two goods (land and location) in only one transaction, and only one payment is made for the

combination. He could buy the same quantity of land at another location, or he could buy more land or less land at the same location. In the analysis, one also encounters a negative good (distance) with positive costs (commuting costs); or conversely a positive good (accessibility) with negative costs (savings in commuting). In comparison with agriculture, the urban case presents another difficulty. In agriculture, the location is extensive: many square miles may be devoted to one crop. In the urban case the site tends to be much smaller, and the location may be regarded as a dimensionless point rather than an area. Yet the thousands or millions of dimensionless points which constitute the city, when taken together, cover extensive areas. How can these dimensionless points be aggregated into two-dimensional space?

Here I will present a non-mathematical overview, without trying to give it full precision, of the long and rather complex mathematical analysis which constitutes a formal theory of the urban land market. It is a static model in which change is introduced by comparative statics, and it is an economic model: it speaks of 'economic men', and it goes without saying that real men and social groups have needs, emotions and desires which are not considered here. This analysis uses concepts which fit with agricultural rent theory in such a way that urban and rural land uses may be considered at the same time, in terms of a single theory. Therefore, we must examine first a very simplified model of the agricultural land market.

## Agricultural model

In this model, the farmers are grouped around a single market, where they sell their products. If the product is wheat, and the produce of one acre of wheat sells for

Source: Alonso W 1960 *Papers and Proceedings of the Sixth Annual Meeting of the Regional Science Association* pp 149–57. Reprinted by permission of the Regional Science Association.

$100 at the market while the costs of production are $50 per acre, a farmer growing wheat at the market would make a profit of $50 per acre. But if he is producing at some distance—say, 5 miles—and it costs him $5 per mile to ship an acre's produce, his transport costs will be $25 per acre. His profits will be equal to value minus production costs minus shipping charges: 100-50-25=$25. This relation may be shown diagrammatically (see figure 6.1). At the market, the farmer's profits are $50, and 5 miles out, $25; at intermediate distance, he will receive intermediate profits. Finally, at a distance of 10 miles from the market, his production costs plus shipping charges will just equal the value of his produce at the market. At distances greater than 10 miles, the farmer would operate at a loss.

In this model, the profits derived by the farmers are tied directly to their location. If the functions of farmer and landowner are viewed as separate, farmers will bid rents for land according to the profitability of the location. The profits of the farmer will therefore be shared with the landowner through rent payments. As farmers bid against each other for the more profitable locations, until farmers' profits are everywhere the same ('normal' profits), what we have called profits become rent. Thus, the curve in figure 6.1, which we derived as a farmer's profit curve, once we distinguish between the roles of the farmer and the landowner, becomes a bid-rent function, representing the price or rent per acre that farmers will be willing to pay for land at the different locations.

We have shown that the slope of the rent curve will

be fixed by the transport costs on the produce. The level of the curve will be set by the price of the produce at the market. Examine figure 6.2. The lower curve is that of figure 6.1, where the price of wheat is $100 at the market, and production costs are $50. If demand increases, and the price of wheat at the market rises to $125 (while production and transport costs remain constant), profits or bid rent at the market will be $75; at 5 miles, $50; $25 at 10 miles; and zero at 15 miles. Thus, each bid-rent curve is a function of rent against distance, but there is a family of such curves, the level of any one determined by the price of the produce at the market, higher prices setting higher curves.

Consider now the production of peas. Assume that the price at the market of one acre's production of peas is $150, the costs of production are $75, and the transport costs per mile are $10. These conditions will yield curve MN in figure 6.3, where bid rent by pea farmers at the market is $75 per acre, 5 miles from the market $25, and zero at 7.5 miles. Curve RS represents bid rents by wheat farmers, at a price of $100 for wheat. It will be seen that pea farmers can bid higher rents in the range of 0 to 5 miles from the market; farther out, wheat farmers can bid higher rents. Therefore, pea farming will take place in the ring from 0 to 5 miles from the market, and wheat farming in the ring from 5 to 10 miles. Segments MT of the bid-rent curve of pea farming and TS of wheat farming will be the effective rents, while segments RT and TN represent unsuccessful bids.

The price of the product is determined by the sup-

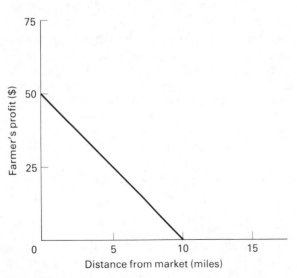

Figure 6.1

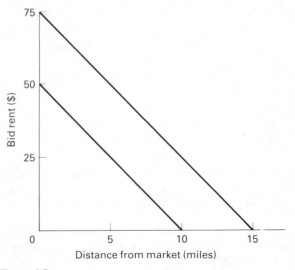

Figure 6.2

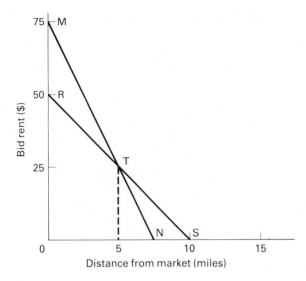

**Figure 6.3**

ply–demand relations at the market. If the region between zero and 5 miles produces too many peas, the price of the product will drop, and a lower bid-rent curve for pea farming will come into effect, so that pea farming will be practised to some distance less than 5 miles.

Abstracting this view of the agricultural land market, we have that:

(1) land uses determine land values, through competitive bidding among farmers;

(2) land values distribute land uses, according to their ability to pay;

(3) the steeper curves capture the central locations. (This point is a simplified one for simple, well behaved curves.)

Abstracting the process now *from* agriculture, we have:

(1) for each user of land (e.g. wheat farmer) a family of bid-rent functions is derived, such that the user is indifferent as to his location along any *one* of these functions (because the farmer, who is the decision-maker in this case, finds that profits are everywhere the same, i.e. normal, as long as he remains on one curve);

(2) the equilibrium rent at any location is found by comparing the bids of the various potential users and choosing the highest;

(3) equilibrium quantities of land are found by selecting the proper bid-rent curve for each user (in the agricultural case, the curve which equates supply and demand for the produce).

## Business

We shall now consider the urban businessman, who, we shall assume, makes his decisions so as to maximize profits. A bid-rent curve for the businessman, then, will be one along which profits are everywhere the same: the decision-maker will be indifferent as to his location along such a curve.

Profit may be defined as the remainder from the volume of business after operating costs and land costs have been deducted. Since in most cases the volume of business of a firm as well as its operating costs will vary with its location, the rate of change of the bid-rent curve will bear no simple relation to transport costs (as it did in agriculture). The rate of change of the total bid rent for a firm, where profits are constant by definition, will be equal to the rate of change in the volume of business minus the rate of change in operating costs. Therefore the slope of the bid-rent curve, the values of which are in terms of dollars per unit of land, will be equal to the rate of change in the volume of business minus the rate of change in operating costs, divided by the area occupied by the establishment.

A different level of profits would yield a different bid-rent curve. The higher the bid-rent curve, the lower the profits, since land is more expensive. There will be a highest curve, where profits will be zero. At higher land rents the firm could only operate at a loss.

Thus we have, as in the case of the farmer, a family of bid-rent curves, along the path of any one of which the decision-maker—in this case, the businessman—is indifferent. Whereas in the case of the farmer the level of the curve is determined by the price of the produce, while profits are in all cases 'normal' (i.e. the same) in the case of the urban firm, the level of the curve is determined by the level of the profits, and the price of its products may be regarded for our purposes as constant.

## Residential

The household differs from the farmer and the urban firm in that satisfaction rather than profits is the relevant criterion of optional location. A consumer, given his income and his pattern of tastes, will seek to balance the costs and bother of commuting against the advantages of cheaper land with increasing distance from the centre of the city and the satisfaction of more space for living. When the individual consumer faces a given pattern of land costs, his equilibrium location and the size of his site will be in terms of the marginal changes of these variables.

The bid-rent curves of the individual will be such that, for any given curve, the individual will be equally satisfied at every location at the price set by the curve. Along any bid-rent curve, the price the individual will bid for land will decrease with distance from the centre at a rate just sufficient to produce an income effect which will balance to his satisfaction the increased costs of commuting and the bother of a long trip. This slope may be expressed quite precisely in mathematical terms, but it is a complex expression, the exact interpretation of which is beyond the scope of this paper.

Just as different prices of the produce set different levels for the bid-rent curves of the farmer, and different levels of profit for the urban firm, different levels of satisfaction correspond to the various levels of the family of bid-rent curves of the individual household. The higher curves obviously yield less satisfaction because a higher price is implied, so that, at any given location, the individual will be able to afford less land and other goods.

## Individual equilibrium

It is obvious that families of bid-rent curves are in many respects similar to indifference-curve mappings. However, they differ in some important ways. Indifference curves map a path of indifference (equal satisfaction) between combinations of quantities of two goods. Bid-rent functions map an indifference path between the price of one good (land) and quantities of another and strange type of good (distance from the centre of the city). Whereas indifference curves refer only to tastes and not to budget, in the case of households, bid-rent functions are derived both from budget and taste considerations. In the case of the urban firm, they might be termed isoprofit curves. A more superficial difference is that, whereas the higher indifference curves are the preferred ones, it is the lower bid-rent curves that yield greater profits or satisfaction. However, bid-rent curves may be used in a manner analogous to that of indifference curves to find the equilibrium location and land price for the resident or the urban firm.

Assume you have been given a bid-rent mapping of a land use, whether business or residential (figure 6.4, curves BRC $_{1,2,3}$). Superimpose on the same diagram the actual structure of land prices in the city (curve SS). The decision-maker will wish to reach the lowest possible bid-rent curve. Therefore, he will choose that point at which the curve of actual prices (SS) will be a

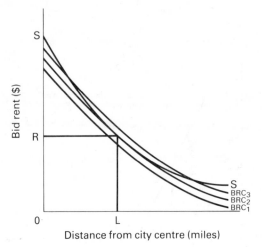

**Figure 6.4**

tangent to the lowest of the bid-rent curves with which it comes in contact (BRC $_2$). At this point will be the equilibrium location (L) and the equilibrium land rent (R) for this user of land. If he is a businessman, he will have maximized profits; if he is a resident, he will have maximized satisfaction.

Note that to the left of this point of equilibrium (toward the centre of the city) the curve of actual prices is steeper than the bid-rent curve; to the right of this point (away from the centre) it is less steep. This is another aspect of the rule we noted in the agricultural model: the land uses with steeper bid-rent curves capture the central locations.

## Market equilibrium

We now have, conceptually, families of bid-rent curves for all three types of land uses. We also know that the steeper curves will occupy the more central locations. Therefore, if the curves of the various users are ranked by steepness, they will also be ranked in terms of their accessibility from the centre of the city in the final solution.

Thus, if the curves of the business firm are steeper than those of residences, and the residential curves steeper than the agricultural, there will be business at the centre of the city, surrounded by residences, and these will be surrounded by agriculture.

This reasoning applies as well within land-use groupings. For instance, it can be shown that, given two individuals of similar tastes, both of whom prefer living at low densities, if their incomes differ, the bid-rent curves of the wealthier will be flatter than those of the man of lower income. Therefore, the poor will tend to central locations on expensive land and the rich to cheaper land on the periphery. The reason for this is not that the poor have greater purchasing power, but rather that they have steeper bid-rent curves. This stems from the fact that, at any given location, the poor can buy less land than the rich, and since only a small quantity of land is involved, changes in its price are not as important for the poor as the costs and inconvenience of commuting. The rich, on the other hand, buy greater quantities of land, and are consequently affected by changes in its price to a greater degree. In other words, because of variations in density among different levels of income, accessibility behaves as an inferior good.

Thus far, through ranking the bid-rent curves by steepness, we have found the relative rankings of prices and locations, but not the actual prices, locations or densities. It will be remembered that in the agricultural case equilibrium levels were brought about by changes in the price of the products, until the amount of land devoted to each crop was in agreement with the demand for that crop.

For urban land this process is more complex. The determination of densities (or their inverse, lot size) and locations must be found simultaneously with the resulting price structure. Very briefly, the method consists of assuming a price of land at the centre of the city, and determining the prices at all other locations by the competitive bidding of the potential users of land in relation to this price. The highest bid captures each location, and each bid is related to a most preferred alternative through the use of bid-rent curves. This most preferred alternative is the marginal combination of price and location for that particular land use. The quantities of land occupied by the land users are determined by these prices. The locations are determined by assigning to each successive user of land the location available nearest the centre of the city after the assignment of land quantities to the higher and more central bidders.

Since initially the price at the centre of the city was assumed, the resulting set of prices, locations and densities may be in error. A series of iterations will yield the correct solution. In some cases, the solution may be found by a set of simultaneous equations rather than by the chain of steps which has just been outlined.

The model presented in this paper corresponds to the simplest case: a single-centre city, on a featureless plain, with transportation in all directions. However, the reasoning can be extended to cities with several centres (shopping, office, manufacturing, etc), with structured road patterns and other realistic complications. The theory can also be made to shed light on the effects of economic development, changes in income structure, zoning regulations, taxation policies and others. At this stage, the model is purely theoretical; however, it is hoped that it may provide a logical structure for econometric models which may be useful for prediction.

## Note
[1] A full development of the theory is presented in Alonso (1960).

## References
Alonso W 1960 A model of the urban land market: locations and densities of dwellings and businesses *PhD Thesis*, University of Pennsylvania

Haig R M 1936 Towards an understanding of the metropolis *Quarterly Journal of Economics* **XL**

——1927 *Regional Survey of New York and its Environs* (New York: New York City Plan Commission)

Hurd R M 1903 *Principles of City Land Values* (New York: The Record and Guide)

von Thünen J 1826 *Der Isolierte Staat in Beziehung auf Landwirtschaft und Nationalekonomie* (1st edn Hamburg)

# 7 Urban Land-use Theory: a Critique

## by D Harvey

Contemporary urban land-use theory is in a peculiar state. Analysis typically concentrates either on use value characteristics (through the study of the life-support system) or on exchange value characteristics (the market exchange system), but there is little or no conception as to how the two may be related to each other.

Geographers and sociologists, for example, have evolved a variety of land-use theories which focus on patterns of use. The concentric zone, multiple nuclei and sectoral 'theories' are nothing more than generalized descriptions of patterns of use in the urban space economy. The tradition of research in factorial ecology attempts the same thing with much greater sophistication (and some enlightenment), while the work of other sociologists such as Gans (1970) and Suttles (1968) brings a certain amount of realism to the somewhat arid statistical summaries of factorial ecology. Various other devices exist for generalizing statistically about the macro-patterns of urban land use. The negative exponential 'model' of population-density (and land-rent) decline with distance from the urban centre has been investigated in some detail. Various models out of the social physics tradition — of which Wilson's (1970) formulations are surely the most sophisticated to date — have also been used to characterize the macro-characteristics of activities and uses in the urban system. All these formulations, however, amount to sophisticated analyses of patterns of use which differ in degree, but not in kind, from those expressed in a land-use map or in a description of daily activity as it unfolds in the life-support system that is the city. A great deal can be gained by such descriptions, but

studies such as this cannot yield up a theory of urban land use.

By way of contrast, land-use theories generated out of neoclassical micro-economics focus upon exchange value, although in so doing they appeal explicitly to the strategy pioneered by Jevons through which use value (utility) is equated to exchange value at the margin. Alonso (1964), Beckmann (1969), Mills (1969) and Muth (1969) presume utility-maximizing behaviour on the part of individuals. In the housing market this is taken to mean that individuals trade off the quantity of housing (usually conceived of as space), accessibility (usually cost of transport to place of employment) and the need for all other goods and services, within an overall budget constraint. It is presumed that consumers are indifferent with respect to certain combinations of space and accessibility. It is also presumed that individuals bid for housing at a location up until the point where the extra amount of 'satisfaction' gained from a move is exactly equal to the marginal utility of laying out an extra quantity of money. From this conceptualization it is possible to derive equilibrium conditions in the urban housing market — conditions which are held to be Pareto optimal. This process can be modelled in a variety of ways. Herbert and Stevens (1960) formulate it as a programming problem in which households seek their best 'residential bundle' of goods out of a general market basket of all goods subject to cost and budget constraints. Muth (1969) provides particularly sophisticated formulations in which he attempts to bring together analyses of the production of housing, the allocation of existing housing stock, the allocation of land to uses and utility-maximizing behaviour on the part of individual consumers with different income characteristics and diverse preferences for housing. Other writers have examined the

Source: Harvey D 1973 *Social Justice and the City* (London: Edward Arnold) pp 160–76.

competition for space and location among different uses (commercial, industrial, residential, etc).

It is tempting to view this corpus of urban land-use theory as providing an adequate framework for analyzing the market forces shaping urban land use. Unfortunately these theories abstract from questions of use value and do as little to bring use and exchange value together as do the formulations of geographers and sociologists, who start with use value as their basic consideration. The fact that utility-maximizing models contain a crude assumption concerning the relationship between use value and exchange value should not deceive us into thinking that real problems have been resolved. This is not to condemn the models derived from micro-economics as useless. They shed light on the exchange-value aspect of urban land-use theory in much the same way that geographers and sociologists have shed light on the use-value aspects. But an adequate urban land-use theory requires a synthesis of both these two aspects in such a way that we grasp the social process of commodity exchange in the sense that Marx conceived of it. This theory will not be easy to construct, particularly in view of the peculiar qualities of land and improvements and the diverse uses to which these may be put.

A critical appraisal of the micro-economic approach will help us identify what the problem is. Kirwan and Martin (1971) have recently reviewed the contribution of this approach to our understanding of residential land uses, and for the sake of brevity I will concentrate on this aspect of urban land-use theory. It should be evident that my remarks can in principle be generalized to all other aspects of urban land use.

The assumptions typically built into the micro-economic approach are obviously unrealistic, and are generally admitted to be so. But then this is true of all micro-economic models of this sort. The question is, how and to what degree is the general conceptualization unrealistic? We can start to answer this by comparing the general nature of the results with the reality we are seeking to understand. The remarkable fact here is that although the theories derived analytically out of the micro-economic framework cannot be regarded as 'true' in the sense that they have been subjected to rigorous empirical testing, these theories of urban land use (although normative) yield results which are not too much at variance with the realities of city structure. Put another way, the case for regarding them as empirically relevant devices may not have been proved, but it has not been disproved either. These theories may thus be regarded as perhaps not unreasonable general characterizations of the forces shaping urban land use. There are, however, grounds upon which this interim conclusion may be criticized: we will now explore them.

There are numerous and diverse actors in the housing market and each group has a distinctive way of determining use value and exchange value. Let us consider the perspective of each of the main groups operating in the housing market.

(i) The *occupiers* of housing consume the various facets of housing according to their desires and needs. The use value of the house is determined by the coming together of a personal or household situation and a particular house in a particular location. Owner-occupiers are basically concerned with use values and act accordingly. But, insofar as the house has a use in storing equity, exchange value may become a consideration. We may alter our house so that we can use it better, or we may modify it with an idea of increasing its exchange value. Owner-occupiers typically become concerned with exchange value at two points — at the time of purchase and when major repairs force them to look to their budget constraints. Renters (and other kinds of tenant) are in a rather different position in that use value provides only limited rationale for action when exchange value goes to the landlord. But all occupiers of housing have a similar concern — to procure use values through laying out exchange value.

(ii) *Realtors* (estate agents) operate in the housing market to obtain exchange value. They realize a profit through buying and selling or through charging transaction costs for their services as intermediaries. Realtors rarely contribute much to the use value of the house (although they may undertake some improvements in certain cases). To realtors the use value of housing lies in the volume of transactions, for it is from these that they gain exchange value. They operate as coordinating entrepreneurs in the housing market, work under competitive pressure and need to reap a certain level of profit. They have an incentive to increase turnover in the housing stock for this leads to an expansion of business. Turnover may be stimulated by ethical or unethical means (blockbusting being a good example of the latter). Realtors can thus play a role on a continuum between passive coordinator of the market through encouraging market activity to forcing it.

(iii) *Landlords* operate, for the most part, with exchange value as their objective. Owner-occupiers who rent out a portion of their house have a dual objective of course, and may be as much motivated by considerations of use value as those who occupy the whole of their property. But professional landlords regard the house as a means of exchange — housing services are exchanged for money. The landlord has two strategies. The first to purchase a property outright and then to rent it out in order to obtain an income from the capital invested in it. The second strategy involves the purchase of a property through mortgage financing: the application of the rental income to servicing the mortgage (together with depreciation and tax allowances) then allows the landlord to increase the net worth of his or her holdings. The first strategy maximizes current income (usually over a short-term horizon), while the second maximizes the increase in wealth. The choice of strategy has an important impact upon the management of the housing stock, the first strategy tending to lead to rapid obsolescence, the second to good upkeep and maintenance. The choice depends upon the circumstances — the opportunity cost of capital invested in housing as opposed to all other forms of investment, the availability of mortgage financing and so on. Whatever the strategy, it still remains the case that professional landlords treat housing as a means of exchange and not as a use value for themselves.

(iv) *Developers* and the housing construction industry are involved in the process of creating new values for others in order to realize exchange values for themselves. The purchase of the land, the preparation of it (particularly the provision of public utilities), and the construction of the housing, require considerable capital outlay in advance of exchange. Firms involved in this process are subject to competitive pressure, and must realize a profit. They therefore have a strong vested interest in bringing into being the use values necessary to sustain their exchange-value benefits. There are numerous ways (both legal and illegal) for accomplishing this, and certainly this group in the housing market has a strong vested interest in the process of suburbanization and, to a lesser degree, in processes of rehabilitation and redevelopment. In much the same way that realtors are interested in increased turnover, so developers and construction firms are interested in growth, reconstruction and rehabilitation. Both these groups are interested in use values for others, only in so far as they yield exchange values to themselves.

(v) *Financial institutions* play an important role in the housing market owing to the particular characteristics of housing. The financing of owner-occupancy, landlord operations, development and new construction draws heavily upon the resources of banks, insurance companies, building societies and other financial institutions. Some of these institutions are locked into financing in the housing market (the Savings and Loan Associations in the United States for example). But others service all sectors and they tend to allocate their funds to housing insofar as housing yields opportunities for profitable and secure investment relative to other investment opportunities. Fundamentally, the financial institutions are interested in gaining exchange values through financing opportunities for the creation or procurement of use values. But financial institutions as a whole are involved in all aspects of real estate development (industrial, commercial, residential, etc) and they therefore help to allocate land to uses through their control over financing. Decisions of this sort are plainly geared to profitability and risk avoidance.

(vi) *Government institutions* — usually called into existence by political processes stemming from the lack of use values available to the consumers of housing — frequently interfere in the housing market. Production of use values through public action (the provision of public housing for example) is a direct form of intervention; but intervention is frequently indirect (particularly in the United States). The latter might take the form of helping the financial institutions, the developers and construction industry to gain exchange values by government action to provide tax shelter, to guarantee profits or to eliminate risk. It is argued that supporting the market is one way of ensuring the production of use values — unfortunately it does not always work out that way. Government also imposes and administers a variety of institutional constraints on the operation of the housing market (zoning and land-use planning controls being the most conspicuous). Insofar as government allocates many of the services, facilities and access routes, it also contributes indirectly to the use value of housing by shaping the surrounding environment.

The operations of all these diverse groups in the housing market cannot easily be brought together into one comprehensive framework for analysis. What is a use value for one is an exchange value for another, and each conceives use value differently. The same house can take on a different meaning depending upon the social relationships which individuals, organizations and institutions express in it. A model of the housing market which presumes all housing stock to be allocated among users (whose only differentiating characteristics are income and housing preferences) through utility-maximizing behaviour appears peculiarly restricted in its applicability. Realistic analyses of how urban land-use decisions are made — dating from Hurd's (1903) perceptive analysis — have consequently led Wallace Smith (1970, p 40), for example, to conclude that 'the traditional concept of "supply-and-demand equilibrium" is not very relevant to most of the problems or issues which are associated with the housing sector of the economy.' It is difficult not to concur with this opinion, for if a commodity depends upon the coming together of use value and exchange value in the social act of exchange, then the things we call land and housing are apparently very different commodities depending upon the particular interest group operating in the market. When we introduce the further complexity generated by competition among diverse uses, we may be inclined to extend Wallace Smith's conclusion to urban land-use theory as a whole.

Another general line of criticism of the microeconomic approach to urban land-use theory stems from the fact that it is formulated in a static equilibrium framework. This would be mere churlish criticism, however, if it were only pointed out that the urban land-use system rarely approaches anything like an equilibrium posture and that Pareto optimality is likely never to be achieved. Differential disequilibrium is everywhere evident and there are too many imperfections, rigidities and immobilities for the market to work well as a coordinating device. But there is a point of substance here that requires examination. The urban area is built up sequentially over a long period of time and activities and people take up their positions in the urban system sequentially. Once located, activities and people tend to be particularly difficult to move. The simultaneity presupposed in the micro-economic models runs counter to what is in fact a very strong process. This indicates a fatal flaw in the micro-economic formulations — their inability to handle the absolute quality of space which makes land and improvements such peculiar commodities. Most writers either ignore this issue or dismiss it. Muth (1969, p 47) for example, holds that:

Many of the features of city structure and urban land use can be explained without reference to the heritage of the past. To the extent that there is any distinction between land, especially urban land, and other factors of production, such a distinction would seem to arise chiefly from the fact of spatial uniqueness. In fact spatial uniqueness is not as clearly a distinction in kind as one might initially suppose. If labour were not sometimes highly immobile, and hence spatially unique, there would probably be no depressed area or farm problems. And the supply of land with certain spatial characteristics is sometimes increased by filling in areas along waterfronts and more frequently through investment in transport facilities.

This treatment of spatial uniqueness (or absolute space) plainly will not do. Spatial uniqueness cannot be reduced to mere immobility nor to a question merely of transport access.

The points raised here are both informal and incomplete but they do provide a useful foil against which to compare the utility-maximizing models of Alonso, Muth, Beckmann and Mills. Insofar as these models are formulated in relative space[1] in a manner which discounts the monopoly characteristics of absolute space, they appear most appropriate as arguments governing what happens to affluent groups who are in a position to escape the consequences of monopoly in space; the formulations are therefore income-biased. The criterion of Pareto optimality also appears irrelevant (if not downright misleading) in any analysis of the urban housing market. The differential distribution of the collective consumers' surplus[2] according to the first-come-first-served principle with the rich at the head of the queue, almost certainly has a differential income effect in which the rich are destined to gain more than the poor in most situations. Sequential occupancy in urban land use of the type we are hypothesizing here does not yield Pareto optimality, but a redistribution of imputed income (which is what consumers' surplus really amounts to).[3] Even if we take into account that new construction is possible (i.e., the housing stock is not fixed), this condition is unlikely to change, for the poor certainly are in no condition to generate activity in the private sector because of the weak effective demand for housing which they are capable of expressing in the market place.

The evidence from the contemporary American city suggests that the dynamics of land-use change remains fairly constant under the capitalist mode of production. The consumers' surplus of the poorest groups is diminished by producers of housing services transforming it into producers' surplus through quasi-monopolistic practices (usually exercised on the basis of class monopoly power). Also the poorest groups generally live in locations subject to the greatest speculative pressure from land-use change. In order to realize an adequate future return on investment in existing commercial urban renewal schemes, for example, financial institutions have a vested interest in expanding commercial development geographically; by this process spatial externalities are created, through which new commercial development enhances the value of the old. New commercial development will usually have to take place over land already in housing. Housing in these areas can be deliberately economically run down by the withdrawal of financial support for the housing market — 'red-lining' by financial institutions is a common practice in the United States, although it is generally explained away as risk aversion: this is but a part of the story, however. Landlords are forced under these conditions to maximize current income over a short-term time horizon, which means a rational business-like milking of a property for all it is worth. The physical obsolescence, generated out of this economic obsolescence, results in social and economic pressures which build up in the worst sections of the housing market and which have to be relieved, at some stage or other, by a 'blow out' somewhere. This blow out results in new construction and the taking up of new land at the urban fringes or in urban redevelopment — processes which are both subject to intense speculative pressure. New household formation and in-migration supplement this dynamic situation.

The same financial institutions which deny funds to one sector of the housing market stand to gain from the realization of speculative gains in another, as land use is subsequently transformed or as suburbanization proceeds. The impulses which are transmitted throughout the urban land-use system are not unconnected. The diversity of actors and institutions involved makes a conspiracy theory of urban land-use change unlikely (which is not to say that conspiracy never occurs). The processes are strongly structured through the market exchange system so that individuals, groups and organizations operating self-interestedly in terms of exchange value can, with the help of the 'hidden hand', produce the requisite result. It is argued by some that this system produces the best possible distribution of use values. This is a presumption which casual observation suggests is wrong: the maximization of exchange values by diverse actors produces disproportionate benefits to some groups and diminishes the opportunities for others. The gap between the proper production and distribution of use values and a system of allocation that rests on the concept of exchange value cannot easily be glossed over.

The diversity of actors operating in a land-use system and the monopolistic quality inherent in absolute space render the micro-economic theories of urban land use inadequate as describers of the mechanisms of allocation when judged against alternative formulations in which exchange value predominates. If we drop the assumption that individuals and groups have homogeneous tastes with respect to housing and allow the great diversity of needs and tastes to play its part, then we stray even further from the framework encompassed in micro-economic theory — and it is thus far that we ought to stray if we are to construct a realistic theory of the forces shaping urban land use. Yet there is something disconcerting about this conclusion, for the micro-economic theories do indeed produce results which are reasonably consistent with the actual results of diverse social processes governing land-use allocation. Alonso (1964, p11) addresses himself directly to this point; he suggests that the micro-economic theories can succeed in doing much more simply what the more elaborate conceptualizations of workers such as Hawley (1950) do more realistically but with less analytic power. We therefore have to consider what it is that makes micro-economic theory so successful (relatively speaking) in the modelling of urban land-use patterns, when it is so obviously wide of the mark when it comes to modelling the real processes that produce these patterns. The solution to this problem may be sought by investigating the meaning and role of rent as an allocative device in the urban system.[4]

## Notes

[1] Space is relative because distance also depends upon the means of transportation, on the perception of distance by actors in the urban scene and so on.

[2] Consumers' surplus is the difference between what an individual actually pays for a good and what he or she would be willing to pay rather than go without it.

[3] Those wishing to consider further the concepts of consumers' surplus and relative/absolute space are referred to pp 168–71 of the original text (Eds).

[4] The role of rent is considered further on pp 176–94 of the original texts (Eds).

## References

Alonso W 1964 *Location and Land Use: Toward a General Theory of Land Rent* (Cambridge, MA: Harvard University Press)

Beckmann M J 1969 On the distribution of urban rent and residential density *Journal of Economic Theory* **I** 60–7

Gans H J 1970 *People and Plans* (New York: Basic Books)

Hawley A 1950 *Human Ecology: A Theory of Community Structure* (New York: Wiley)

Herbert J and Stevens B 1960 A model for the distribution of residential activities in urban areas *Journal of Regional Science* **2** 21–36

Hurd R M 1903 *Principles of City Land Values* (New York: The Record and Guide)

Kirwan R M and Martin D B 1971 Some notes on housing market models for urban planning *Environment and Planning* **3** 243–52

Mills E S 1969 The value of urban land *The Quality of the Urban Environment* ed. H Perloff (Baltimore, MD: Johns Hopkins)

Muth R 1969 *Cities and Housing: The Spatial Pattern of Urban Residential Land Use* (Chicago: University of Chicago Press)

Smith W F 1970 *Housing: The Social and Economic Elements* (Berkeley, CA: University of California Press)

Suttles, G D 1968 *The Social Order of the Slum* (Chicago: University of Chicago Press)

Wilson A G 1970 *Entropy in Urban and Regional Modelling* (London: Pion)

# 8    Issues in Retailing

*by R L Davies*

The business of shopping has changed dramatically in the last twenty years, from the point of view of both the retailer and the consumer. There have been enormous upheavals in the methods and organization of retailing, especially since the introduction of self-service techniques on a wide scale and the proliferation of supermarkets. There has been the development of cash-and-carry forms of wholesaling, leading in turn to the formation of voluntary trading groups for small independent retailers (such as Spar and VG), the expansion of mail-order trading, the emergence of superstores and discount houses, and most recently the inception of trade centres and trading marts. Collectively, these innovations have led to an increasing integration of the distributive trades and a blurring of the traditional distinctions between wholesaling and retailing. There has been a general escalation in bulk-merchandising practices to achieve economies of scale, and an increase in the corporate sizes of multiples or chain companies as well as the physical sizes of stores. The corollary of such growth, of course, has been a decline in the number and trading strength of the smaller traditional activities, particularly the retail markets and corner stores.

On the consumer side, there have been commensurate changes in shopping habits and travel patterns. Most noticeable has been the reduction in frequency of short daily shopping trips for small quantities of convenience goods and the greater emphasis on once- or twice-weekly bulk-buying expeditions. This has occurred partly because of the growth of female employment and partly because of the increase in personal mobility and the acquisition of freezers. The general rise in disposable incomes has also allowed for a greater total expenditure on shopping and particularly on durable goods and luxury items. Clear preferences have been expressed for the new forms of precinct and enclosed shopping centres which have further encouraged a greater trend towards family participation in shopping, especially on Saturdays. Perhaps the most significant factor affecting retailing itself, however, has been the overall shift in the pattern of population in the post-war years. The movement of people from the inner city to the suburbs has left behind a legacy of out-moded trading conditions in the older urban areas while creating vast new reservoirs of purchasing power in outlying housing estates.

In view of these spatial and other changes one might have expected to find some clear areas of conflict or controversy between consumers on the one hand and retailers on the other. In fact, however, the instances of direct and serious confrontation between them have been relatively few. This is not to deny that there have been periodic outbreaks of dissatisfaction, particularly over the prices and quality of goods, but, in the main, there has been considerable public acquiescence to even the most undesirable consequences of retail change. The main reason for this can probably be attributed to government intervention, which has contained the main forms of abuse and worst excesses to relatively low levels. In acting so vigilantly on the consumer's behalf, however, various government agencies have also tended to be identified as working against the whole process of retail change. It is therefore in the relationship between these government agencies and certain retail firms that most of the confrontation over retail change has occurred.

There is no other country in the western world which exercises such stringent planning controls over the retail system as Britain, particularly in terms of curtailing market pressures for a greater amount of decentralization of trade. While other countries have provided

Source: Davies R L 1978 *Issues in Urban Society* Ch 5, eds. P Hall and R L Davies (Harmondsworth: Penguin) pp 132–60. Reprinted by permission of Penguin Books Ltd.

similar levels of consumer safeguards and in some cases given stronger powers to consumer protection agencies, their influence over the spatial distribution of retailing activities has been markedly less than in Britain. It is this spatial intervention that has led to most conflict between retailers on the one hand and government agencies (or 'planners') on the other; and hence it is essentially a set of spatial issues that forms the subject of this essay. There are three particular questions that we might most profitably consider:

(1) To what extent have various planning authorities been justified in concentrating new retail investments in traditional city centres rather than permitting the development of new forms of suburban shopping facilities?

(2) What have commercial redevelopment programmes achieved with respect both to central-area reconstruction and to the revitalization of other parts of the inner city?

(3) How can the problems of smaller businesses and the weaker elements of the distributive trades be alleviated while accepting that planning policies must provide for an efficient and modern system of retailing?

## The debate over decentralization

The process of suburbanization is a world-wide phenomenon, but at its most conspicuous in North America. There the massive shifts in population concentration have been accompanied by extensive changes in the retail pattern, such that considerably more shopping is now undertaken in the suburbs than in the traditional centres of cities. Since there has been little planning control over this decentralization of retail trade, vast new 'out-of-town' shopping facilities have been allowed to compete with the central area and collectively their impact has led to its deterioration and decline. Some serious problems have thus emerged over the spatial inequalities in quality levels of shopping facilities. In essence, a highly attractive convenient and modern set of shopping centres has grown up in suburban areas to serve the needs of a relatively wealthy middle-class population; while in the inner cities, a poorer older and predominantly coloured population has been left with a legacy of outworn and inferior shopping facilities (Simmons 1964, Cox and Erickson 1967).

The experience of North America represents an extreme case of what happens when there is uncontrolled growth, and its importance here lies in the fact that it is often taken as showing the sorts of consequence that might follow from an expansion of suburban retail trade in Britain. By contrast, however, the experience of other countries, particularly in western Europe, provides examples of how a more limited amount of decentralization can be accommodated without such deleterious effects. In various planning climates, from France to West Germany and Sweden, large numbers of superstores and hypermarkets have been allowed, although large new suburban shopping centres have generally been restricted. Essentially, a distinction has been made between those new shopping facilities providing convenience goods and those that might compete with the more specialized functions of city centres. The result has been a considerable increase in overall suburban shopping with little decline of traditional centres except in the case of surrounding local shops (MPC and Associates 1973, NEDO 1973, International Geographical Union 1975).

The debate over decentralization in Britain therefore needs to be considered against the background of the types of shopping development that are likely to be involved. A wide assortment of proposals has been made both for centres and individual types of store. For simplicity we will distinguish here between three kinds of centre: 'out-of-town' centres, 'edge-of-town' centres and district or neighbourhood centres; and three kinds of store: hypermarkets, superstores and discount houses. It is the out-of-town centres and hypermarkets which have drawn most controversy because these are both large developments that tend to occupy greenfield sites and are most likely to have radical effects on traditional patterns of trade. The district or neighbourhood centres and discount houses are generally the most acceptable forms since these can be integrated into the existing hierarchy of shopping provisions. In between are the edge-of-town centres and superstores which in many ways constitute compromises between the extremes. Edge-of-town centres are more akin in their functional role to district centres in providing mainly convenience goods, but their location will not necessarily be confined to the middle of housing estates and could include accessible points alongside main roads in the outer parts of cities. Similarly, superstores are more like giant supermarkets than hypermarkets, although their trade need not necessarily be in foods but could include the cheaper range of household goods as well.

*The main arguments*

There are essentially three main arguments involved in the debate over decentralization at the present time:

(1) There is a set of economic considerations wherein lower operating costs for retailers and hence potentially lower prices of goods for consumers have to be measured against the erosion of trade in other localities and the possible creation of blight conditions. Suburban locations are clearly attractive to large-scale retailers because of the greater amount of space available, the ease of access for both consumers and delivery vehicles, the generally lower building costs incurred and the close proximity to other elements of the distribution system (namely warehouses and redistribution centres) that can be achieved. It is argued that the greater business efficiency that can be attained can lead to price reductions on goods of up to 10% (on national averages), besides providing shoppers with a greater choice of goods and a more congenial shopping environment. The establishment of a large new suburban facility will inevitably lead to the capturing of trade from existing and mainly smaller shops, however; and in certain cases, may even lead to loss of trade from a traditional town centre. Many councils and planning departments are particularly sensitive to this because of the amount of public investment that might already have been channelled into providing local shopping needs and redeveloping the traditional town centre. The counter-argument is therefore essentially one of protection rather than antipathy to a modern form of shopping.

Not unexpectedly perhaps, claims over price advantages and over trading effects have both tended to be exaggerated and often disputed by various research findings. The real benefits in lower prices have to be considered against the additional costs which may be incurred in visiting a suburban store, where these include not only journey costs but also the outlay in purchasing storage facilities for bulk produce, together with their running costs. It has been found that the impact of a new superstore or hypermarket falls hardest on the small branch outlets of the multiples rather than independent corner stores (Lee and Kent 1974, Thorpe and McGoldrick 1974); and many retailing firms are so confident of having only a minimal effect on town-centre trade that they continue to retain their own branches there even when opening a larger suburban facility.

(2) A second area of contention concerns the threat to the environment which large-scale suburban developments might pose. This involves not only the premises themselves but the amount of traffic which they would be likely to give rise to. Traffic considerations are important because of the possibility of overloading suburban roads in the course of alleviating bottlenecks in other parts of the city. Without a considerable road widening and improvement scheme, bottlenecks might simply be shifted from one point to another. While much has been made of this argument by planners, however, the same degree of opposition does not seem to have been met by other categories of land use (namely wholesaling and industry) which have been more actively encouraged to decentralize. There are clearly differences in the volume and character of flows to be considered here, but the possibility of linking certain kinds of retail development to new wholesaling and industrial areas does not appear to have been fully explored. Potential conflict between shoppers and heavy lorries has been raised as one form of objection, but this is a problem that can surely be resolved in the way it has on a much larger scale within the central area.

The possible links that might be drawn between certain kinds of retail development and outlying wholesaling and industrial areas are worth examination in a further context, that of the extent to which suburban shopping facilities can or should be allowed to penetrate green-belt land. This is an issue which has been vigorously argued in planning inquiries, where planners have taken the view that both hypermarkets and out-of-town shopping centres are likely not only to take up much valuable space because of their large size but also to spoil the character of the countryside by their (mainly poor) physical appearance. Developers have then tried to counteract these points by suggesting that many pockets of dereliction within the country side could actually be improved by the presence of shopping facilities, especially if strict standards in appearance were laid down, and that the employment opportunities created by these developments might well lead to a general upgrading of the locality. There is no doubt that in many conurbations excessive urban sprawl should be checked; but at the same time it seems that green-belt protectionist policies have been more strictly adhered to where retailing is in question than for most other categories of land use.

(3) The provision of new suburban facilities on a much larger scale than is now usual would, however, be interpreted by many planners and various consumer

pressure groups as discriminating against the poor, the elderly and other minority sections of the population. Since these people are predominantly confined to the inner areas of cities, they would not have the same benefits of access as younger, more mobile, middle-class consumers. Moreover, the growth of extensive suburban facilities might lead to a serious deterioration in their own local provisions. These potentially inequitable effects, however, need to be considered against the background of other retail changes. The inner areas, by virtue of their proximity to the city centre, are likely to suffer proportionately more from the effects of redevelopment schemes and the building of new 'in-town' shopping centres than they are from out-of-town or edge-of-town developments. Even in the middle and outer parts of cities, local shops remain just as much in danger of closing because of the competition of large new stores coming into the existing urban market as they do because of developments in the suburbs.

Local authorities have rarely worried about the social inequalities resulting from their own in-town shopping developments, but continue to use these arguments as a major reason for curtailing decentralization. Recently the principal criterion for refusing to allow a Tesco superstore on the edge of Swansea was that it would 'result in a material reduction in trade in local shops and some would be forced to close. People without cars, especially old people, would be seriously inconvenienced' (*Planning Newspaper* October 1975, Lee and Kent 1976). Research into the shopping patterns of the elderly, however, has indicated that these are by no means a homogeneous group; and, indeed, in some cases evidence has been found that they are often more likely to travel further than younger people because of their greater amount of free time and the availability of free or cheap bus passes. There is no doubt that superstores appeal more to the younger, family-based households, but the elderly are often prepared to foresake their local-store loyalty for visits to the central area. Within the older age band, however, there is obviously a difference between those people who remain fit and well and those who are seriously incapacitated. It is the minority who are effectively housebound who are most in need of care and attention. They have difficulty visiting even local shops, but few suggestions have been made about improving their situation.

### The way ahead

The debate over decentralization in Britain has been going on for more than ten years and unfortunately has not been as constructive as one might have hoped. Instead of a healthy exchange of ideas and the reconciliation of different needs, there has been too much antagonism between retailers and planners, with the taking up of firmly entrenched positions. The debate has not been helped by a surfeit of emotion and confusion over the semantics of the new developments involved. As in many other aspects of urban society in Britain, there has been too much confrontation and not enough action — at least in a forward or positive direction.

The loser has ultimately been the consumer. The bulk of the population remains heavily constrained in its choice of shopping facilities and is effectively restricted to two kinds of centre: the city centre for specialized or comparison goods and the neighbourhood centre or local parade for convenience items (Davies 1973). In certain localities, of course, superstores and other new developments have been introduced, but these are not always in the most ideal locations and still represent the exceptions to the general case. For most suburban residents, the main form of modern provisions has been the open precinct district centre, the equivalent of older inner-city ribbons but designed to be safer and easier to approach. While this modern form of traditional concept has no doubt made travelling to shops easier, its material effect on the actual economic and social functions involved has been small. Prices of goods remain high because of the difficulty of achieving scale economies in small separate shop units; the variety or range of goods remain equally limited, often with too much duplication; goods have to be hauled around without the use of trolleys or store buggies; there remains the discomfort of shopping in bad weather conditions (except in the cases of enclosed district centres); untidiness and vandalism are difficult to control when the upkeep falls to the local authority; the main visual impact on the shopper is usually one of concrete and bricks rather than of the internal decor of the actual shops.

It needs to be made clear, however, that we are not advocating here a complete relaxation of planning constraints such that new types of suburban development can be established at will. We are concerned with providing a greater degree of choice in situations where this can be sensibly accommodated. There are certain rules of thumb that may be used in this decision-making. First, there are only a limited number of sites throughout the country as a whole in which large new out-of-town shopping centres could conceivably be built. These are around the major cities and conurbations and ideally should be linked to potential growth areas. The commitment already made to city-centre redevelopment schemes, however, and the prohibitive costs involved in this scale of development mean that it is likely that few if any will actually be built in the 1980s. Secondly, the case for a greater decentralization of convenience trade is stronger than that for more specialized needs, hence applications for edge-of-town centres should primarily be adjudicated on the basis of their intended functional role. In many cases, these might well resemble in form the conventional district centres; but in others, a small cluster of independent shops may well be dominated by two or even three very large establishments. Thirdly, the most worrying features of hypermarkets, by comparison with other types of superstores, are their preferred location in greenfield sites and their enormous size. The two features are inextricably linked, such that the potential penetration of rural areas looks far more damaging in the case of a large proposal than a small one. Restrictions on the overall size of permissible developments would reduce not only the environmental threat but ensure competition within only one part rather than the whole of an urban market. Fourthly, superstores and discount stores need to be considered much more in terms of their compatibility with, rather than difference from, traditional neighbourhood shopping centres. In functional terms, many of them provide the same kinds of goods as a series of small independent stores, but with these arranged under one roof rather than in separated units. A deliberate integration of these into the hierarchy of shopping facilities within the city would ease the pressure on alternative sites which is now being felt. This positive arrangement in the provision of more special-purpose sites could then be matched with more restrictive policies towards the use of vacant buildings in inner-city areas.

## Commercial redevelopment programmes

In contrast to the procrastinations over the amount of decentralization of retail trade that should be allowed to take place, there has been a quiet but radical transformation of shopping facilities in the inner areas of British cities. This has involved the redevelopment of city centres and of the myriads of local clusters of stores in the so-called twilight zone. Curiously, despite the enormous sums of public money used, there has been little argument over the desirability and aims of these

programmes. This is due partly to the general recognition that these areas have long been in need of renovation and partly to the fact that those changes which have been made have often been incorporated into other redevelopment schemes (involving the clearance of houses and warehouses and the building of new roads, etc) which in themselves have commanded more attention or drawn more controversy. Even so, the record of achievement in terms of the improvements which have been made is not very impressive and a number of issues can be identified which need more thoughtful attention in future.

### City-centre shopping schemes

The initial stimulus to city-centre redevelopment came from the bomb damage of the last war when many cities, most notably Coventry, Plymouth and Southampton, were in serious need of physical reconstruction. The complete obliteration of large parts of the central area in these places allowed for experiments in the separation of vehicles from pedestrians and the planning of a more orderly arrangement of retail land uses. As economic conditions improved and the precinct concept of shopping evolved, more and more cities embarked on comprehensive redevelopment programmes, prompting the publication of a series of guidelines (Ministry of Housing and Local Government and the Ministry of Transport 1962) on how these policies should be carried through (see Holliday 1973). Subsequently, in the late 1960s, new types of enclosed shopping centre began to be built instead of the open-air precinct forms; and the construction of inner ring-roads and urban motorways allowed for the banning of traffic from the traditional High Street.

The general principles embodied in these city-centre shopping schemes are difficult to fault. They provide a much safer and usually more congenial shopping environment for the bulk of the population. What can be criticized, however, is the rather arbitrary way in which many new centres have been built and their frequent lack of attunement to local business needs. These shortcomings are manifested in a variety of ways, but most noticeably through their poor design and the difficulties developers have faced in getting units let. From the consumer's point of view, there has usually been a significant increase in the number of specialist chain shops, but shops which offer a rather standardized range of products and lack the individuality and character of those in unplanned shopping streets.

The most contentious developments are the new large regional shopping centres which add considerably to the specialized retailing stock of the city. These have been questioned not so much in terms of quality of their design, which generally speaking has been quite high, but for the rationale behind their development, given the enormous impact they have. The impact can be felt in three different ways. First, because of the overall increase in the size and importance of city centres in which they are found, surrounding town centres at lower levels in the regional hierarchy find themselves depleted of trade and likely to suffer from closures and commercial blight. Secondly, within the city centre in which the regional shopping centre is built, there are radical shifts in the centre of gravity of trade, leading to a considerable upgrading of certain shopping streets but also a substantial decline in those which suddenly find themselves in an off-centre location. Such shifts in the centre of gravity of trade are most severe when a regional shopping centre is built on the edge of the central area away from the High Street and traditional axis of pedestrian flow, and the effect is compounded by the movement of many of the larger stores into the new centre itself and even in some cases the relocation of a department store. The third type of change that causes concern is the growth in the dominance of chain stores or multiples at the expense of smaller independent stores. Smaller stores are effectively penalized in two ways: through the loss of sites when an area is cleared to make way for a new regional shopping centre, then through inability to go into the new scheme because of the high level of rents that must be paid. This leads to fewer personalized stores and a reduction in the variety of services and unusual trades.

Given the fact that most of the in-town regional shopping centres have been developed by the local authorities themselves (in a partnership arrangement with a property company) it seems remarkable that they have been allowed to be built with a minimum of public scrutiny and fuss.

The lack of a regional policy towards redevelopment schemes has meant that most towns and cities have tried to promote their own shopping importance and growth, leading to the general over-provision of retailing stock. At the same time, the larger centres have gained in commercial strength at the expense of smaller ones. Much of this is due not to the single effect of a large new regional shopping centre but to the cumulative effect of smaller precinct and enclosed centres. The net benefits of many of these schemes are dubious because of duplication of existing stores and a lower level of environmental standards. Some schemes have been abject failures (such as the Elephant and Castle Centre and The Tricorn Centre in Portsmouth) combining a poor design with an inappropriate location. The majority, however, have emerged as rather uniform monotonous concourses with the same types of stereotyped chain store, adding little variety and pleasure to the actual business of shopping. The slight advantages in safety and weather protection (in enclosed centres) are probably counteracted by the loss of small specialist traders forced out through their competition. Exceptions include several of the Arndale Centres, which, whilst perhaps lacking imagination and finesse, usually look bright and cheery and have considerable consumer appeal.

*Redevelopment in the inner city*

Redevelopment programmes outside the city centre have mainly been concerned with removing commercial slums and reducing the surfeit of small marginal businesses to be found in the inner parts of cities. Few people would quarrel with the underlying need to modernize the retail system in areas where there has been a substantial decline in trade. The arguments arise over the scale and apparent impartiality of clearance schemes and the form which new shopping centres take.

It is impossible to quantify the amount of change which has taken place in Britain. Surveys undertaken in the US, however, indicate that more than 100 000 small businesses have been dislocated through redevelopment schemes in that country since 1950. Since only about one third of these have managed to start up again, we must question whether the rate of enforced closure has been excessive. Detailed case studies undertaken in Chicago suggest it has not, given the high rates of liquidation to be found in areas unaffected by redevelopment (Berry *et al* 1968). As many as one third of all new businesses in the inner parts of American cities fold within the first two years of operation and the average life-span for all new businesses is only about seven years. However, even if redevelopment simply intensifies what is already happening, it is inevitable that certain healthier businesses become casualties when they might otherwise have survived. It is this inequality in localized situations and the lack of compensation (both in monetary terms and the provision of appropriate new sites) that is of most concern.

## The problems of local shopping

There has been a substantial reduction in the number of small local shops throughout the urban area, primarily because of the competitive effects of chain stores and the inability of independent traders to generate sufficient profits in an era of increasing operating costs and shrinking markets. These difficulties have been exacerbated by planning interventions in two ways: through redevelopment programmes in the inner city, with the sorts of consequence we have already discussed; and through restrictions on the number and type of new locations permitted in outlying areas, which effectively consist of only local parades and neighbourhood shopping centres confined to the middle of housing estates.

The overall reduction in numbers of small shops is not in itself of major concern to the majority of consumers for they have been primarily responsible for this state of affairs. Apart from a sentimental attachment to such shops, what worries them most is the disappearance of certain more specialized or unusual trades which they would like to have available for occasional needs. These businesses often provide a range of goods that the multiples will not supply because of limited demand. There are certain sections of the population, however, who are clearly more dependent than others on the presence of a wide assortment of local shops; and it is with these groups in mind that various policies to prevent any further significant decline are now being discussed. Such groups include the elderly concentrated in the inner city and women with small children on outlying housing estates.

Two different schools of thought appear to be emerging, nevertheless, on how the market conditions for small independent traders can be improved. The first is in favour of a greater degree of local authority support, whereby certain types of shop might be given rent subsidies in new shopping centres or preferential treatment over chain stores in the allocation of sites. An extension of the change-of-use orders might also prevent more desirable and essential local shops from being converted to service establishments and non-retailing activities. The second school argues for a reduction in State interference and the promotion of an atmosphere much more conducive to self-help within the business community itself. Changes envisaged would include a relaxation of planning and building controls, allowing smaller businesses greater freedom of choice over new locations and the premises they use, and a revision of the tax laws, particularly regarding

VAT. These diametrically opposite viewpoints obviously have political undertones and are difficult to reconcile; but it may be that compromise solutions can be worked out at least with respect to the spatial considerations. There could, for example, be a general loosening of planning constraints as a general principle throughout the city as a whole, but a tightening of controls in specific localities where protective measures are clearly called for.

These possibilities need to be examined in the context of the different requirements of different kinds of shop, however. To date, most sympathy has been shown for the problems of the corner shop or the traditional counter-top establishment engaged in the convenience trades. This is because they are felt to play a valuable social as well as economic role within a community. There are a number of other businesses, however, which could justifiably claim that they face more intractable operating difficulties and have been systematically discriminated against by local authorities. These include many of the services found in ribbon developments that we have mentioned before and also a set of more dubious traders concentrated in the 'twilight' zone.

The risk that goes with a general relaxation of planning controls, of course, is that there will be a lowering of environmental standards. Given the current conditions to be found within most new local authority shopping developments, however, we would judge these effects to be relatively slight. A further consideration is that any significant locational shifts that emerge might hamper those who are relatively housebound or strongly dependent on nearby local facilities. This problem could and ought to be met by an increase in home-delivery services.

## Conclusion

There have been two predominant areas of concern in retail planning during the 1960s and 1970s: the need to redevelop the outworn parts of the inner city; and the need to provide new shopping facilities in the rapidly growing suburbs. These two requirements are clearly inter-related, for an overall balance has got to be created between the total amount of new floorspace added and the total population which is being served. Instead of pursuing these problems together, however, each one has been made the subject of a separate policy, with the result that they have been viewed as a choice between alternatives: either we have a continued concentration of retail investments in city centres or a wide-spread dispersal into suburban locations. The consensus response within planning circles has been to bolster up the traditional system. It could hardly have been different given the fact that a choice had to be made.

Now that the main period of central area reconstruction is over, however, and the economic climate makes it unlikely that there will be any more large in-town shopping centres built in the near future, it seems an appropriate time for a more comprehensive planning approach to be adopted giving more scope to the decentralization at least of the convenience trades. This has clearly got to take account of the needs of small businesses as well as the larger chain companies and provide a modern and attractive local shopping environment for both the mass of middle-class consumers and more disadvantaged minority groups. It should also recognize the claims to new locational settings of a wide variety of retail services, particularly those which have been the major casualties of redevelopment schemes. The coordination and integration of these different requirements into a single commercial plan may need some new concepts both in terms of the spatial organization of shopping for the future and in terms of the individual forms that shopping centres or store clusters take. We need a design framework which enables us to relate hypermarkets and superstores to the traditional hierarchy of shopping facilities, and some imaginative blueprints for what the modern counterparts of ribbon developments and street parades should be.

What consumers have really lacked in the emerging retail system to date is an air of enjoyment and comfort in shopping, particularly for convenience goods. They have been provided with too limited a range of dull and stereotyped shopping facilities that do little to remove the drudgery and boredom involved in the routine aspects of shopping. Now that the planners have expanded their ranks, have more control over wider territories and have almost completed the task of removing the worst of the commercial slums, let us hope that we shall see more enlightenment in their policies and greater flair in the physical design of their schemes.

## References

Berry B J L, Parsons S J and Platt R H 1968 *The Impact of Urban Renewal on Small Business: The Hyde Park-Kenwood Case* (Chicago: University of Chicago, Center for Urban Studies)

Cox E and Erickson L G 1967 *Retail Decentralization* (Bureau of Business and Economic Research, Michigan State University)

Davies R L 1973 Patterns and profiles of consumer behaviour *University of Newcastle, Department of Geography Research Series No. 10*

Holliday J 1973 *City Centre Redevelopment* (London: Charles Knight)

International Geographical Union 1975 Urbanisme commerciale et renovation urbaine *Commission on Applied Geography Symposium Papers*, University of Liège

Lee M and Kent E 1974 *Caerphilly Hypermarket Study, Year 2* (London: Donaldsons)

——1976 *Planning Inquiry Study* (London: Donaldsons)

Ministry of Housing and Local Government and Ministry of Transport 1962 *Town Centres: Approach to Renewal* (London: HMSO)

MPC and Associates 1973 *The Changing Pattern of Retailing in Western Europe* (Worcester: MPC and Associates)

NEDO 1972 *The Distributive Trades in the Common Market* (London: HMSO)

Simmons J W 1964 The changing pattern of retail location *University of Chicago, Department of Geography Research Paper No. 92*

Thorpe D and McGoldrick P J 1974 *Carrefour Caerphilly: Consumer Response to a Hypermarket* (Manchester Business School, Retail Outlets Research Unit)

# 9 Housing Conditions in the United Kingdom: Who Got What at What Cost, 1919–77

*by B Headey*

## Introduction

A basic assumption of policy-makers in the UK has been that separate provision needs to be made for middle-class and working-class housing. Middle-class people live in suburbs of owner-occupied houses bought with building society mortgages and built by private enterprise firms. An increasing number of working-class people are also becoming owner-occupiers but most live either in council estates (i.e. housing owned by local authorities) which are immediately recognizable and geographically distinct from the suburbs, or rent accommodation in central-city areas from private landlords. The British public undoubtedly perceives government policy as being concerned to 'help people who cannot help themselves' — a social welfare approach designed to improve what nineteenth-century Royal Commissions and Acts of Parliament plainly called 'the housing condition of the working classes'. The live issues in the housing field have always related to council housing, rent control and protection of the security of tenure of those who rent in the private market. In reality, owner-occupiers in the UK (as in Sweden and the US) benefit very substantially from tax concessions but these have attracted almost no publicity — they are simply not recognized as subsidies — and middle-class demands have generally been confined to urging government to reduce its subventions to the public sector.

In presenting evidence of the changing distribution of housing services and costs in the UK we shall generally be comparing the position of tenure groups:

owner-occupiers, local authority tenants and private tenants. This is the way statistics are usually presented in government documents; a method of presentation which reflects both the official view that different types of housing are appropriate for different social classes and the observed fact that far fewer households move between than within tenure groups (Murie *et al* 1976). It is broadly true to say that most people with higher incomes and socio-economic status choose to be owner-occupiers, that local authority tenants occupy an intermediate position in the social hierarchy and that people with the lowest incomes and status are disproportionately found in the private rental sector (table 9.1). In the UK, as elsewhere, rising incomes in the post-war period, coupled with favourable tax regulations, have drawn more and more families into owner-occupation. Before the building boom of the 1930s only a quite small proportion of families (perhaps 10%; exact figures are not available) were owner-occupiers. In the 1950s and 1960s owner-occupation increased very rapidly, so that by 1975, 53% of households (including about half of those headed by skilled workers) were buying or had already bought their own house. Table 9.2 shows the trend towards owner-occupation and also the decline of the private rental sector relative to the local authority sector.

## Housing requirements

Compared with their counterparts in other countries, policy-makers in the UK in this century have been fortunate in not having to cope with rapid demographic changes. From 1921 to 1971 (the date of the last census) the population increased only from 44 million to 55.7

Source: Headey B 1978 *Housing Policy in the Developed Economy* (London: Croom Helm) pp 100–14.

**Table 9.1** Income groups by tenure, 1974 (after *Social Trends* **7** 1976)

| | | Income (£) | | | | | | |
|---|---|---|---|---|---|---|---|---|
| Tenure | Under 1000 | 1000 to 1499 | 1500 to 1999 | 2000 to 2499 | 2500 to 2999 | 3000 to 3499 | 3500 to 4999 | 5000 and over |
| Owner-occupied | 39 | 41 | 43 | 47 | 50 | 60 | 64 | 75 |
| Local authority tenants | 30 | 43 | 40 | 39 | 37 | 32 | 28 | 18 |
| Privately rented and unfurnished | 31 | 16 | 17 | 14 | 13 | 8 | 8 | 7 |
| | 100% | 100% | 100% | 100% | 100% | 100% | 100% | 100% |
| *N* | 113 | 228 | 399 | 606 | 631 | 609 | 1162 | 655 |

million, and the proportion living in urban areas stayed constant at around 77%. Regional migration was also on a small scale; the much deplored drift to southeast England increased that region's share of population only from 12.3% in 1921 to 17.3% in 1971. This was a very minor shift compared with the drift away from Norrland in Sweden and from the South to the northern cities and California in the United States. Immigration has affected things very little: there has been a small net outflow for most of this century, except from 1958 to 1962. Immigrants moved mainly to London (which, with 4.7% of its population foreign born, is much the most cosmopolitan city) and the West Midlands, though there are concentrations in some northern cities too.

The demographic change which has made most impact on housing demand has been the increasing proportion of married people and old people. These two groups generally form separate households and their requirements have boosted demand ever since the 1930s. This can be seen from the fact that between 1931 and 1951 the population increased by just under 10%, while the number of households increased by 28.2%

**Table 9.2** Dwellings by tenure, 1950–75 (after *Social Trends* **7** 1976)

| Tenure | 1950 (%) | 1960 (%) | 1970 (%) | 1973 (%) | 1975 (%) |
|---|---|---|---|---|---|
| Owner-occupied | 29 | 42 | 49 | 52 | 53 |
| Rented from local authority or new town | 18 | 27 | 30 | 31 | 31 |
| Rented privately | 45 | 26 | 15 | } 17 | } 16 |
| Other tenures | 8 | 6 | 5 | | |
| | 100 | 100 | 100 | 100 | 100 |

(Murie *et al* 1976). In the 1960s and 1970s demand has been further stimulated by more single people and divorcees forming their own households. A further 10% increase in population from 1951–75 was accompanied by an increase of over 30% in separate households. The position now is that only 40% of households include children, and the number of one- and two-person households continues to rise rapidly.[1] These smaller households are not well catered for, due to the tendency of both local authorities and private enterprise to concentrate on building five-room (i.e. three-bedroom) houses. Such houses are cheaper to construct per square metre than smaller units but involve smaller households in 'wasting' a great deal of space.

**Resources devoted to housing**

The UK's relative economic position has deteriorated throughout this century. At the end of World War I per capita income was considerably less than in America, though probably higher than in Sweden. In the postwar years the UK was rapidly overhauled by Sweden too, so that by the 1970s British per capita income was about half the Swedish and American. The slow growth of the resource base, coupled with the priority given to macro-economic policy, has meant a relatively slow building rate measured in terms of new dwellings per thousand population. The highest building rates ever achieved in the UK were in the years 1934–9 when construction was mainly by private enterprise. Other boom periods were the early 1950s when the Conservatives achieved their election pledge to build 300 000 houses a year and the 1965–8 period when Labour tried but failed to achieve its pledge to build 500 000 houses a year. (A high of 413 700 was achieved in 1968.) In general though, except for the 1930s, the UK's building rate has been well below that of Sweden and somewhat below the American level. Investment has simply not been available and in fact since 1950 has averaged only 3.2% of the gross national product (GNP).[2]

While not achieving a high rate of new building, British government has managed to knock a lot of old houses down. Local authority slum clearance plans, designed to remove the worst of nineteenth-century jerry-built housing, were initiated on a large scale in the UK in the 1930s when other countries still relied wholly on haphazard private enterprise clearance and replacement of old dwellings. It can be seen from table 9.3 that slum clearance has moved in cycles. Government policy has oscillated between concentrating on increasing the total stock while neglecting old houses (1919–33, 1945–55) and concentrating on slum clearance and improvement of old houses (1934–9, 1955–64) while hoping that private enterprise would take up the slack in new building. One reason for the failure to pursue a more balanced strategy is that it is politically more dramatic — it makes for better slogans — to set a high annual building target or, alternatively, to declare that the years of shortage are over and that all that is required now is to refurbish neglected but basically sound dwellings.

## The quantity of housing available in relation to requirements

Relative to all other western European countries, the UK has long had a relatively plentiful supply of housing. Significant shortages arose in the wake of the two world wars but these were less serious than in other countries because no invasion occurred and the destruction of housing stock was therefore less serious.

It is clear that in most of the country overcrowding is no longer a problem; by 1971 only 1% of households were living at a density of more than 1.5 persons per room. There is, however, still overcrowding in London (2.4% of households lived at more than 1.5 persons per room in 1971) and in the Scottish cities, where housing conditions are generally worse than in the rest of Britain.

In recognition of the diminishing value of persons per room as a measure of housing inadequacy the government now uses a 'bedroom standard'. This decrees that a bedroom is required by each married couple, each other household member aged 21 or over, and each two members aged less than 21, except that those aged 10 to 20 should share only with a person of the same sex. On this (strict) definition 6.8% of households

were shown by the 1971 General Household Survey to be inadequately provided for (Office of Population Census and Surveys 1973). 11.3% of unskilled manual workers' families were below the standard and, at the other end of the social scale, only 1.3% of professional families.

Lastly, as a measure of psychological overcrowding, we may take the number of households sharing accommodation. This would most commonly occur, of course, when a newly married couple moved in with the husband's or wife's parents for a few years before finding their own place. The proportion of households sharing has in fact reduced from 19.8% in 1921, to 18.3% in 1951 to 7.3% in 1966 (Rollett 1972).

## Housing quality

No reliable estimates of housing quality exist for the pre-war years. Local authorities were first asked to declare how many 'unfit' houses there were in their area in 1934, but on that occasion (and subsequent ones) they tended simply to state the number they intended to demolish. If we were to believe these returns, we would have to say that 'unfit' houses had increased from 266 851 in 1934 to 820 000 in 1965. Since all other indicators show improving conditions it is clear these figures are valueless.

As a preliminary measure of quality we may take the age of the current stock. Here we assume that newness correlates with quality and we find that, compared with Sweden and the US, the UK has very old housing. As of December 1975, 18% of dwellings dated from prior to

**Table 9.3** Slum clearance, 1930–73: England and Wales (after Rollett 1972 and Department of the Environment quarterly *Housing and Construction Statistics*)

|  | Total houses demolished or closed | Parts of buildings closed | Number of people moved |
|---|---|---|---|
| To March 1934 | 27 564 | — | 91 109 |
| April 1934 |  |  |  |
| —March 1939 | 245 272 | — | 1 001 417 |
| 1940–4 | nil | — | — |
| 1945–9 | 29 350 | 3 850 | 98 950 |
| 1950–4 | 60 532 | 5 913 | 211 090 |
| 1955–9 | 213 402 | 8 571 | 682 228 |
| 1960–4 | 303 621 | 5 514 | 833 746 |
| 1965–9 | 339 419 | 4 253 | 896 352 |
| 1970–5 | 367 381 | 4 668 | 821 076 |

1891, 15% were of 1891–1918 vintage, 22% were inter-war units and 45% were post-war (*Social Trends* **7** 1976). The oldest units were preponderantly in the privately rented sector which serves the lowest-income families.

The most reliable data on housing quality relates to plumbing facilities. Table 9.4 reports the proportion of households revealed by the 1955–71 census to be sharing or lacking hot water, a fixed bath and a WC. The figures show a steady improvement, though with 8.4% of households in 1971 still lacking their own hot water supply, 12.3% lacking a fixed bath and 4.4% lacking exclusive use of a WC there is no reason for complacency. Conditions have always been worst in privately rented accommodation. In this sector in 1966 only 35.3% of households (as opposed to 81.1% of owner-occupiers and 87.8% of local authority tenants) had exclusive use of hot water, a fixed bath and an indoor WC (Rollett 1972). By 1972 conditions had improved somewhat, but even then about 50% of tenants living in furnished accommodation and a third of those in unfurnished accommodation lacked one or more basic amenity (*Social Trends* **4** 1973).

It bears repeating that these basic amenities are indeed desperately basic. If we raise our standards a bit, the picture looks considerably worse. The very thorough Housing Condition Survey of 1967, undertaken by the Government Social Survey in England and Wales, reported that $2\frac{1}{2}$ million dwellings lacked one or more of four amenities — an internal WC, a fixed bath, a wash basin and hot and cold water at three points — and that another $3\frac{3}{4}$ million needed repairs which would cost more than £125. In other words $6\frac{1}{4}$ million units — 40% of the total stock — were found to be somewhat

defective (Rollett 1972, Berry 1973). (And these figures exclude Scotland where conditions are known to be worse.) The government's gift for understatement in these matters can be judged from the fact that, on the basis of this survey, it declared only 1.8 million units 'unfit'.

UK housing continues to be deficient in design and finish and to lack such facilities as roof insulation, built-in closets and central heating which are standard in new houses in other northern European countries. In these respects local authority housing is often better than owner-occupied housing particularly since new standards (Parker Morris standards) were introduced in 1969. By 1974, 90% of new local authority housing was built with central heating,[3] certainly a far higher percentage than in the private sector which, as a Swedish housing official remarked to the author, 'showed its traditional determination not to infringe the civil rights of Eskimos.'

## Housing choice and neighbourhood quality

It is often said that British people lack effective choice over the location and tenure of their housing. While it is true that most middle-class people in fact live in owner-occupied suburbs and most working-class people live in council housing, choice is not as restricted as might appear. Quite a lot of older dwellings at relatively low prices are available for working-class people who want to become owner-occupiers and in fact over 20% of unskilled, 30% of semi-skilled and 40% of skilled workers are owners. For middle-class people most cities offer a wide range of inner and outer suburbs which vary according to status and the cost of housing but which are by no means so segregated by income as, say, American suburbs. This is because local authorities, unlike their American counterparts, have not used their planning powers systematically to enforce social segregation. The main restrictions on housing choice in fact arise as a result of the contraction of the private rental sector (as we shall see, the effect of successive government measures have been to force landlords out of business) and the absence of any equivalent of the (middle-income) cooperative housing sector found in Scandinavian countries. Also serious, as a restriction on labour mobility as well as on personal choice, is the fact that council tenants are often reluctant to move from one city to another because they would lose their houses. They would have to get on their new city's waiting list and, since length of resi-

**Table 9.4** Housing quality: possession of amenities in England and Wales (after Rollett 1972 and *Social Trends* **6** 1975 Crown Copyright).

|  | 1951 | 1961 | 1966 | 1971 |
|---|---|---|---|---|
| Percentage of households without amenities: |  |  |  |  |
| hot water tap | n.a. | 21.9 | 12.5 | 6.5 |
| fixed bath | 37 | 22.0 | 14.9 | 9.1 |
| WC | 8 | 6.9 | 1.8 | 1.1 |
| Percentage of households sharing: |  |  |  |  |
| hot water tap | n.a. | 1.8 | 2.1 | 1.9 |
| fixed bath | 8 | 4.6 | 4.3 | 3.2 |
| WC | 13 | 5.8 | 6.0 | 3.1 |

dence is usually one criterion in the 'points' systems used by local authorities, their chances of getting a house quickly would be slight.

More persistent than complaints about the choice of housing available are complaints about neighbourhood quality. Most council-house tenants are reasonably satisfied with their dwellings, but there is considerable dissatisfaction about lack of amenities and services and (even harder to assess) unsatisfactory community life on council estates. The charge is made that by building satisfactory houses but not providing amenities government in some sense treats council tenants as second-class citizens. The provision of amenities on council estates — shops, transport, entertainment and recreation — was undoubtedly skimped in the inter-war years and in the drive to remove the post-war housing shortage. Councils could justify their austerity on the grounds that it would be wrong to use limited funds to provide 'extras' at a time when many families were still living in appallingly overcrowded or slum conditions. Impressionistic evidence suggests that by the 1960s many councils had relented to the extent of providing adequately for shopping and transport and also for recreational activities deemed 'worthy': play areas and swimming pools for children, youth clubs and community centres.[4] They were still unwilling, though, quite largely for fear of an electoral backlash, to build and rent cinemas, dance halls, bingo halls and pubs. In a study of Castlemilk, Glasgow — for many years the largest council estate in Europe — the author found that the lack of such amenities locally, and the cost of going into the central city to find them, were the most commonly cited sources of dissatisfaction by residents (see Headey 1972, Ministry of Housing and Local Government 1966a, 1969). Castlemilk is a well established area: in newer council estates surveys have generally found that more distress is caused by loss of friends and break-up of the three generation family[5] (especially of the relationship between married women and their mothers) arising from movement out of the central city to estates in which everyone is newly arrived and friendships and community life develop slowly.[6]

One group whose grievances have received a great deal of publicity are residents of high-rise flats. These flats were system-built on a large scale from 1964 to 1968, when the Ronan Point disaster led to drastic curtailment of plans.[7] Some councils placed families with small children in such flats and the accidents which resulted, the phenomenon of children playing in corridors because official play areas were too far away for effective parental supervision, and the alleged anomie engendered by high-rise living, produced an outcry against the inhumanity of bureaucrats and planners. The general revulsion against high-density high-rise living has now died down however, partly because surveys have shown that many households — old people and those without young children — find it reasonably satisfactory (Stevenson et al 1967, Jephcott 1971, Ministry of Housing and Local Government 1969).[8] It appears that anomie and insecurity result not from high-rise living as such but from what Newman (1972) calls lack of 'defensible space', of an external area close to the home which an individual family or a small group of families feel they can watch over and call their own (see also Department of Environment, Housing Development Group 1972).

The saddest disadvantage of council-estate living, which nobody knows how to overcome, is the social stigma attached to it. Owner-occupiers in their suburbs generally want to be as clearly differentiated from council tenants as possible and have induced authorities to put up huge barriers separating them from their presumed social inferiors. The most notorious barrier was erected in Oxford, where the seven feet high Cutteslowe Walls topped with revolving spikes divided a council estate from a suburb (Collison 1963).[9] 'Social grading' of a peculiarly pernicious kind occurs within council estates and when comparisons are made among different estates (see Tucker 1966). Local authorities typically send 'housing visitors' to inspect the homes of prospective tenants and it is the task of these visitors to decide whether their clients are Grade 1 (most respectable, assign to a recently built area), Grade 2 (moderately respectable, assign to an older post-war estate) or Grade 3 (problem tenants, assign to a pre-war or dilapidated estate) (Tucker 1966). Interviewers have found that council tenants themselves feel that, as recipients of subsidized housing, they are in some sense downgraded and fully understand the desire of owner-occupiers to be physically and socially separate from them (Tucker 1966). It is also clear that within their own estates residents soon distinguish relatively prestigious areas inhabited by families who obtained their houses as a result of being on the waiting list from inferior areas inhabited by 'problem families', or families rehoused due to slum clearance.

Years ago, Labour party spokesmen feared that social stigma would attach to council housing if size and quality standards were set too low. Mr Jimmy Wheatley in 1923 accused a Conservative Minister who proposed reducing standards of 'stereotyping poverty in housing for half a century to come and giving Parliamentary acceptance to the permanence of class distinction in this country. Why do you propose these boxes for our people? Are they inferior people to you?' What no one foresaw was that even when the quality of council building reached a high level (as under the current Parker Morris standards it now does), the social stigma would remain.

### Housing cost-to-income ratios

The proportion of its income a household spends on housing generally varies a good deal over the housing 'life-cycle'. Young married couples in which the wife is still working may be able to spend large sums without undue hardship. When children arrive and the wife is tied to the home the burden may become almost intolerable, but the couple may be willing to bear it for the sake of decent housing, for status reasons and in the expectation that the husband's income will rise while house payments remain more or less constant. Typically, both for owner-occupiers and renters, housing cost-to-income ratios are at their lowest when unmarried children are also bringing in incomes. It is people at this stage of the housing life-cycle, living in council accommodation, who are regularly assailed by right-wing commentators for living in subsidized housing when (allegedly) they could well afford owner-occupation. In reality though, these tenants, and their counterparts in other types of accommodation, in many cases again find themselves in difficulties in old age. Poverty in old age, the most common and shameful type of poverty in the UK today, is reflected in housing as in all other material conditions.

Reliable data on housing cost-to-income ratios are only available for recent years and come from the family expenditure surveys carried out by the Department of Employment. For earlier years A A Nevitt collected evidence on the ratio of manual workers' earnings to the cost of a new local authority house. She found that between 1931 and 1962 such houses cost about 2.3 times a skilled worker's earnings and 3 times an unskilled worker's earnings. However, since the standard of council housing during the period improved enormously, the information can only be regarded as of general interest rather than as giving an indication of the housing services people could actually afford. The latest available figures, broken down by income, are given

in table 9.5. It can be seen that wealthier households spend considerably lower proportions of their income, though more absolutely on housing and fuel than poorer households. As compared with an average expenditure for all households of 19% of income, households in the four lowest brackets spend over a quarter of their income on housing and those in the four highest brackets spend around 15–17%. On the basis of these figures it would certainly seem — and this is on the face of it surprising — that the nationwide introduction of rent rebates in local authority housing in 1972 and of rent allowances in the private sector has not removed the class of people whom Americans term 'house-poor'.[10]

As well as facilitating comparisons of income groups, the data collected by the family expenditure surveys also enable us to compare households by age of head, by family composition, by region and by tenure groups. The plight of the old is revealed with appalling clarity. In 1974 women over 60 living by themselves (usually widows) spent 37.4% of a total expenditure of £16.18 on housing and fuel, single old men spent 32.7% of an average expenditure of £18.02, and elderly couples spent 36.8% of £16.10 (Department of Employment 1974). The next most disadvantaged type of household are those headed by single parents who on average spent 24.5% of £33.06. Turning now to regional differ-

**Table 9.5** Housing expenditure by income, 1974 (Department of Employment *Family Expenditure Survey* 1974, Crown Copyright)

| Household weekly income | Costs as percentage of total household expenditure | | |
|---|---|---|---|
| | Housing (1) | Fuel (2) | (1) + (2) |
| Under £12 | 16.8 | 11.8 | 28.6 |
| £12–15 | 23.6 | 12.2 | 35.8 |
| £15–20 | 20.7 | 10.4 | 31.1 |
| £20–25 | 18.3 | 8.1 | 26.4 |
| £25–30 | 17.3 | 7.0 | 24.3 |
| £30–35 | 16.2 | 7.0 | 23.2 |
| £35–40 | 16.2 | 6.4 | 22.6 |
| £40–45 | 14.8 | 6.0 | 20.8 |
| £45–50 | 14.2 | 5.8 | 20.0 |
| £50–60 | 13.1 | 5.2 | 18.3 |
| £60–70 | 12.8 | 5.0 | 17.8 |
| £70–80 | 12.7 | 4.4 | 17.1 |
| £80–100 | 12.6 | 4.4 | 17.0 |
| £100–120 | 11.1 | 3.8 | 14.9 |
| £120+ | 13.1 | 3.4 | 16.5 |
| All households | 13.8 | 5.2 | 19.0 |

ences, the housing problems of Londoners can be judged from the fact that a family of two adults and two children typically spent 21% of total expenditure on housing and fuel, compared with 17.7% spent by similar families in provincial conurbations, 17.8% by such families in other urban areas and 17.3% in rural areas.

All the comparisons we have just made are flawed in two respects. First, they do not take account of the tax allowances — on mortgage interest, capital gains and imputed net rent — available to owner-occupiers but not to renters. Since, as we know, owner-occupiers are concentrated among higher-income groups, the figures cited above understate by a considerable margin the disadvantaged position of lower-income groups. Secondly, no consideration is given to the amount of housing service received by different tenure groups and types of household, and hence no attempt can be made to calculate the value-for-money (or rate of return) people are getting for their housing expenditure.[11] If we assume that a similar balance of advantage/disadvantage has prevailed in the past (though the data are simply unavailable to confirm this), owner-occupiers have been in a much more favourable position than local authority or private tenants.[12]

## Notes

[1] Such households were 38.1% of the total in 1951 but by 1971 had reached 49.3% (see *Social Trends* **4** 1973).

[2] For comparative data on building investment see United Nations *Yearbook of National Income Statistics* and United Nations Economic Commission for Europe, *Annual Bulletin of Housing and Building Statistics*.

[3] In 1975 the number of units built with central heating declined.

[4] Evidence on these points has to be impressionistic since councils are not required to report on the amenities of their estates.

[5] Typically, in working-class central-city areas, grandparents live close to married children and hence their grandchildren.

[6] The first study to bring out these family and social difficulties was Young and Willmott (1968, originally published 1957). Their analysis of changing relationships within the family is presented in Young and Willmott (1973, originally published 1960). An excellent series of 'before' and 'after' studies was conducted by the Ministry of Housing and Local Government (1970a, b, 1971). See also Ministry of Housing and Local Government (1966b).

[7] A gas explosion at the Ronan Point flats in Canning Town, London, on May 16 1968, led to the collapse of part of the building, indicating serious design faults.

[8] For a summary of this research see Department of the Environment (1975).

[9] The walls were pulled down in 1959.

[10] The take-up rate, particularly in the private sector, is appalling; in some areas it is as low as 20%.

[11] In Chapter 1 of the book that this passage was taken from it was calculated that (on the basis of some fairly heroic assumptions) the rates of return that local authority tenants, private tenants and owner-occupiers might expect to get over the next forty years would be 7%, 7–9% and at least 10% respectively.

[12] The position of owner-occupiers may well have further improved relative to tenants because, while many ordinary forms of saving and investment lost value due to interest rates not keeping up with inflation in the 1970s, house prices (i.e. people's investments in their own homes) kept ahead of inflation.

## References

Berry F 1973 *Housing: The Great British Failure* (London: Knight)

Collison P 1963 *The Cutteslowe Walls* (London: Faber)

Department of Employment 1975 *Family Expenditure Survey 1974* (London: HMSO)

Department of the Environment, Housing Development Group 1972 *The Estate Outside the Dwelling: Reactions of Residents to Aspects of Housing Layout* (London: HMSO)

Department of the Environment 1975 *The Social Effects of Living Off the Ground* (London: HMSO)

—— 1976 *Housing and Construction Statistics* **16** (4th Quarter 1975) (London: HMSO)

Headey B 1972 Indicators of housing satisfaction: a Castlemilk pilot study *University of Strathclyde Survey Research Centre Occasional Paper No.* 10

Jephcott P 1971 *Homes in High Flats* (Glasgow: Oliver and Boyd)

Ministry of Housing and Local Government 1966a *Home and Environment: A Pilot Study of Reading* (London: HMSO)

—— 1966b *The Deeplish Study: Improvement Possibilities in a District of Rochdale* (London: HMSO)

—— 1969 *Estate Satisfaction on Six Estates* (London: HMSO)

—— 1970a *Living in a Slum: a Study of St Mary's, Oldham* (London: HMSO)

—— 1970b *Moving out of a Slum* (London: HMSO)

—— 1971 *New Housing in a Cleared Area* (London: HMSO)

Murie A *et al* 1976 *Housing Policy and the Housing System* Ch 2 (London: Allen and Unwin)

Newman O 1972 *Defensible Space* (Washington, DC: American Institute of Planners)

Office of Population Census and Surveys 1973 *The General Household Survey* (London: HMSO)

Rollett C 1972 *Housing Trends in British Society Since 1900* ed. A Halsey (London: Macmillan)

*Social Trends* **4** 1973 (London: HMSO)

*Social Trends* **5** 1974 (London: HMSO)

*Socail Trends* **6** 1975 (London: HMSO)

*Social Trends* **7** 1976 (London: HMSO)

Stevenson A *et al* 1967 *High Living: A Study of Family Life in Flats* (London: Cambridge University Press)

Tucker J 1966 *Honourable Estates* (London: Gollancz)

United Nations *Yearbook of National Income Statistics* (New York: UN)

—— Economic Commission For Europe *Annual Bulletin of Housing and Building Statistics* (New York: UN)

Young M and Willmott P 1968 *Family and Kinship in East London* (London: Routledge and Kegan Paul)

—— 1973 *The Symmetrical Family: A Study of Work and Leisure in the London Region* (London: Routledge and Kegan Paul)

# Section III
# Social Processes in Cities

**Introduction**

Cities are not just a collection of urban–regional relationships and an amalgamation of land uses but they are also the arena of a large proportion of the country's lives, the place where leisure and work activities take place, where decisions are made and implemented, where most people are born and grow up. Although a growing proportion of the population now take an annual holiday away from home, often travelling to other parts of the country or abroad, and more and more people migrate between the country and the city or between cities when changing jobs or houses, the largest part of most people's life is spent in the familiar surroundings of one small part of a particular town or city. Many people develop a strong attachment to their own local area and have strong feelings about it. Indeed for newcomers to the city, particularly those from other countries and other races, certain small areas have become identified with their homeland or at least with members of their own nationality who now live there. Many cities have their 'Chinatown' or 'Soho', an identifiable Italian area, a Jewish area, an Asian or West Indian area. Similarly, other parts of the city acquire a reputation for certain characteristics — in most towns, the residents when asked would easily be able to identify 'a nice neighbourhood', 'a solid suburb', an area that is 'going up' or 'coming down' and it is generally expected that certain types of people, of a particular social class or income level, would live there.

This social differentiation of cities — the extent to which individuals and households with varying characteristics live in different areas of the city and engage in different types of life styles — has long been a major focus of interest for social scientists and for the emergent area of urban studies. The six articles in this section of the reader have been selected to illustrate the diversity of approaches and range of issues that have been tackled under the broad heading of social processes in cities.

The first two extracts, from books by Raban and Thorns, represent a long tradition (which has been particularly important in sociological studies of urban areas) of documenting and describing in detail the differences between small areas in cities. Together the two extracts provide both a description of and a contrast between areas that are commonly found at the present time in the towns and cities of market economies — the inner city and suburbs.

*Raban*, by describing in detail the occupants of a single square in Islington in inner London, draws attention to the diverse characteristics of the inhabitants of inner areas. The operation of general processes of invasion, succession and domination identified as characteristic of the inner areas of North American cities by sociologists working at the University of Chicago in the 1920s is illustrated by Raban's specific example. Although the extract is descriptive rather than analytic in tone, certain of the processes that lead to population change and turnover and urban decay, later to be followed in this particular example by overall improvements in living conditions, in inner areas are identified by Raban. The role of the local authority in designating the square as a conservation area, the impact of redevelopment and rehousing policies, the greater purchasing power of the middle class are all mentioned as factors leading to social change in the area. These are some of the social processes that the Chicago school subsumed under the general heading of the 'hidden hand of competition' regarding competition as a 'natural' urban phenomenon similar to competition among plant, rather than human, communities.

In the second extract, *Thorns* is concerned with a contrasting area of the city where the social heterogenity of inner areas is replaced by the social homogenity — at least in terms of social class and life-cycle stage — of the suburbs. Thorns' study is more analytical than that of Raban. He is concerned to challenge the popular myth, which is also to some extent found in sociological writing about suburbs, particularly in North America, that suburban residence leads to a particular life style. He subjects the general features of 'the suburban myth' to a critical analysis based on a number of studies of suburban life styles.

Both suburban and inner areas, as well as being the subject of academic investigation, have been the focus of social concern. Indeed in Great Britain the two have been fused in an investigative sociological tradition that Halsey (1980) has labelled 'political arithmetic' because of the desire of its practitioners not only to describe social differences in cities but to use the results of their work for definite political aims — usually based in practical urban reforms. The processes of social differentiation and residential segregation, that have led to areas as different as the inner city and the suburbs in the cities of market societies, stem from the vast social and spatial upheavals that were wrought by rapid industrialization and urbanization in the nineteenth century — first in Great Britain and then in other countries in western Europe and North America. Before that time residential segregation — the spatial separation of households and families of different social classes, races or ages — was much less pronounced. The changes that occurred in Victorian England, in particular the increasingly visible numbers of the working and unemployed poor concentrated in certain areas of the city, gave rise to concern: not only among philanthropists and social observers such as William Booth, Beatrice Webb and Seebohm Rowntree but also among politicians and legislators who endeavoured to introduce legislation to improve housing standards and living conditions. Support was often forthcoming from industrialists who recognized the heavy burden that ill health and poor living conditions placed on industry through lost days and low productivity. The concern of British philanthropists and reformers was manifested in a number of ways including the erection of model dwellings and even entire towns.

Of greater influence on the intellectual development of social science in this country was the beginning of a tradition of describing in detail the main features of daily life in particular areas, often predominantly working-class, of the city. This type of work is commonly labelled 'community studies'. In some of these case studies surveying, mapping and counting areas of poverty or of social disadvantage remained the focus of attention, continuing the earlier tradition of combining academic and political interests. The classic study of working-class life in Bethnal Green, for example, was undertaken by Young and Willmott (1962) in an attempt to persuade the local housing authority not to demolish the nineteenth-century housing of the area and rehouse its occupants elsewhere, but to engage in more selective policies which would enable those residents who wished to do so to remain in Bethnal Green. The later work of Young and Willmott has also been undertaken with specific political ends in view rather than purely as an academic exercise.

The 1950s and 1960s was the period when 'community studies' were most influential and many studies of small areas in British and North American cities were published, influenced both by the Chicago tradition and by the British 'social concern' tradition. Common patterns of association between a range of social characteristics and problems and particular areas of the city were identified in many towns and cities, in both Britain and North America. In some cases this led to the mistaken assumption that living in a particular part of the city caused social problems — known as the 'ecological fallacy'. This confusion of correlation with causation has been subjected to wide reaching criticism and it is now recognized that the concentration of high unemployment rates, poor housing and families with low incomes in certain parts of the city results from the operation of more general social and economic processes.

In his demolition of the 'myth of suburbia' Thorns makes clear that the particular family and home-centred life style of the suburbs is not caused by living there but is already a desired feature of the life style of households who move to suburban locations. In the third extract, *Abrams* expands this theme in a more general critique of the concept of 'community' arguing that sociologists must separate location from life style. Local small-scale analysis still has its place in urban studies, not least because of recent governmental concern with inequalities in the distribution of goods and resources within cities, but more recently academic effort has turned to studying the links between small-area social and spatial differences and more general social and economic forces. In particular, the influence of the capitalist mode of production and state intervention policies on urban form and structure have begun to be analyzed by researchers approaching urban questions from backgrounds in a number of disciplines as well as sociology. The extract from *Cockburn*'s book is one example of this approach. She attempts both to locate existing definitions of 'community' in the operation of broad social forces and to redefine it in such a way as to make clear what she sees as the most important links between small area differences in cities and the organization of an advanced capitalist economy. She defines 'community' in terms of 'collective reproduction', analyzing the key role of women as workers and clients in this sphere and as providers of services in the family. In this way, she is able to make links between domestic life and work in the labour market, which are neglected in the earlier 'traditional' community studies, and also to argue that such a linking makes the way clearer for the combination of class struggles in the workplace and what were previously seen as less important community-based struggles about issues that primarily affect families' living conditions.

Cockburn's analysis has a great deal in common with Castells' identification of collective consumption as the central theme of urban studies. Dunleavy outlines Castells' argument in Section 1 (§1.2) and includes a short extract which defines collective consumption. As he uses the term, it is virtually identical to what Cockburn calls collective reproduction, although she goes further than Castells in explicitly recognizing the importance of women as providers, recipients and workers in this sphere.

Although the neo-marxist arguments about the importance of wider social forces in influencing urban form and structure are becoming more widely accepted, the question still remains as to whether there are specific urban processes that act to reinforce the differences between small areas in the city and between the people living there. In an extract from a book specifically concerned with the educational system in the inner areas of British cities, *Quinton* tackles this question by comparing families living in inner London and in a small town on the Isle of Wight. Although he rejects the hypothesis that living in cities in itself affects children's achievements, the evidence is less certain for adult behaviour. Women living in cities, in particular, appear to suffer more psychological and social problems than their counterparts in small towns. Quinton's article is included not only for its specific content but also for its methodological approach. His comparison of a matched sample of children with behaviour problems living in different sized urban areas is an excellent example of a clear research design to explicitly test his hypothesis that the size of cities affects social behaviour. He also demonstrates the inadequacies of official statistics and other studies that purport to demonstrate an area effect on behaviour.

The re-orientation of studies of social differentiation and segregation in cities that has taken place in more recent years has led to a shift of emphasis from the characteristics of individuals and of households to the structure of urban institutions, particularly those of the local government system, that are responsible for the management and allocation of goods and resources within the city. The final extract in this section, from a book edited by *Coates et al* is included as an example of this focus on urban institutions. Here the authors attempt to assess the extent to which the distribution of resources by decentralized systems of local government counteracts urban or spatial inequalities that arise in capitalist cities from the unequal distribution of income. Thus, although the focus is on city-specific institutions, the explanation for the persistence of the striking differences found between small areas in cities is ultimately located at the more general level of income and wealth inequalities.

The six extracts included here present a range of approaches to understanding social differences in cities. They tackle a variety of issues from different perspectives and disciplinary backgrounds, including sociology (Raban, Thorns and Abrams), psychology (Quinton), geography (Coates *et al*) and political economy (Cockburn). They all fall within the ambit of urban studies, which in itself has developed from diverse disciplinary perspectives, and cover the range of approaches identified by Dunleavy in his introductory article.

## References

Halsey A H (eds) 1980 *Origins and Destinations* (Oxford: Clarendon)

Young M and Willmott P 1962 *Family and Kinship in East London* (Harmondsworth: Penguin) (First published 1957)

**Linda McDowell**

# 10     The Importance of Place

*by J Raban*

## The urban patchwork

It is, perhaps, a symptom of our superstitious habitation of the city that we map it, quarter by quarter, postal district by postal district, into a patchwork quilt of differently coloured neighbourhoods and localities. For in fact urban man is less of a robin, less tied to rigid territorial boundaries, than his country cousin. The 'Italian' or 'Jewish' or 'Black' quarters are not exclusively inhabited by Italians, Jews and Blacks; they are more or less arbitrary patches of city space on which several communities are in a constant state of collision. A colourful and closely knit minority can give an area its 'character', while its real life lies in the rub of subtle conflicts between all sorts of groups of different people, many of whom are visible only to the denizen. Conversely, a strong cultural community can exist quite outside the geographical definition of a quarter. I live in a community whose members are scattered piecemeal around London (some of them live outside the city altogether); the telephone is our primary connection, backed up by the tube line, the bus route, the private car and a number of restaurants, pubs and clubs. My 'quarter' is a network of communication lines with intermittent assembly points; and it cannot be located on a map.

Yet place is important; it bears down on us, we mythicize it — often it is our greatest comfort, the one reassuringly solid element in an otherwise soft city. As we move across the square to the block of shops on the street, with pigeons and sweetpapers underfoot and the weak sun lighting the tarmac, the city is eclipsed by the here-and-now; the sight and smell and sound of place go to make up the fixed foot of life in the metropolis. Place, like a mild habitual pain, reminds one that one is; its familiar details and faces — even the parked cars which you recognize as having been there in that spot

for months — assure us of a life of repetitions, of things that will endure and survive us, when the city at large seems all change and flux. Loyalty to and hunger for place are among the keenest of city feelings, reverenced and prized precisely because they go against the grain of that drift towards the formless and unstable which the city seems to encourage in us. My two quarters are parts of London where I have lived; a square in the northeast of the city on the border of Islington and Holloway, and a square in Earl's Court, on the scruffy western fringe of Chelsea and Kensington. They are not ghettos, and to prise them apart from the rest of the city entails a considerable amount of surgery on the veins and arteries which, in real life, keep them organically joined to the messier and much larger entity of London. (This is not merely a problem of piety. The outstanding weakness of much classical sociological study of the city, from Louis Wirth's (1928) *The Ghetto* to Willmott and Young's (1962) *Family and Kinship in East London*, is its insistence on the neighbourhood as a culturally self-sufficient community — as if it really were a village inside the city. My quarters are both chronically dependent places, and when they look most like villages or ghettos it usually means that some people are trying very hard to play at being villagers or persecuted refugees, which is not at all the same thing.)[1]

## Living in Islington

### The square

The northern square is not actually a square at all. It has three sides, each with its own rather odd cheapskate style of Victorian architecture — one grand with balconies and rotting porticoes, one derelict gothic, and one orotund fake Georgian terrace. It looks as if it had been put together by a shady and squabbling trio of

Source: Raban J 1975 *Soft City* Ch 8 (London: Fontana) pp 184–95.

speculative builders, while their fourth friend found a more profitable racket: the south side of the square is a cutting, in which the goods trains of the North London Line from Broad Street to Richmond rattle through where the slaveys ought to have been steaming in a line of basement washrooms. In the centre of the square, there is an ornamental garden; mostly grass and flowering shrubs that never seem to be in flower, a children's playground, the superstructure of a bomb shelter which has been turned into a pair of public lavatories and a padlocked tennis court whose gravelled surface has cracked and subsided as if it had had its own earthquake. It is a square that architectural writers about the area leave conspicuously alone; it shows its own history, of unimaginative enterprizes, both private and civic, all too well. Lots of people have had ideas for it at one time or another; none of them have been very bright ones. On the Georgian side, someone thought he would smarten his house up by slapping a coat of gravel stucco on it three inches deep and painting it nightdress pink; and just by the railway line, on the central space, a truck owner built a large wooden and corrugated-iron hangar to unload his lorries in. Floodlit, on a winter evening, it looks as if he is putting on a passion play, a ghostly Islington Oberammergau.

*The pattern of population change*
Yet in the late 1960s the council put a conservation order on the square — perhaps because it was, almost, a square and squares were in fashion; perhaps because no-one from the council had ever seen it (or maybe because they all lived in it); perhaps because Islington Council has a perversely sophisticated taste in architectural curiosities — and it was this, the most recent idea, which dominated the life of the square while I lived there, skunklike, in a newly converted basement on the west side. For eighty years the square had been deserted by the middle classes for whom it had originally been built. They fled north, scared of their proximity to the East End and its suppurating brew of cholera and typhoid germs. By Edwardian times, their only relics were the kept women of gentlemen in the City, who sat out, so people remembered, on the sunny balconies of the north side, waiting for the clank of carriage wheels coming up Liverpool Road from the Angel. But the carriages stopped coming. The houses were broken up into flats. It was a cheap place for immigrants to get a peeling room; and it bulged with Irishmen who had come to London to burrow tunnels

for the underground railways, with Greeks who were in the tailoring trade as machinists and hoped to open cafés, with Poles, and West Indians and the more feckless members of the Cockney working class for whom the square was a vague stopover, a place they had landed up in on the way to somewhere else.

Since 1900, the square, like so many metropolitan districts, had been a shabby *entrepôt*, steadily declining as more and more shipments of people washed up in it for no special reason except that it was cheap and close to the centre of things. That was how I found myself there, too; I didn't feel I had to love it or be excited about its local colour — it was just a place to tie up until something better happened, it commanded no particular loyalty. But the conservation order changed that; it slapped a sudden value on all the fungoid stucco fronts, and young couples with a little money to put down on a deposit took to patrolling the square on Sunday afternoons. They stared at it squint-eyed and, after their first blank shock at the mean disorder of the place, they learned to look at it with affection. They saw a uniform line of white fronts, with brass knockers in place of the tangle of electric bells; they put leaves on the trees; they stripped the pavements of fish-and-chip papers; they looked at the people on the streets, thought for a while, and came up with the word cosmopolitan. Tennis in the evenings, dinner on the balcony in the sun, the rippling laden branches of the elms, the clicking wheels of smart prams parading on the shady walk between the trees — this vision was far removed from what one could actually see, but the city is a stage for transformation scenes, and the square was a challenge to the imagination. It had sunk so low that it could only be rescued, restored like one of General Booth's street-harlots to rosy cheeked prosperity by evangelism and a change of diet.

From then on, estate agents' boards began to sprout from the basements, and builders and interior decorators swarmed round the square, carrying whole walls away from inside houses, pointing the brickwork, painting the fronts, taking long speculative lunch hours in the pub, while Nigel and Pamela, Jeremy and Nicola, made flying spot-checks in their Renaults and Citroens. The leaves on the trees grew greener; old absentee landlords suddenly started to take an unprecedented interest in the lives of their tenants, shaking their heads gloomily at the absence of bathrooms and the damp patches and the jags of falling plaster, and suggesting that the tenant would be better off by far in a spick and span council flat in Finsbury Park. For every signboard advertising 'Vacant Possession' there had been some degree of harassment, alternately wheedling and vicious. Some landlords locked their tenants out of their lavatories; some hired thugs in pinstripe suits; some reported their own properties to the council as being unfit for habitation; many offered straight cash bribes — the going rate for eviction was £200, a large sum to tenants but one that was a fraction of the rising monthly value of the property. (In three years, a typical house on the terrace rose in price from £4000 to £26 000). The tenants were winkled out; most of them were innocent of their rights, and those who tried to stay were subjected to long elaborately mounted campaigns of harassment. Alongside the builders' skips on the pavement, there were handcarts and aged minivans, with a transistor radio squawking pop tunes on top of the roped together pile of tacky furniture. The local barber told me about the girl-graduates who came to sweet-talk him out of his flat, flashing their miniskirts and going on about vinyl bathrooms and 'conveniences'. He had already lost most of his customers, since the council was tearing down several acres of neighbouring streets to make way for glass-and-concrete apartment blocks.

The people went, to Finsbury Park, to the darker reaches of Holloway Road and to brazen GLC estates on the Essex border. The people who came in (a disproportionately small number since a single middle-class couple and their baby can displace anything up to a dozen sitting tenants) were from Kensington, or the northern outer suburbs, or the Home Counties; and they settled avidly on the land, taking over its shops and pubs, getting up little campaigns to preserve this and demolish that, starting playgroups and giving small intense dinner parties for other new pioneers. They took tutorials in local gossip from their chars, and talked knowledgeably about 'Ron' and 'Cliff' and 'Mrs H' and 'Big Ted', as if the square and its history were their birthright.

## The new occupants

### Ethnic diversity

But it was by no means a complete takeover; people still ran car repair businesses from their homes along the square, and on the unreclaimed east side, the houses seemed to actually swell with an influx of the displaced and the very poor. This part of London responds like a needle-gauge to disturbances across the world, and the

expulsion of Asians from Kenya showed up on the square in a sprinkling of faces who looked bitterly unaccustomed to such want and cold, and were determined to move out before they got bogged down in the rising damp, and learned to shiver with fortitude before the popping gas fire with half its plaster columns cracked or gone. On the far side, the sunny side, of the square, these unluckier migrants could see a future of a sort; a future of Japanese lampshades, *House and Garden*, French baby cars, white paint, asparagus tips, Earl Grey tea and stripped pine stereo systems — the reward of success is the freedom to choose a style of elegant austerity. It is hard to guess how such conspicuous rejection of the obvious fruits of wealth must have looked to the Asians, who had so recently been deprived forcibly of their money and its attendant powers.

Both groups of immigrants were innocent of the improvised network of rules which had evolved around the square. The smart young things trespassed casually, arrogantly, secure in the conviction that the future of the square lay in their hands. The Asians were shy, beadily knowing, and usually wrong. Both groups made themselves unpopular.

On the nearby block of shops there were two opposite, nearly identical, grocery stores — one run by a Greek Cypriot and one by a family of African Asians who had come over some years before. Each shop had its band of regular customers. The Cypriot was liked by the West Indian and Irish women, and kept their local foods in stock. The shop opposite was mainly patronized by the respectable Cockney working class; it was full of notices about not asking for credit as a refusal might offend, and piled high with cans of petfood. It was the chief gossip centre of the quarter, and sent calendars and cards at Christmas to its regulars. Jokes were the main currency of both shops; facetious insults, usually of a bawdily intimate or racial kind, which buzzed back and forth over the racks of breakfast cereals. These jokes kept insiders in and outsiders out; they reflected — harmlessly would not be strong enough, cathartically would give them too importantly theatrical a function — the tensions in the composition of the quarter. But the immigrants could never get them quite right.

Both the Asians and the young pioneers wandered promiscuously into both shops. Once there they laughed too loudly or not loudly enough at the ritual jokes; they asked for foods which the shop they were in did not stock. When they tried to initiate jokes themselves they were too elaborate, too humorous — for humour was not the point at all — and they were received with wary tolerance. Had they been able to joke about their own peculiar tastes for white paint and thrifty ecologism it might have been different; but they presumed to a matey all-men-are-equal tone — and the visible untruth of that ploy was all too clear to their auditors, whose own lives were day-to-day struggles with a dramatic inequality (made harsher by the presence of these well heeled young couples), and hard bitten compromises with the other people in the quarter who were their territorial competitors, if not their enemies.

The local pub was carefully demarcated into symbolic territories. The public bar was the West Indian province, with a smattering of white girls of catholic tastes and inclinations. The saloon was for the Irish. There was one black, in a shiny felt hat, who ambled leggily round the saloon bar picking up empty glasses; and in the spade bar of the pub there was a single, very sodden Irishman. These two hostages strengthened the division. In the public bar, the juke box hammered out Reggae records; in the saloon on Friday nights Bridie the Singing Saxophonist carolled about the Mountains of Mourne and the Rose of Tralee through an amplifying system which made her sound like Frankenstein's monster.

But the pioneers went everywhere regardless. They stood on the pavement beside the West Indians, pretending they were at a garden pub in Hampstead; they slummed in the public bar, and put the occasional sixpence in the juke box; they investigated the churches; they joined the villainous Golden Star club, a black dive where you drank rum or beer in paper cups and shuffled on the tiny dance floor to records of amazing rhythmic crunch and volume. Wherever they went, they spread money and principled amity. They were interested in everybody, as temporary and ubiquitous as secret policemen. The Polish tailor who had a workshop in his front room in the block of shops grimaced at them as they went by; he, presumably, had prior experience of people who invade places armed with good intentions.

### Social class change
All round the square, more working-class streets came down. Windows were boarded up, and the London Electricity Board put LEB OFF in swathes of paint across the front doors when it disconnected each house's supply. It looked like a sign of a plague or a primitive curse: 'Leb off!' As the demolition men moved in, a cloud of old brickdust hung over the streets, and more frightened cats migrated southwards into the square. Soon after I stopped living there, Lesly Street residents refused to go; they barricaded both ends of the road and tried to sit it out as the bulldozers rumbled closer, taking down their garden walls and flattening flower borders under their tracks. The residents were supported by some rebellious social workers and one sympathetic Labour councillor, and their 'No Go Area' got into the national newspapers. But Lesly Street has gone too, now. It wasn't perhaps a class war, but there has certainly been something suspiciously like a class victory.

In the converted house where I lived, privilege and choice, like a thick belt of insulation, kept the destruction of the surrounding streets to little more than an annoying noise that interfered with one's reading. The garden in the centre of the square was replanted; the trees, now we were in a smokeless zone, looked less sorry; the careening streams of afterschool kids thinned; the front area was easier to keep clean. It was regarded as a small victory for conservation that the wall at the back of the garden was kept high, so that the tenants of the new council low-rise flats would not be able to watch the sunbathing Brahmins leafing through the *New York Review of Books* on their breakfast patios.

The off-licence which I used came down. The betting shop was served with a surprize notice of demolition. Little backstreet businesses shifted or died. Every big city, even at times of desperate property shortage, has crannies into which the feckless and the transient and some, at least, of the dispossessed, can lose themselves. No-one knows exactly where everybody goes in these upheavals. Municipal housing helps some; but the unlucky have to search out those remaining areas which have not yet been picked out by the spotlight of wealth and fashion or the dimmer beam, which often accompanies it, of urban renewal: ruinous terraces backing on to the railway tracks in Camden; grim apartment houses in east Holloway; derelict streets running alongside disused wharves and warehouses south of the river . . .

### A new respectability?
But the pioneers, the new Brahmins, are there to stay; their money firmly invested, their place assured. The trees grow greener for them; they outlive cats, charladies, sitting tenants, they ride the tides of inflation

and depression and mobility just as their parents did, though they have loftier spirits than their parents, or more innocence. The square is not — will not be — as 'real' as it was; but it is coming to be a place of substance, which, on the economic scale of things, is perhaps more important. If the frontier spirit with which it was colonized is fading, it is being replaced by a sense of imminent history — it is growing a pedigree from scratch, as if the departed tenants had no more claim on it than the delivery drivers who take cars out to their new owners and put the clock back to zero. The tenants — 'temporary people' in the words of a lady Brahmin — will always be on the road somewhere, being passed by.

## Note

[1] In this extract the first of these two quarters is described (Eds).

## References

Wirth L 1928 *The Ghetto* (Chicago: University of Chicago Press)

Young M and Willmott P 1962 *Family and Kinship in East London*, (revised edition) (Harmondsworth: Penguin Books)

# 11    Suburbia: Myth or Reality?

*by D Thorns*

## Aspects of the myth

*The North American stereotype*

The suburban way of life is a complex mixture of fact and fantasy which is extremely difficult at times to unravel, but it is essential for clear understanding that these two parts of the suburban way of life are clarified.

The growth of the myth of suburbia has been coincident with the physical growth of the suburb and has been most extensively analyzed in America where it reached its height in the late 1950s. There are eight main elements of this myth in American society (Berger 1960, Dobriner 1963, Donaldson 1969). These are:

(1) The suburb as a transient centre, the population undergoing a constant turnover with the average length of residence between four and five years. This is due to the composition of the suburb being predominantly young, upwardly mobile individuals (age span between 25–35 years).

(2) The suburbs are uniformly middle-class areas with a well educated population: both husbands and wives are usually college trained.

(3) The population is homogeneous. This results from the one-class nature of the population, the similarity of house design and the similarity of interests, with all of the families being at the same stage of the family cycle, at the same stage of their careers, all having just bought new houses and all busy home-making.

(4) There is extensive social activity in both formal and informal participation, the latter taking the form of the Kaffeklatsches for the women during the day at which the conversation revolves around child-rearing and home improvements. In the evenings, the suburb is seen as the scene of parties and dropping-in of couples for drinks etc, as a place with a considerable level of social activity.

(5) Child-rearing is important and the latest authorities are taken as the guides rather than those of the extended family group. The fact that the husbands are mostly commuters means that suburban society is female dominated and that child-rearing is very much the preserve of the woman.

(6) There is a return to religion, the church performing social as well as strictly religious functions in the community.

(7) Traditional Democratic voters become Republican.

(8) The suburb is seen as the second 'melting pot' in American society where the upwardly mobile can learn the new norms of behaviour of the class to which they aspire in a community where the criterion of judgement is how the individual acts within the community rather than his social background.

In the 1960s this particular myth of suburbia was shown to be erroneous by the studies of Berger and Gans. These studies of a working-class and lower-middle-class suburb have shown just how far divorced this myth was from the reality of many suburban areas. However, despite these efforts at myth-breaking, the myth has shown a great measure of resilience, continuing to dominate the thinking of many. The continued existence of the myth is due as much as anything to the fact that the critics of the suburb have become thoroughly caught up in their own creation. Donaldson (1969) writes: 'The curious thing is that much of the criticism which has been levelled against the suburb has been written by commentators whose thinking is far more powerfully influenced by this same myth than is that of the normally non-ideologically oriented suburban home owner.' The preservation of the myth, however, despite the contrary evidence provided by sociological studies, shows the extent to which the myth forms part of contemporary American culture. It

Source: Thorns D 1973 *Suburbia* Ch 9 (London: Granada Publishing Limited) pp 147–55.

also supports the view that it is not based so much upon residential changes in American society as upon modifications and adaptations within the culture of the middle classes, of which Park Forest and Crestwood Heights are good examples.

## The British version

In Britain the myth of suburbia has never received the same level of prominence nor the same volume of discussion but there does nonetheless exist a clearly held set of popular ideas regarding what constitutes the suburban way of life. This view, as in America, has often been somewhat derogatory with some viewing the suburb with disdain as the centre of the middle-brow, conformist, respectable, uninspiring members of society who are quite content to potter around in their own rather limited world. This view is again somewhat divorced from the realities of suburban life. The suburb is a term which does for the majority conjure up a particular view of a residential neighbourhood with its rows of semi-detached houses cast in an identical, or nearly identical, mould and associated with this physical structure is a particular way of life. In both cases these tend to be stylized and often bear only a vague resemblance to the reality of life in the suburbs. This rather stylized view was illustrated by some of the respondents in the suburbs studied. In the Bristol suburb, for example, one respondent did not consider that he lived in a suburb as his house was not part of a new building development. Such a development, however, was taking place across the road from where he lived which he did term suburban, using the term in a somewhat derogatory way as if it in fact provided a well understood and neat summary of undesirable features which were associated with new developments.

In the study of the two suburbs some attempt was made to investigate this particular aspect of the suburban controversy and discover what were the elements of the myth. This was attempted by presenting the respondent with thirteen elements which have formed part of the myth and finding out the degree to which the sample expressed agreement or disagreement with them. From their replies it was possible to build up a picture of what the majority considered to be the main elements of the suburban way of life. This way of life had seven elements. These were:

(1) That it was a way of life which had a prevalence of and gave importance to do-it-yourself activities.
(2) It gave a central role to the family and child-

rearing. These first two give a picture of the central importance of familism as a major social value within the suburban image.

(3) It was a middle-class way of life.
(4) Great importance was attached to the possession of status symbols.
(5) It was a life which was marked by a high degree of social activity.
(6) It was strongly conformist in character.
(7) Suburbs were politically conservative.

## Reality

### The importance of family and home-centred activities

The picture which emerges appears to have a considerable degree of correspondence with what Riesman has described as the values of 'suburban peace and domesticity' with the consequent emphasis upon the work of suburbia rather than upon the work involved in the individual's occupation. The importance of home improvements is related to the familism of the suburban dweller but it is also related to another issue — that of status competition. It is claimed that people are motivated in their home improvements by a desire to have all the latest gadgetry to 'keep up with the Joneses'. However, in the survey, only a small proportion did, in fact, consider that suburban life was a 'keeping up with the Joneses' activity. An alternative explanation of the emphasis on home improvements is that this is a sphere in which the individual can, in fact, stress his individuality and seek to bring variety into his environment. The fact that the suburban dweller does not believe that the suburbs are depressing because they are identical in house design and estate layout suggests that it is an important aspect of the suburban life to produce variety through alterations to both interior and exterior design. One of the most striking things about new suburban estates is the speed at which variety is introduced. Again the conformity is perhaps more strongly in the mind of the critic than in the actual dweller in the suburb. The two popularly held notions or myths of suburbia depicted here for Britain and America show both similarities and differences. There is agreement, for example, on five of the elements: those of familism, the active social life, the homogeneity of the population, the middle-class nature of the population and political conservatism. There was, however, divergence on one key issue, associated strongly with the suburban way of life in America, that of the extent of transience. This was held to be a characteristic, if not the characteristic, feature of the American myth. However, only 24% of the British sample cited it as an element.

The analysis of popularly held beliefs is only one aspect and one which often bears little resemblance to the actual activities of suburban dwellers. This gap between myth and reality is well demonstrated by Gans's essay on the man who frantically searches for the mythical social whirl of suburbia. In the suburbs studied at Bristol and Nottingham the important point which stood out was how little activity appeared to be significantly related to the degree of suburban identification held by the individual. The most significant of these relationships concerned the lack of support for a non-instrumental view of work and an indication that this was replaced by involvement in the family and home centred activities rather than widespread participation in social organizations. The picture which emerges does not provide evidence of massive and widespread social activity of either a formal or an informal nature nor does it show a tremendous amount of interest in such supposed suburban activities as religion, parent–teachers associations or conservative political involvement.

*The suburban way of life*
From this examination of the actual activities of the suburban dweller and of the popularly held notions or myths it is possible to draw certain conclusions and inferences about the origin and perpetuation of the suburban way of life. It can firstly be seen that there is little real evidence that the simple fact of living in a suburb leads the individual to adopt a particular style of living. This can, of course, be broadened to say that any residential situation, be it urban, rural or suburban, does not of itself produce a way of life. The production of the way of life is dependent upon other factors but it may be that these are related to particular residential areas. The second thing which is clear is that the failure of many writers in both Britain and America to define clearly and to investigate the actual behaviour of people in the suburb has led to the growth and perpetuation of erroneous ideas. These ideas have been nurtured and developed for the most part because they correspond to the view of reality that some groups in society would like to exist, largely so they could then attack it as inadequate and in the case of America, as a betrayal of the 'American Dream'. Hence the whole edifice of the suburban myth is unsound and needs replacing by a carefully developed picture of what life is actually like in suburban areas. There is also the necessity to seek a more reasonable basis for the examination of the way of life in which people living in the suburbs share. Gans (1967) in his study of Levittown, is concerned to show how the way of life of the people is a product of social class rather than the residential situation in the suburb. This is also the conclusion of Berger's (1960) study of Militipas where he found, within his working-class suburb, an adaptation of their previous working-class pattern of living rather than any move to adopt a suburban style as it is popularly believed to exist. The popular myth of suburban living consequently is seen by these writers as clearly associated with the middle class, hence it is correctly seen as not about suburban living but about middle-class life. The importance of this insight is considerable and when social class as a key variable was examined in relation to the evidence from the Bristol and Nottingham suburbs, it was found that for a majority of the questions relating to the three central themes examined, social class group either in objective occupational terms, or by self-assigned placement did show a much more significant relationship than did suburban identification. Hence this lends considerable support to the views of Gans and Berger that social class is a more significant variable in the determination of the style of life adopted by an individual than suburban residence. This relationship between social class and suburban imagery is further demonstrated by the significantly higher degree of suburban identification in the subjectively defined middle class. The popular image of the suburb in Britain, therefore, more correctly should be seen as a middle-class style of living rather than as a suburban style in that it is based upon occupational, educational and status criteria as much if not more than upon those of residence, e.g. neighbourhood and house type. The adoption within the middle class of this particular style of life is not, however, universal. For example, there was found to be a negative relationship between the desirability of suburban life and occupational status, with the routine white-collar group showing the strongest support for the view that the suburban way of life is desirable. This lends some evidence to support, though by no means conclusively, the view that it is the middle-middle class to lower-middle class, those who have perhaps not been successful at getting to the top of their occupational ladders, who have moved to adopt the essentially family centred ethic which has become, more or less, synonymous with the suburb; but the

question of cause and effect here is difficult to determine as it could be that the adoption of familism leads to lack of success rather than vice versa. The essential role and hence value of the analysis of the suburb is that it is the place where the effects of these other changes can be clearly seen. The mistake of previous writers has been to confuse the causes of changes which lie outside the suburb with the effects of these changes which are increasingly manifest in the suburb and suburban way of life.

## The future role of the suburb

The suburb is rapidly becoming a world-wide phenomenon with the devolution of the major cities of the world as they grow increasingly congested. The continued importance of the suburb as a physical entity is assured for the foreseeable future within, for example, Britain, the US, France and Japan; although in both Britain and the US there are those who argue that there is evidence of the beginning of a return to the city, but this has in neither country reached sizeable proportions. This is quite clear from the evidence of growth figures for the areas around the metropolises and the continuing population decline in the major cities of the world. What are the implications of this likely growth of the physical form of the suburb? The consequences for urban government and planning which are currently affecting many of the cities of America could well become the problems of Britain, if the cities decline to such an extent that they produce a marked imbalance in the population structure of the city. The suburban migration has been predominantly one of middle-class, higher-income earners, coupled with the local authority rehousing of the urban working class. This has resulted in the income of the city dwellers as a whole shrinking in times of rising expenses. This situation has led to considerable financial problems in American cities where the city's income is fast becoming inadequate to pay for the services the city has to provide. The other result of the selective character of migration is to leave pockets of particular ethnic or socio-economic groups in the city. In many English cities the migration to the suburb of the middle class from what were substantial and prosperous Victorian, middle-class residential areas has produced areas of large houses which have increasingly been taken over for a number of purposes, the most important of which is as multiple-occupancies for immigrant groups. The development of this kind of an area and its social implications are well documented in a study of Sparkbrook in Birmingham (Rex and Moore 1967). The creation of urban ghettos at their most extreme and local government financial difficulties are both in part related to the decentralization of the city and the corresponding growth of the suburbs. Neither of these has yet reached serious proportions in Britain but both could, before the end of the century, unless a constructive effort is made to counteract the existing trends.

Turning to the suburb itself, and its problems of planning and growth, Mumford's (1961) attack upon the suburban way of life rested primarily on the grounds that what had begun as a valuable thing when it was confined to only a small group, has been ruined by its own popularity. This popularity has destroyed the ideal and any virtues it may have originally possessed. However, in the two suburbs studied, of those who considered the areas to be suburbs, 56% thought suburban life, i.e. presumably life as they lived it, to be desirable and only 22% considered it undesirable. Hence, whatever the critics may think, the people who actually live in the suburbs do not appear to be too dissatisfied with the quality of their life. The suburbs do, in fact, offer an alternative arena to that of the individual's occupation and work organization in which he can find some kind of satisfying social role. This would appear to be all the more necessary where there is an increasing proportion of non-manual workers who are sharing the manual workers' instrumental view of work. For these people work is simply a source of funds and satisfaction is sought outside paid employment, not often in organizational activity of another type, but in informal social relationships based upon the family and in some cases the neighbourhood. The suburb, hence, could increasingly become the arena where people play their important social roles. If the analysis here is correct then the demands of the suburban dweller are not particularly complex: what he does not need are many extra-familial facilities as most of his time and energy will be devoted to the establishment and improvement of his house, home-making and family life in general. Hence community facilities which require an organizational base, such as community centres, are not likely to prosper.

The confusion and contradictions which have bedevilled this area of urban sociology have, to a great extent, been due to a failure of diagnosis by the investigators, who have consistently failed to isolate the causes of the changes they have discussed from their effects. Once this is done, then the whole area of debate can immediately be clarified and progress can be made in the further investigation and elaboration of the varied styles of life which are present in each of the types of suburb distinguished. This means that just as there is not only one type of suburb, there is not only one type of suburban way of life.

## References

Berger B 1960 *Working-class Suburb* Ch 1 (Berkeley and Los Angeles, CA: University of California Press)

Dobriner W M 1963 *Class in Suburbia* Ch 1 (Englewood Cliffs, NJ: Prentice-Hall)

Donaldson S 1969 *The Suburban Myth* (New York: Columbia University Press)

Gans H J 1967 *The Levittowners* (London: Allen Lane)

Mumford L 1961 *The City in History* (London: Secker and Warburg)

Rex J and Moore J 1967 *Race and Community* (London: Oxford University Press)

# 12     From Analysis to Facts: the Problem of Community

*by P Abrams*

**Industrialization and loss of community**

The apparent weakening of the family indicated in many descriptive studies is as nothing compared to the loss of community predicted by the whole tradition of sociological theory. Because sociology came into being as an attempt to apprehend the nature and dynamics of the transition to industrialism, it tended to identify the emergent industrial world in terms of a series of stark polarities, or contrasts of type, between pre-industrial and industrial societies. The most characteristic of these centred on the idea of community. Virtually without exception the pioneers of sociology understood industrialization to involve a disintegration of the bonds of community. Community became, as Nisbet (1966) puts it, 'the most fundamental and far-reaching of sociology's unit ideas.' The term referred not only to attachments grounded in locality but, much more generally, to an inclusive moral cohesion believed to have characterized pre-industrial society. As Nisbet remarks again: 'the word encompassed . . . all forms of relationship which are characterized by a high degree of personal intimacy, emotional depth, moral commitment, social cohesion and continuity in time. Community is founded on man conceived in his wholeness rather than in one or another of the roles, taken separately, that he may hold in a social order'. Since the essential nature of industrial society involved an explosion of the division of labour which splintered people's lives into separated roles, industrialization logically threatened community. The theme is endlessly reiterated, becoming one of the commonplaces of sociological wisdom. The loss of community was associated especially with the growth of industrial cities which were contrasted not just with the past, but with the countryside as well. It is in the early urban sociology of Simmel and Wirth that these ideas are most directly presented. But this theme was not the sole preserve of sociologists. It was Benjamin Disraeli who gave it one of its crispest statements (quoted in Nisbet 1966):

> There is no community in England; there is aggregation, but aggregation under circumstances which make it rather a dissociating than a uniting principle. . . In great cities men are brought together by the desire of gain. They are not in a state of cooperation but of isolation, as to the making of fortunes; and for the rest they are careless of neighbours. Christianity teaches us to love our neighbour as ourself; modern society acknowledges no neighbour.

Ruth Glass has suggested that this tendency to contrast urban present and rural past in terms of community, and to the disadvantage of urban present, is a peculiar tendency of British intellectuals. While there is some truth in this, it is also the case that this particular contrast pervades the whole of sociology, European and American as much as British. A representative passage from Wirth (1938) may speak for the whole sociological tradition, both in its assertion of the loss of community and in its belief in the vital importance of the city as a distinctive social form of industrial society.

> The multiplication of persons in a state of interaction under conditions which make their contact as full personalities impossible produces that segmentalization of human relationships which has sometimes been seized upon by students of the mental life of the cities as an explanation for the 'schizoid' character of urban personality. Characteristically, urbanites meet one another in highly segmental roles. They are, to be sure, dependent upon more people for the satisfactions of their life-needs than are rural people and thus are associated with a greater number of organized groups, but they are less dependent upon particular persons, and their dependence upon others is confined to a highly fractionalized aspect of the other's round of activity. This is essentially what is meant by saying that the city is characterized by secondary rather than primary contacts. The contacts of the city may indeed be face to face but they are nevertheless

Source: Abrams P (1978) *Work, Urbanism and Inequality* (London: Weidenfeld and Nicolson) pp 10–14.

impersonal, superficial, transitory and segmental. The reserve, the indifference and the blasé outlook which urbanites manifest in their relationships may thus be regarded as devices for immunizing themselves against the personal claims and expectations of others.

## Theory and evidence

### The paradox

The most remarkable thing about the body of doctrine and inference which such a passage represents is that it is so deeply rooted in the logic of sociological analysis that it has successfully resisted modification, despite a glaring absence of confirmatory evidence and the uncomfortable presence of a good deal of evidence pointing in a contrary direction. To begin with the analysis as a whole can not be fully validated as we lack thorough empirical studies of the supposed pre-industrial 'base-line' from which the loss of community is believed to have developed. We simply don't know what community in eighteenth-century England was like and it is unlikely that we shall ever be able to gain such knowledge. For the most part early sociologists treated the presence of strong communities in pre-industrial societies as axiomatic; and subsequent advocates of the loss of community thesis have been increasingly constrained to do the same. More disturbingly, quite a large number of descriptive studies of contemporary industrial societies, including the UK, have suggested that the various elements of community are in fact quite strongly present in such societies, even in their most highly urbanized areas (see Bell and Newby 1971). The paradox of the sociology of community is the coexistence of a body of theory which constantly predicts the collapse of community and of a body of empirical studies which finds community alive and well.

### Neighbouring patterns

In practice the problem is not quite as clear cut as that, nor is the difficulty all on one side. Nevertheless, if we take one theme from the field of community studies as an example, that of neighbouring, we can see that this is a field where applied sociology has been quite severely hampered by the general predominance of theory over description. In line with the tendency of community theory as a whole, the prediction emerging from sociological analysis about neighbouring has been that it too will be eroded with the development of industrialism. For Wirth (1938) again states 'the bonds of neigh-

bourliness are likely to be absent or, at best, relatively weak' in urban industrial milieux. In a comprehensive and perceptive review of the whole literature on neighbouring, Keller (1968) was able to condense conventional wisdom into a number of quite sharply focused propositions explaining why attachment and interaction among neighbours must be expected to be less important and less intense in the typical settings of industrial societies (cities) than in the typical settings of pre-industrial societies (villages). From this point of view the UK is one of the most 'industrial' of nations. Its population density, 601 people per square mile, is one of the highest in the world; and that population is predominantly urban. The ratio of those living in urban areas to those living in the countryside is about 7/2:1, with approximately a third of the population living in seven great conurbations. Keller (1968) summarizes the analysis as follows:

(1) As crises diminish in number and kind, where, that is, self-sufficiency increases, neighbour relations will diminish in strength and significance.

(2) As new forms of social control arise, the significance of neighbouring as a means of social control will recede in importance.

(3) Where neighbouring is a segmental activity in an open system rather than an integral part of a closed system, it will be a highly variable and unpredictable phenomenon.

(4) Since all three conditions are more true of urban than of pre-urban or suburban areas, neighbouring should diminish in extent, significance and stability in cities.

More recently and in very much the same spirit, Key (1965) has claimed to have shown that 'there is a negative relationship between the size of the community in which people reside and the frequency with which they participate in social relationships involving other people residing in portions of the community contiguous to their residence'.

One difficulty with this sort of argument is that a large number of empirical studies have documented the existence of intensive and extensive neighbourliness in highly urbanized areas; the best known being the work of Young and Willmott (1957) in Bethnal Green. A further difficulty is that even when the effects predicted for urban industrial societies *are* observed it is not at all clear that the effects that occur *in* cities or in industrial societies are in fact caused *by* cities or by industrialism. The implicit causal relation asserted in the theory of community has proved extraordinarily difficult to demonstrate explicitly. That is one reason why modern urban sociology has increasingly moved away from the idea of the city as a distinct social entity, found in the work of Simmel and Wirth, towards a type of analysis in which cities are treated merely as settings in which the characteristic relationships of some larger social structure and some more inclusive division of labour work themselves out. It is a move away from urban sociology towards the political economy of towns. Alternatively, one could say that it is an attempt to place the analysis of urbanism within the broader sociological analysis of the division of labour and social inequality, instead of allowing it to remain a seemingly independent and isolated branch of the discipline.

## Redefinitions

### Community

But to return to community and to neighbouring: the concept of community for its part is slowly being evicted from British sociology, not because there is agreement on the empirical collapse of community, but rather because the term has come to be used so variously and different relationships, identified as those of community, have been discovered in so many different contexts that the word itself has become almost devoid of precise meaning. In particular there has been a determined effort to detach the study of social relationships from the study of spatial relationships — two themes which are hopelessly jumbled together in the traditional idea of community. The problem of solidarity is, recent authors have suggested, one that should be separated as a matter of principle from the problem of the effects of spatial arrangements on social relationships. As Pahl (1966) says in an important article developing this theme: 'any attempt to tie particular patterns of social relationships to specific geographical milieux is a singularly fruitless exercise.' And Stacey (1969), in a parallel argument about the 'myth of community studies', concludes that 'our concern as sociologists is with social relationships; a consideration of the social attributes of individuals living in a particular geographic area is therefore not sociology, although it may well be an essential preliminary to sociological analysis'. In sum, there is now a prevailing scepticism about the possibility of studying spatial relationships sociologically, and a resulting insistence that insofar as the concept of community implicitly prejudges the nature of such relationships it should be expelled from the dictionary of sociology.

### Neighbouring

Where does that leave the study of neighbouring in contemporary industrial society? Obviously, unlike the concept of community, the concept of neighbouring *must* have spatial connotations: neighbours are defined as people who live close to each other. So the sociological problem involving neighbours has to be one that includes a question about the effects on social relationships of different ways of living close to others. More generally, the issue is to determine what type of social and/or spatial circumstances are associated with what types of relationships between neighbours. And here again such evidence as there is tends to belie the predictions of sociological theory. Two findings emerge especially strongly from the whole body of research. One concerns prevailing social norms about neighbourliness; the other involves the conditions under which members of our society are likely to be 'good' neighbours. So far as the norms of neighbourliness are concerned it seems clear that neighbourliness is very widely understood, in both town and country, and for that matter in both present and past, as a three-dimensional relationship composed of friendliness, helpfulness and distance (Robinson and Abrams 1977). Ideally, normatively, a neighbour is someone who is agreeable when casually encountered, there when you want them in an emergency and yet who does not 'live in your pocket'; who, while being both friendly and helpful, also respects your privacy. Not only is this norm widely diffused in contemporary UK society, but the circumstances under which such a norm can actually be realized in relationships between neighbours are also widely understood.

Thus, although the urban–rural contrast does not stand up to scrutiny, it seems that the nature of the locale in which people live does affect patterns of neighbouring in some quite definite ways. For example, the social homogeneity of a neighbourhood and the average length of residence of its inhabitants are both factors which seem to vary positively with the intensity and extensiveness of neighbouring (Gans 1967, 1968, Willmott 1963). There are also well established social class variations, characterized by Keller (1968) as 'working-class solidarity, middle-class selectivity and suburban sociability'. But these are typically much finer, subtler and more complex than the familiar generalizations would suggest. For example, one of the most recent English studies reports 'no clear dichotomy' in either behaviour or attitudes to neighbouring between classes but, within this 'general similarity of response', notes that working-class people tend to 'see' their neighbours more often, and that the street is a more common meeting place for them than for middle-class people — a difference which would seem to spring from variations in built-form, mobility and occupation rather than from class as such (Kingston Polytechnic 1976). More generally, class differences in neighbouring would appear to be rooted in the ways in which the meaning of class is bound up with and affected by (for example) mobility, income, kinship, age, residential arrangements, social service provision and occupational patterns; in other words more specific influences operating within the worlds of class. A typical instance is the finding of one study that while working-class people express rather more positive attitudes about the desirability of contact with neighbours than middle-class people, this difference is not reflected in any degree of greater neighbouring activity. The authors explain this absence of variation by pointing out that the working-class people in their study *did* have a great deal more to do with their relations than the middle-class people (Kingston Polytechnic 1976). For practical purposes kinship replaced neighbouring in the functions through which strong relationships with neighbours might have been constructed. Similar observations could be made about the relationship between neighbouring and built-form, mobility, stages in the life-cycle and numerous other factors. What emerges consistently throughout is that neighbourliness is a social relationship facilitated or impeded by a wide array of social-structural influences but determined by none of them. Disraeli was wrong: modern society *does* acknowledge neighbours and the social analysis of neighbouring is slowly moving away from the sweeping application of deterministic social theory towards the detailed empirical rediscovery of the conditions under which it does so.

## References

Bell C and Newby H 1971 *Community Studies* (London: Unwin University Books)

Gans H 1967 *The Levittowners* (Harmondsworth: Penguin)

—— 1968 Urbanism and suburbanism as ways of life *Readings in Urban Sociology* ed. R Pahl (Oxford: Pergamon Press)

Keller S 1968 *The Urban Neighbourhood* (New York: Random House)

Key W H 1965 Urbanism and neighbouring *Sociological Quarterly*

Kingston Polytechnic 1976 *The Buxton Report*

Nisbet R 1966 *The Sociological Tradition* (London: Heinemann Educational Books)

Pahl R E 1966 The rural-urban continuum *Sociologia Ruralis* **6** 299–329

Robinson F and Abrams P *What We Know about Neighbours* (Rowntree Research Unit, University of Durham)

Stacey M 1969 The myth of community studies *British Journal of Sociology* **20** 134–47

Willmott P 1963 *The Evolution of a Community* (London: Routledge and Kegan Paul)

Wirth L 1938 Urbanism as a way of life *American Journal of Sociology* **44**, July, 1–24

Young M and Willmott P 1957 *Family and Kinship in East London* (London: Routledge and Kegan Paul)

# The New Terrain of Class Struggle

*by C Cockburn*

## Community action: redefining the concept

In this extract I look first at the category of *community action* and the weaknesses that experience has shown it to hold, and demonstrate on the contrary the strengths of strategies for collective action that are developing around the concept of *reproduction*. I then go on to look at the three places where such action is developing: the point of *collective reproduction*[1] (where we are 'clients' of State 'services'); the point of *employment in reproduction* (where we are the workforce of the local State); and the point of *privatized reproduction* (our family life).

### *'Community' belongs to capital*

The phrase that has come to be used to describe almost any collective action going on outside the workplace is 'community action'. It rings with implausibility. Why? It is not the activity, so much as the *category* that needs questioning. It is not just a question of using or not using a form of expression, but of *thinking* with it. I'll suggest four (all related) reasons — three of which are fundamental and one tactical.

First, to think in terms of community action places struggle on ground prepared, over a long historical period, by the State. It takes a shape that is expected, anticipated and even proposed by the State. In a sense (to use an ecological metaphor) the State is the environment that offers a vacant 'niche', a milieu that will reward and foster a certain kind of behaviour, and the fledgling initiatives of struggle step in to fill it.

The local electoral representative system has been, since its inception, based on territorial definition of interest group. The local councillor represents the ward, wards are grouped into constituencies, constituencies into boroughs. More recently the local services of councils, too, have decentralized themselves partially, into area teams and area offices, again using

territory as the basis for this organization. In recent years, and in some boroughs, not only is there a ward councillor and an area housing manager and several other officials on the spot to relate to whatever 'community' may arise — there is a community worker too. They have even revamped the local bobby into the community policeman. When territorial working-class community groups arise there is a set of officers and councillors, in a sense waiting for them, to whom the community group is of vital relevance and who have their own preconceptions which they will bring to bear on its activities.

One effect of the presence of a ward councillor and a community worker in the territory is to encourage the idea that problems arise in 'officialdom' — because this is something that they both know something about and are ready to tackle. 'The expression that workers and peasants initially give to their discontent is generally diffuse and fragmentary and it often moves into a simple anti-authoritarianism such as "dislike of officialdom" the only form in which the State is perceived' (Gramsci 1976). If discontent is addressed against the local bureaucracy and the top politicians (the Labour leadership)[2] that join with them in urban management, the answer is too easily seen as lying with the ward councillor or left-wing or populist sympathies, and this, as I'll suggest below, has its dangers for collective action.

### *'Community' and consumer protection*

The second, related, argument against relying on a concept of 'community action' is that it has been closely connected since its rise to popularity in the late 1960s with consumer protection. It tends to cast us in the role of consumers (of capital's products and the State's services), a position that is economically and politically weak. What has been called community action has been rationalized as something that arises *not* from capitalism itself, but from some of the more unfortunate but

From Cockburn C 1977 *The Local State* (London: Pluto Press) pp 158–84.

curable effects of the current stage of technological development. 'We all suffer at the hands of large and insensitive organizations and we are all emotionally stunted by the amorphous uniformity of the cosmopolitan culture to which we belong' — that is the way it is commonly put (Greaves 1976). James (1973) put her finger on what is happening: 'we have inherited a distorted and reformist concept of capital itself as a series of things which we struggle to plan, control or manage, rather than as a social relation which we struggle to destroy.' Community action points not to deficiencies in the mode of production but in the products: the goods or services.

### 'Community' and class

Thirdly, community action is all too often defined as classless. In common usage it is a populist formulation, open to all classes, groups and interests. Where it *is* defined to exclude 'the middle class' it is nonetheless normally focused not on the working class as such but on 'the deprived', 'the poor' or even the 'poor-poor'. In other words, it bites off what the Victorians called the residuum, the problem fraction of the people, and distinguishes it from the 'real' working class. This splintering is reflected in the bourgeois ideology of pluralism and participatory democracy, the essence of which is that no one group in society should be too *big*. There is in the idea of community action the idea of smallness up against bigness. We are asked to think of the David of one small council estate taking on the Goliath of the town hall: 'small is beautiful'. It is an image which totally rules out the reality of class struggle, in which huge and powerful forces are ranged against each other, not momentarily, but over centuries. It imposes blinkers which stop one working-class group looking to another with similar problems as its natural ally and leads to a situation where groups in neighbourhood territories struggle in competition for the limited resources offered them — a situation often exploited by a local council.

### 'Community' and electoral politics

The fourth (tactical) point is essentially an illustration of the first. The national political parties are now using community action and community politics for competitive *electoral* purposes. Even the Communist Party recognizes its political uses. 'Community Groups Fill Gaping Vacuum' wrote Dave Cook, national election agent of the CP. 'The growth of community action is very much in line with the perspectives outlined in the CP

programme. Greater participation by Communists and others on the left is necessary to ensure that community action draws closer to traditional working-class organization' (*Morning Star*, 10 and 11 February 1975). It is interesting to note that the CP was included with the three bigger national parties in the invitation by the Department of the Environment to comment on the Neighbourhood Councils scheme.

The Liberals too adopted 'community politics' as a shot in the arm for the Party after a crushing electoral defeat in 1970. At the next Party conference a resolution was put forward by Young Liberals that the Party see its role as helping 'organize people in communities to take and use power'. Failing any other new idea to shore up the Party in its collapse, or even of any concerted opposition to this one, the resolution was accepted and grassroots goings-on became a plank in the Party's platform — to the rage of Tories and Labour alike who called it mindless opportunism and marketing technique. It led the Liberals to success in Liverpool Corporation and to big votes in Manchester and other urban areas.

If 'community action', then, is not a helpful concept to describe working-class struggle outside the job situation, what is? Since many of the issues arising in communities arise in the State services, there is a case for defining our action as anti-State struggles. But to do so deflects attention away from the mode of production, which is the real cause of exploitation. It tends to give the State too much importance and apparent detachment from the economic base.

## An alternative definition

### Shifting the emphasis to reproduction

What then? There is no similar catch phrase. I believe we have to rely clearly and simply on the analysis of the local State that we are attempting; we have to see that what we are involved in is struggle in the field of capitalist reproduction. We have to recognize that alongside struggle at the point of production, in the mines and factories, there is struggle at the point of reproduction, in schools, on housing estates, in the street and in the family.

This definition is a strong one because it immediately brings into the field of action, alongside 'client' action by residents on housing, patients' action on health services, etc, the workers employed by the State, both professional and manual. Furthermore, it identifies the significance of the action of women in the home, in privatized reproduction.

A number of other insights follow. Now we see the inseparable nature of production and reproduction in capitalism we can also see the inseparability of industrial struggle and action over reproductive services such as housing. It also shows us ways of making the connections that are so badly needed between the two. Even the male worker now recognizes that, when he went home with his wage packet formerly he was expected to 'reproduce' his own labour power with it — pay his rent, pay his doctor, insure himself through his provident fund. If now the State takes on some of these responsibilities, it is because he has established this right through struggle. The social services have become an intrinsic part of his wage: the social wage. In part, too, the cost of them has been recouped by the State in the increased tax and insurance contributions that he pays. Struggles around housing or benefits or schools are *economic*, as well as 'merely' political. Those things too must be protected against the erosions of inflation and the pressure of profit.

Secondly, it underpins changes in the nature of workplace struggles too. These are coming to be extended to take account not just of wages and conditions but of that part of life where reproduction goes on — life at home. Working mothers are never likely to make the mistake of seeing higher pay as the only, or even the main, demand. For them the most important thing is that the conditions of paid work respond, when necessary, to the needs of children, to take account of sickness and so on. It will force onto the agenda of workplace action the reproductive issues that should be there.

For years revolutionary parties of the left have neglected all forms of action but those of the factory floor: because in the employer you come up against the real class enemy in person. Not only is this crude, it is wrong. Many workers in their place of work are up against no fanged demon of capitalism but a benevolent head teacher, a petty minded park superintendent or a mild professor, whose status as class enemy is difficult to realize. At home, on the other hand, they are up against the landlord and the hire purchase credit chaser whose part in capitalism is clear enough. The analysis has to be more comprehensive.

To organize at the workplace alone leaves out half the worker's own experience of exploitation — speaking as it does of the cash wage but not of prices or of the social wage. More important, it excludes all wageless people

from organization. Pensioners, women doing unpaid domestic work, students at school and college, the unemployed and the invalid collecting State benefit — such people are a political resource, needed in struggle and needing it too. Reproduction, whether it be the practical reproduction of labour power or the ideological reproduction of our class system, our relationship to capital, is something in which everyone is involved.

Industrial struggle in the key industries is still the heart of the labour movement and will remain so. But, if the movement is to continue to gather strength in the contemporary phase of capitalism, others must be drawn into it and their different struggles recognized and joined by industrial militants.

*Collective action about services*

We are used to talking about struggle at the point of production. We know what it means — it is easy to envisage the factory floor, the building site or the office. One reason it is so easy to see is that it nearly always implies a place. The point of collective reproduction is not so easily pinned down. In some of its aspects it is of course a place too: housing takes place in a distinct location, and a council estate is a territorial unit, but private tenants renting from the same landlord may live miles, even cities apart. Education takes place in identifiable schools and colleges, but social casework and social security are delivered to people as individuals, sometimes at home, sometimes across a counter.

In spite of the intangible nature of the process of collective reproduction, many struggles *have* been developing around it in recent years. Some are easier to organize than others. The fact that all people claiming social security benefits in a certain area obtain them through one office has enabled claimants to contact each other and form claimants' unions, of which there are now very many in Britain. Action around demands for nurseries and other forms of child care have been organized by local Working Women's Charter and other groups. Schools have presented a problem of organization, because parent–teacher associations, even in working-class schools, are often dominated by more well-to-do parents. But the scope for change is there — aided in primary schools by the fact that mothers meet each other outside the school gates. Health services are specially difficult to organize because patients are usually preoccupied with personal problems to the point that collective action is inconceivable to them. Besides, the catchment area of hospitals is very large. Some women, however, have been forming health groups, often starting by learning how to administer their own pregnancy tests and understand their own (physical) reproductive problems.

The last few years have seen a rapid growth in collective action of this kind and two conclusions have been drawn and are beginning to be acted upon. One is that trades unions and rank and file groups have been for too long unconscious of the importance to them of the social wage and the need for them to go out of the way to contact and support such struggles. The second is that a natural connection exists between the interests of State workers and State 'clients'.

What of *elected members* of local councils? Much liberal democratic literature and practice makes a sharp distinction between bureaucrats and politicians. It is implied: if one could just put more power into the hands of the elected member and weaken the officers' grip on policy, all would go very much better for working-class interests. This is at one and the same time a misinterpretation of the position of members, and of that of officers. Many 'officers' are also 'workers' and identify as such. Conversely, it is far from evident that all elected members are politically distinct from senior officers in the bureaucracy. They are drawn even closer through the mechanisms of corporate management. As far as the public can see they often stand shoulder to shoulder. Besides, even should they wish to do so, local Labour leaderships do not in reality have the power radically to change the circumstances of the working-class population.

But if the leadership is inexorably caught up in the procedures of the State and the management of the economy, what, all the same, of the backbencher? Many ward councillors are people who have stood for election, even taken the step of joining the Labour Party, just for the advantages they hope it may give them as advocates or representatives of local working-class interests. Once in the Party and the council chamber, however, the impediments standing in their way are many. Corporate management has meant that the backbencher, the ordinary ward councillor, is further from the sources of decision and power (such as they are in the town hall) then ever before. He is excluded from the high-level partnership between leadership and senior officers and takes little part in policy planning processes. It is seldom that a left-wing captures power on a majority group in a local council and when they come near to it they do not remain for long undiluted by the right.

The isolated backbench councillor can take part directly in the action — working with squatting groups, tenants associations, women's centres and so on. Such a councillor has the advantage over an outsider (though not necessarily over a council worker) of having access to council information which he can pass on to groups. Councillors' intervention can however have bad effects on both action and organization; sometimes it delays direct action, defuses energy, blocks the view of the problem — limiting its definition to matters the councillor can or is willing to see as within his scope, or that of local government. It increases the tendency to believe that someone else, specifically someone in authority, alone can solve the problems of a situation that in reality is a reflection of an entire mode of production and balance of class power.

Such difficulties faced by elected members will be confirmed by many who've experienced them — but who may nonetheless insist that it is possible to find ways of furthering working-class struggle in matters of collective reproduction from within the council chamber.

### On the local State's payroll

The second place of reproduction struggle is in the State's own workplace: offices, swimming baths, parks, building sites and day centres. This includes not just local council departments but other (non-elected) local State agencies that play in whole or in part a reproductive role. Examples are the urban transport services, the hospitals and so on. We can also include national State jobs of a reproductive kind where they take a local form, such as work in employment exchanges, social security offices and the post office (Giro, family allowances).

The State, like industry, is hierarchical. At the top are senior officers, who take a big hand in making State policy (though convention has it that they subordinate themselves to politicians in this respect). In the middle ranks are the professional-style jobs of social workers, teachers, public health inspectors, rent officers etc, and at the bottom are the low-paid jobs of clerks and manual workers. To take social service department staffs as an example, in 1974 there were 5000 senior bureaucrats; 20 000 field workers and no less than 82 000 'dishwashers, residential, etc' in the children's homes of Britain (Case Con No. 2 1976).

It has been a characteristic of reproductive jobs of the local State, jobs in social services, health services, housing, education and so on, that they have become increasingly routine, disciplined and de-skilled. This is what people mean by 'the social factory'. Employees of the State, as well as those who receive its services, have felt less and less in control of the processes of welfare. In education the creativity of both teacher and child is subordinated to streaming, grading and exams. In colleges and universities education is lost in training geared to the needs of employers. In the hospitals patients often feel they are treated as social equipment, rather than human beings, and for their part doctors, nurses and auxiliaries feel they are cast as maintenance engineers. The institutions of the Welfare State have been 'established closer to the production system. The schools and colleges, the welfare system, the new housing estates were all made subordinate to the needs of capitalist production. This was achieved by applying the mode of that production — standardization and profitability — to social life outside the factory' (see Big Flame).

In the State, as in modern industry, growth and development have necessarily led to more 'proletarianizing' of employees. They are more clearly *workers*. In the nature of capitalism, though, its gains bring about its losses. The State's professionals also *feel* more like workers, and so there has been a rapid growth of union membership and activism during the 1960s and 1970s. Before World War II NALGO (the National Association of Local Government Officers) was a Conservative-led professional club, to whose members the idea of trades unionism, let alone strikes, was anathema (White). Now it is a mass-membership union and since the NALGO Action Group was formed in 1970 it has a rank and file of growing strength organized around shop stewards. NALGO cost employers about 100 000 days of work by industrial action in 1974. The 1970s also saw a growth of militancy in unions representing State manual workers with long national strikes of dustmen, postal workers, etc.

The dramatic cutbacks in public expenditure imposed by the Labour government in 1975 and 1976[3] had the effect of impressing upon many State workers the unity of interest of the worker as employee of the State and as client of the State. It is starkly clear that the existence of 50 000 posts in local councils standing unfilled and a threat of redundancies to come means both fewer jobs and deteriorating services.

The high proportion (56%) of women workers in State jobs sharpens this awareness. The perception that many women have of the connection between State job and service 'client' is made more acute by the fact that they themselves are often both worker and client. A national women's organization is emerging within NALGO. NALGO women in Hackney have linked up nursery workers with mothers in a campaign over nursery provision. In other boroughs social workers have taken direct action over homelessness, refusing to put families into bed-and-breakfast hotels and squatting them, instead, in the council offices. On a bigger scale, NUPE have been in the forefront of the fight against private pay beds in the hospital service. Maybe the next step will be (as in Italy) bus drivers and conductors refusing to collect fare increases, rent officers taking action in support of rent striking tenants and so on.

There are two strong reasons why, in their own interest as *workers*, local State employees are finding it necessary to include issues of reproduction (the nature of the service they are asked to perform) with their wage demands. These reasons are quite apart from the fact that, when at home, they are also clients. Firstly, unlike workers in private firms, they cannot point to the company's growing profits when asking for more pay. Their pay cheques come from taxes, not profits. Secondly, when public employees strike it is often of real inconvenience to 'the public', who in one guise of course are workers in private industry. Unless State workers include along with their demands on pay and conditions demands relating to the quality of service to the working class, private industry workers may feel such action to be directed against them as much as against capital (O'Connor 1973).

### Making over the family

We have no choice but to look to women, and to the family, as soon as we begin to examine the State. 'The State in its welfare aspects begins and ends with the family' (Wilson 1974). A striking feature of the instances of working-class housing action is the key role that women played in them. This is in strong contrast to industrial action, where among union and rank and file activists women are as a rule greatly underrepresented. In tenants' associations and street groups in many towns and cities women were among the organizers. Their involvement sprang direct from their experience in the home. The engagement of mothers in employment is usually provisional, being placed second to care of children and home. There is thus little time or energy left for organizing at work, but hit by intolerable housing conditions, or by actual homelessness, a woman becomes the most likely member of the family to take defensive political action.

Struggle inside the family and outside of necessity go together. It was clear from interviews with women in Lambeth that for involvement in struggle in the street or on the estate, a shift in attitude to housework, home and husband was both a precondition and a result. 'As I began to go out, things indoors seemed more trivial. Oh, things are never like they used to be here. I let things go now. It used to be well turned out, all clean.' It also had a profound effect on the confidence of the women involved, the way they thought of themselves. 'In those days I had no confidence, not for that sort of thing. When you start getting involved you find you're not a cabbage any more.' 'When it comes to it — now I know I'll fight.'

As Dalla Costa said, in a wider context, 'struggle demands time away from the housework and at the same time it offers an alternative identity to the woman who before found it only at the level of the domestic ghetto. In the sociality of struggle women discover and exercise a power that effectively gives them a new identity. The new identity is and can only be a new degree of social power' (Dalla Costa and James 1973).

In many ways the State and capital compete with us for definition of the 'family'. What they need it to be is not necessarily what we want it to be. Look at the State's manipulation of family in Lambeth. On the one hand the council depended upon women's self-definition of family to keep children from getting into 'care' and becoming a burden on the rates, but the State was also using 'family' as an administrative device to place some limit, even if an arbitrary one, on its housing responsibilities. Squatters were placed in the ludicrous position of having to find or beget a child (or get a doctor's certificate to prove that one of them was six months pregnant) in order to become a 'family' and qualify for rehousing — or indeed even to qualify as a recognized squatter. At no time did the council have a policy with regard to the housing problems of the single, i.e. non-family, people in the borough.

Again, family was necessary to the council as a measure of control, to avoid antisocial behaviour particularly on council estates. Where this failed and tenants were causing administrative problems to the council, they were invariably known as 'problem *families*', emphasizing the breakdown in the family control function required by the State. The opportunist nature of the local State's approach to family, however, is illustrated by the fact that when it suited it, as in the case of the Acre Lane reception centre for the homeless, fathers were quite callously separated from mothers and children. The cohabitation clause in social security also, of course, in Lambeth as elsewhere, forces men wherever possible into a 'family' role and into the financial maintenance of children of the women with whom they have a relationship.

The State sometimes uses the word community when family, or more strictly, women, is meant. 'The government's policy of unloading the burden of its cuts in the Welfare State onto women, of keeping them at home through necessity, is a means of preventing them from seeking work and figuring in the unemployment statistics. Instead local authorities talk of progressive sounding ideas such as returning the old to care within 'the community' rather than in old people's homes and day centres. The 'community' they have in mind is women' (Case Con No. 2 1976). Both 'family' and 'community' thus have a strictly utilitarian function for the local State and the terms of its support for neither necessarily coincide with the particular need that the individual has of them.

From a slightly different perspective it is possible to see some of the things that local *capital* wants of the family in Lambeth. In the first place, capital has actually defined the very shape of the family. The geographical mobility of people in Lambeth, especially in the poor central and northern areas, is extraordinarily high. People do not move for the fun of it. They move because industrial capital (that provides or fails to provide jobs), finance capital and property interests (that provide or fail to provide housing) push and pull them. The capitalist economy has stripped the family down to its bare nuclear essentials. A sociologist taking a strictly conventional view of families writes that those which are 'highly integrated into a kinship network or into a community inhibit the social mobility of their members. They hinder geographical mobility (a necessity for the smooth functioning of the economy) for whole kingroups are too cumbersome to respond sensitively to the changing demands of the economy (Farmer 1970). The result is that families in Lambeth tend to be strictly nuclear families, stripped of their older or unmarried members. There is also a high proportion of unsupported mothers with their children. Because of their mobility it is reasonable also to assume that few of them have relatives living nearby on whom they can call for support. The deterioration of the position of women with this paring down of the family has been well documented by the women's movement.

So the family is different things to capital, to the State and to the working class. Capital sees the family as market and labour force. The State needs a residential institution for workers and their young — with a strong element of social control over such matters as workshyness, truancy and delinquency. Ideally we should be coupled — though it will settle if necessary for the one-parent family. State and capital combine, too, of course to design the physical accommodation that moulds our family shape. Building societies, property developers and local councils have reduced our housing possibilities to the nuclear family carton.

## The future struggle

But what do WE want? We want personal space — people to be near us but not oppressively so. We want a possibility of genuine growth and learning through association that is fairly continuous over time. We want a base from which to link ourselves into many different collective groups, especially class groups, without being splintered between them all. We want supportive, practical working relationships, not predetermined by sex, nor by age, nor constrained by the physical space available to us. All these things conflict with the 'family' that the State and capital tend to make us. One requirement is shared: a family that is useful for capitalism is warm, caring and attached. And we want to be warm, caring and attached. This is the raw nerve of our struggle. In trying to make the family dysfunctional for capital, how do we avoid further hurting ourselves?

To struggle against dominant values in this field of culture and ideology may seem at first sight like a cost-free game, as though we have nothing to lose but our blindfold. The truth of course is that capitalist relations are not a separate field of struggle. We cannot change our ideas and our beliefs without changing our practice. The two are knit into one: each makes the other. Which comes first is as impossible to answer as whether the chicken comes before the egg. And to change practice is a painful and costly process. There are many women who will bear witness to the fact that the ideas implicit in women's liberation only become real when they are no longer just a set of ideas to be considered, toyed with, but are *one's own ideas*, matching in all respects one's experience of life. As that happens the nature of relationships invisibly changes too, and there is no going back. The costs have been paid before we had time to measure them against the gains. That is why struggle in this and all fields of experience has to be collective: otherwise the individual may pay too high a price.

The situation in which we practise is a set of contradictions. We need the family, capitalism needs it too: the two uses of the family are incompatible. We need services: capitalism needs us serviced. We need jobs: capital needs the work done. On the other hand, capital needs our participation, yet we use these openings in a way that can threaten capital. Capital can benefit from having us grouped into associations; but collectively we are strong. That our struggle seems to take place in contradiction should encourage us because we are seeing dialectically something that is real. The contradictions are not so immobilizing as they seem, because in their particular shape and form they are always changing and so opening up new possibilities for action. Uncovering truth by stepping outside the conventional ideas of family, school, local government or electoral democracy makes it possible to see the present situation for what it is *and at the same time* to roll it onward so that new terrain comes into sight.

## Notes

[1] The term 'collective' is used to mean that the financial provision is no longer an individual and family responsibility. It does not mean that people necessarily undertake the activity together. Council housing is an example — it is State provision but families live in separate dwellings.

[2] Cockburn's case-study was undertaken in the Labour controlled London Borough of Lambeth between 1973 and 1976 (Eds).

[3] And also the cutbacks imposed by the Conservative Government since 1979.

## References

Big Flame *What is Big Flame?* Information paper of the socialist revolutionary group Big Flame, Liverpool

Dalla Costa M and James S 1973 *The Power of Women and the Subversion of the Community* (Bristol: Falling Wall Press)

Farmer M 1970 *The Family* (London: Longman)

Gramsci A 1976 in Boggs C *Gramsci's Marxism* (London: Pluto Press)

Greaves B 1976 Communities and power *Community Politics* ed. P Hain (London: John Calder)

James S 1973 A woman's place, in Dalla Costa M and James S *The Power of Women and the Subversion of the Community* (Bristol: Falling Wall Press)

O'Connor J 1973 *Fiscal Crisis of the State* (London: St. James Press)

White B Whitleyism of Rank and File Action *NALGO Action Group Pamphlet*

Wilson E 1975 Women and the Welfare State *Red Rag Pamphlet No. 2*

# 14 Family Life in the Inner City: Myth and Reality

*by D Quinton*

## Introduction

When we consider the problems of the inner city we are part of a long tradition among professionals and administrators of concern with the conditions of life of urban and metropolitan families. This concern has always been prompted by a mixture of compassion for the disadvantaged and fear of social disorder and breakdown. The statistics persistently and consistently point out that there are higher concentrations of adverse living conditions in urban areas. There is more crime, vandalism and delinquency; there are more low-status and single-parent families; there is more psychiatric disorder, and so on. Many studies have shown the relationship between these factors and various educational and behavioural problems in children. Thus, low social status, large family size and over-crowded conditions have consistently been shown to relate to delinquency, to poor attainment and to behavioural difficulties (Davie *et al* 1972, Rutter and Madge 1976). Since the concentration of certain of these adverse inner-city characteristics (the proportion of poor or low-status households, for example) is increasing in many areas, it seems logical to suppose that behavioural and educational problems will also increase. But the relationship between various disadvantages and the behaviour and development of children are not as straightforward as they sometimes seem. Families with problems are not necessarily problem families. Disadvantaged areas do not necessarily have high rates of delinquency. Schools in similar areas can have markedly different amounts of difficulty (Power *et al* 1972, Rutter *et al* 1979). This does not mean that the roots of these problems do not lie primar-

ily in social disadvantages, or that we should not remove these, but it does mean that we need to be careful about our assumptions concerning the connection between disadvantage and children's difficulties or capabilities. We may mistakenly blame social disadvantage for school problems when on occasions the blame should lie with the school. We may expect lower attainments from children from disadvantaged homes. We may presume that adverse effects may follow from setting our expectations for these children too high.

There have been many studies of inner-city family life: some concerned with the problems of multiply-disadvantaged families (Tonge *et al* 1975, Wilson and Herbert 1978) or with abnormal development in deprived areas (West 1969, West and Farrington 1973) and others with social class differences in patterns of child-rearing (Newson and Newson 1963, 1968). However, in order to look at the contribution of the city itself to various family difficulties, comparative studies of urban with other environments are necessary. Our research comparing one inner London borough and the Isle of Wight was designed to examine family, social and environmental factors associated with children's behaviour problems in schools and with psychiatric disorder. The two areas were chosen because they were radically different environments: the one, part of a 'decaying' inner city; the other, a settled area of small towns.

## A comparative study of two urban environments

*Inner London and the Isle of Wight*

It is often pointed out that although such data may reflect the lives of representative samples of families, they deal with just two geographical areas. The London

Source: Quinton D 1980 in *Education for the Inner City* ed. M M Marland (London: Heinemann) pp 47–67.

borough investigated is in the middle range of inner London areas in terms of various social problems. It has a number of features which parallel the circumstances of families in the inner areas of other cities but there are also clear differences. London, for example, does not have the problems of adult or youth unemployment to anything like the extent that some provincial cities do. The Isle of Wight similarly has a number of features which distinguish it from other small-town areas. The importance of the holiday industry in the island's life and economy is an obvious example. From this it is sometimes argued that the value of such a comparison is limited because the areas are atypical. It is true that these considerations mean that generalizations about city and small-town life cannot be made, but it is quite incorrect to imply either that the nature of the area affects the conclusions that can be drawn concerning the inter-relationships between social factors and children's problems, or that other areas might be chosen which are somehow more representative. This is the first myth or assumption we need to dispose of. There is no such thing as a typical inner city or a typical inner-city family. The same is true of small-town areas. Cities have more of certain problems than other areas and more families are affected by these, but the patterns of association between disadvantages are different in different urban areas and the rates of particular adverse circumstances vary quite substantially between cities.

*The stereotypical view of inner-city family patterns*
Nevertheless the accumulation of statistics sometimes leads to a stereotype of a city family which might be summarized as follows. Inner-city families are often large, poor and of low social status. Many of them are headed by single mothers or unemployed fathers. They live in poor quality or overcrowded conditions or on anonymous new housing estates. Stable, supportive working-class communities have been decimated by disruptive housing policies and by the out-migration of the more skilled and more competent. The families that remain are either the disadvantaged and feckless remnant of these communities, in-migrating problem families or minority ethnic groups forced by discrimination into the decaying areas. In these circumstances children are growing up poor and unstimulated with resulting low educational attainments and disruptive behaviour. School problems are exacerbated by the alienation of parents and children from the school system. As the children get older they drift into vandalism and petty crime, find it hard to get work, have children early and often, and create a cycle of disadvantage.

*The comparative study*
The inner London–Isle of Wight study was designed to compare family patterns in different environments and to investigate how these relate to a variety of educational, behavioural and psychiatric problems. A two-stage strategy was followed. First, the total populations of 10-year-old children in each area were screened for possible behavioural problems at school, using questionnaires filled in by teachers. In addition, all children were assessed using group tests of reading and non-verbal intelligence. On the basis of this screening, samples of children were randomly selected to represent both the general population in each area and groups of children with particular educational and behavioural problems. The parents of these children were then interviewed on a wide range of topics concerned with the children's behaviour, family interaction and relationships, social contacts, and so on.

Both official statistics and data from the study confirm that the London borough has more of the adverse characteristics associated with family disadvantage. Thus there are more low-status, one-parent, large and overcrowded families; more children in care; more in-migrants from outside the UK; much less home ownership and many more homeless families. Consistent with expectations arising from this concentration of disadvantage, the London children are twice as likely to have psychiatric problems and to have behavioural difficulties at school. The case therefore seems clear: more social problems, more children with disorders. Paradoxically, however, differences between the areas are not strongly reflected in a range of family patterns which might be expected to connect these two sets of facts. Thus, within the family there are no differences in the frequency of parent–child interaction or communication, no differences in maternal warmth towards the children, no differences in the amount the parents expect the children to do in helping round the house, or the amount of control they exert over what they are and are not allowed to do. In much the same way the similarities in marital relationships and family patterns between the two areas are much more striking than the differences. Thus, a similar number of mothers are working; the average amount of warmth expressed by wives about their husbands is the same; the number of wives critical of their husbands is the same; couples share daily conversations to much the same extent; husbands help around the house to much the same extent; families on the island tend to go out slightly more together but on the whole the level of joint leisure in both areas is quite high. There is a small but significant difference in the quality of marital relationships but the great majority of marriages in both areas are without noticeable problems (see table 14.1).

In addition, two common assumptions concerning the reasons for problems in the city are not supported by these data. First, although it is true that the London borough is one which has been subject to much redevelopment and to a fall in skilled job opportunities, the idea of a consequent excess of socially isolated or unsupported families is not true. The amount of contact with, or support from, relatives is exactly the same in both areas and the vast majority of mothers have supportive relationships either with relatives or friends or both. Community change and environmental turnover may have something to do with increased family difficulties in London but the connection is not simply through an increase in social isolation due to the destruction of neighbourhoods. There are many more social isolates in the cities but isolation is not characteristic of families. Multiple-problem families have often been shown to lack support or social contacts or to have poor relationships with kin (Schaffer and Schaffer 1968, Wilson and Herbert 1978, Quinton 1981). However, this seems to be linked to psychological problems associated with adverse personal histories rather than to lack of potential support as such. Secondly, there is no evidence that selective in- or out-migration is directly responsible for the increased rates of children's problems. It has been suggested that data such as ours can be explained by the fact that 'happy families move out' (Ward 1978). This is not the case. In a separate four-year follow-up of families in this area it has been shown that the selective out-migration of families with

**Table 14.1**  Marital relationships

|                                              | ILB | IOW |
|----------------------------------------------|-----|-----|
| Mother working in past two years (%)         | 63  | 62  |
| Warmth of wife to husband (mean)             | 3·4 | 3·3 |
| Wife moderately or markedly critical (%)     | 20  | 21  |
| Daily conversations (%)                      | 60  | 59  |
| Husband helps little in the house (%)        | 26  | 29  |
| Frequent joint leisure (%)                   | 42  | 56  |
| Marriage rating (mean)                       | 2·3 | 1·9 |

less disturbed children has not occurred. Nor is the in-migration of problem families an explanation. The rate of disorder in children of in-migrants from other parts of Britain is no higher than that of children born and bred in London of London parents. Moreover, the parents of indigenous children received into care are not significantly more likely to be in-migrants than parents in the general population (Quinton 1981). Finally, it is known that the semi- or unskilled are not over-represented among in-migrants to this area from other parts of Britain (Gilje 1976).

In many ways then the patterns of family relationships in this part of London are remarkably similar to family relationships on the Isle of Wight. A number of common assumptions concerning the explanations for the higher rates of children's problems have been shown to be inadequate. Yet the London borough has many of the characteristics of disadvantage common to inner-city areas, and these are known to relate to children's educational and behavioural problems. Moreover, the children in this London area do have more of these problems. We need, therefore, to explain the nature of the connection between area indicators of disadvantage, family life and children's development.

## Area-based disadvantages

### The limitations of official statistics

The first point that needs to be made concerns the limitations of the conclusions that can be drawn from area-based statistics. It is often assumed that because a particular neighbourhood is shown to have high rates of various disadvantages, such as poor housing, individuals of low social status, one-parent families, etc, that there will be a high proportion of low-status single-parent families in poor housing. A number of intervention policies have been based on the identification of deprived areas using such statistics in this way. However, the assumption that such areas contain multiply-deprived families has generally been shown to be wrong. For example, the inner-area study of Lambeth (Department of the Environment 1975) examined the relationship between housing provision and family income. It is commonly assumed that the poorest families are the poorest housed but this proved not to be the case. Overcrowding was related to family size but not to income. Other indicators of basic amenities showed no relationship between income level and housing provision, nor were single parents worse housed than two-parent families. The overlap of housing problems with other disadvantages varies from area to area. In this case it is probably particularly affected by the high proportion of local authority housing.

The evidence suggests that even in multiply-disadvantaged areas the proportion of multiply-disadvantaged families is relatively small. Conversely, a number of multiply-disadvantaged families live in areas which are not deprived on the evidence of area statistics. This is recognized by the authors of the Liverpool inner-area study (Department of the Environment 1977) who conclude after a careful analysis of census data that

> even in the worst areas the proportions of the resident population in the age, family, social or racial groups most at risk of deprivation are quite small. In virtually no area is the majority of the population actually deprived according to any of the indices of personal, family and social character. The only common form of deprivation shared by the majority of the population is their physical environment and, possibly in some instances, the quality of their housing. In particular the proportions are too small and the areas too large for any conclusions to be drawn about the incidence of multiple deprivation meaning the extent to which a person or family is experiencing a range of different problems. Finally, a large proportion of even the most concentrated indices are to be found not in the worst areas but scattered in other parts of the city.

Two points arise from this. First, that identifying concentrations of families with problems on the basis of area statistics will generally not be successful. Secondly, statistical association between area indicators of disadvantage and such problems as delinquency or child disorders can lead to premature causal assumptions. Errors of this sort can arise as Robinson (1950) has shown in his discussion of the ecological fallacy when the area indicators of disadvantage do not apply to the circumstances of the individuals who are exhibiting the particular problems.

Nevertheless, is it not the case that numerous studies show direct relationships between all kinds of inner-city problems and children's behaviour and attainments? Yes they do, but these findings need to be put in perspective. Although we now know a good deal about the general correlations between multiple disadvantage and the various problems, we know much less about the processes linking particular disadvantages with the various difficulties of children. Moreover, the kind and extent of overlap of adversities can vary quite considerably between different populations and areas and we know little about the effects of particular combinations of disadvantages. We are certainly not in a position to draw direct causal conclusions from correlations between disadvantages and problems.

### Results from the London and Isle of Wight comparison

Some of these points can be illustrated with our own data. Firstly, in terms of examining general relationships we were able to see whether the adverse factors highlighted in other studies also occurred in our two populations. Secondly, we could examine whether the differences between the two areas in the rates of children's problems could be explained in terms of these adverse factors, or whether some other city processes seem to be operating.

First, our data do confirm many of the relationships found in other studies. Thus, children's behaviour problems in both areas are related to a number of family adversities such as low social status, overcrowding, large family size, the child having been in care, parental psychiatric disorder and criminality, and marital discord. The second point concerns the extent to which the differences in the rates of children's problems between the two areas can be explained by the excess of family adversity in London (see figure 14.1). A psycho-social adversity index was used for this analysis calculated by combining scores on the six variables listed in the figure. Over half of the London children were exposed to two or more of these adverse circumstances, compared with less than one-fifth of those on the island. A standardized comparison of the two areas

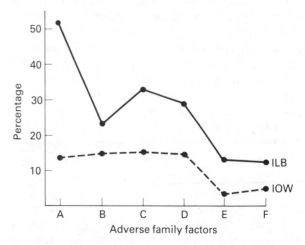

**Figure 14.1**  Adverse family factors in the Isle of Wight (IOW) and inner London borough (ILB). A, homes overcrowded; B, fathers with semi-skilled and unskilled jobs; C, mothers with psychiatric disorder; D, fathers with any offence; E, with severe marital discord; F, children in care (number per 1000).

was made: that is, the question was asked if the amount of disadvantage was the same in both areas, would any difference in the rate of children's disorders remain? As figure 14.1 shows, when such a standardized comparison is made, the difference between the two populations virtually disappears and is no longer statistically significant.

The important conclusion from this stage of the analysis is that behaviour problems in primary-school children in inner London can be largely explained by certain adverse factors operating through the family and, conversely, these factors have similar implications in other reputedly more favourable environments. It is not necessary to invoke any notion of the operation of special direct city processes to explain the differences between the areas.

## Two remaining questions

Two questions remain. Firstly, why are there more disadvantages in inner-city areas? Secondly, what contributions can schools make to ameliorating their effects?

*Why is there more family adversity?*

As has been shown the adverse effects of inner-city life on children's behaviour can almost entirely be explained in terms of factors operating through the family. No distinct city processes need to be invoked. This, however, is not the case for the adults.

Certain factors such as the greater proportion of semi- or unskilled parents or the differences in housing circumstances are clearly due to various processes in the development of cities. Whether these are inevitable characteristics of city life has not been demonstrated. Similarly, the greater number of large families may be explained through the association between family size and lower socio-economic status, although the reasons for this association remain obscure. However, certain other factors, including ones which may affect family interactions and child development directly, seem to be attributable to adverse city processes, although the ways in which these adverse processes operate remain remarkably elusive.

Maternal psychiatric problems may be taken as an example. Such problems are twice as common in the inner London area as on the Isle of Wight. Now it is generally the case that studies of samples of psychiatric patients have shown a strong association between depression or neurosis and marital and family prob-

lems. This is also true in our study. Where marital relationships are poor, the majority of mothers have psychiatric problems both on the island and in inner London. However, a more important fact is that there is a considerable excess of such psychiatric problems in London mothers in the absence of marital difficulties or less satisfactory family relationships. Thus 24% of women in this area in satisfactory marriages nevertheless have psychiatric problems. A clue concerning the possibility of additional city stresses acting upon adults is given by the association of disorder with social class. On the Isle of Wight there was no association between social class and psychiatric problems in women. Thus increased psychiatric problems are not necessarily a correlate of low social status. Further, there was no area difference in the prevalence of depression in women whose husbands had non-manual occupations. Thus depression is not an inevitable consequence of city life. In London, however, depression was much commoner in working-class women and it was only in this social group that the rate of psychiatric disorder was higher than on the Isle of Wight. The reasons for the increase in disorder in working-class mothers are not yet known but the data suggest that some additional city factors may be operating. What these factors may be is unclear at present. A number of common assumptions about family life in the city have been considered earlier. These fail to explain maternal disorder, just as they fail to explain children's problems. Thus the increase in difficulties for mothers is not related to patterns of selected migration or to lack of community support.

However, I have yet to discuss the most obvious way in which the city is different from the small town and that is the environment itself. It is necessary to ask whether differences in this explain the differences in individual family problems. The assumption that such environments are somewhat more stressful is commonly made. For example, Milgram (1970) has hypothesized that many city behaviour patterns are due to greater information overload in cities and Ward (1978) has suggested that a higher level of competence may be needed to handle the city. The fact that inner-city environments are often seen as unpleasant is frequently taken as evidence that they are harmful, but it is important that these two notions should be kept distinct. An unpleasant environment should be changed regardless of whether it has adverse effects or not, but an unpleasant environment does not necessarily cause social problems.

There are a number of problems in investigating the

effects of the built environment on the individual and the family. These include the difficulty of taking previous housing and personal history into account as well as considering the ways in which cultural factors affect vulnerability to stresses (Quinton 1980). Nevertheless, the evidence that the built environment itself is a markedly harmful feature of city life is not clear-cut. For example, psychiatric problems in adults have not generally been shown to relate to environmental circumstances, and do not do so in the present study, nor do they change when environments are improved (Hare and Shaw 1965, Taylor and Chave 1964). Housing styles do not generally appear to affect psycholgoical functioning in the absence of other stresses. It is true that certain environmental features, such as overcrowding, are related to various family problems but this is usually in the context of other overlapping disadvantages. Thus living in flats is not consistently related to adult psychiatric problems but rates are higher for mothers in high-rise blocks (Moore 1974). We have yet to produce convincing evidence of direct links between wider environmental variables and the increase in adult psychiatric problems or marital difficulties in urban areas. However, the possibility remains that other features of housing are more important than the nature of the dwellings themselves. Thus, Brown et al (1975) have shown that stresses, such as threatened eviction or loss, have an effect on maternal psychiatric state. The critical environmental variable may thus be a perceived lack of control over individual circumstances rather than the physical characteristics of the area itself. This feeling may have been exacerbated in recent years by the rapid change in the environment both in the physical sense and in terms of visible population changes, especially with the influx of black families. However, it is extremely unlikely that any straightforward relationship between environment and developmental problems exists. For example, wide variations in rates of deviance are often found in similar physical environments (Jephcott and Carter 1954, Edwards 1973, Power et al 1972). Nevertheless certain environments such as the Piggeries in Liverpool do seem to have become inimical to habitation. Situations such as these have been linked by Newman (1976) and others to features of design such as the lack of defensible space or of semi-public areas where safe, informal contacts might be established (Yancey 1971).

At present the evidence suggests that there are stresses in life in inner London which relate to the increase in maternal psychiatric problems but the nature of these stresses is not well understood. The relationship with social class indicates that city stresses only have an effect in the absence of adequate resources. Material resources are clearly important here but they are not the only ones. For example, Brown's study shows that working-class women in a similar inner-city area are subject to more acute and more chronic stresses than middle-class mothers. However, they are only more likely to succumb to these by developing psychiatric problems if they are further stressed by the presence of pre-school children, or especially if they lack an intimate, confiding relationship with a husband or boyfriend.

From the point of view of the child, however, family values and relationships seem to be the variables which most strongly mediate the effects of the physical environment and the child's view of it.

### The school as a positive factor in ameliorating social disadvantage

The question of ameliorating factors brings me to the last point I wish to consider and that is what the role of the school might be in attenuating the effects of social disadvantage. It should be made clear that I am dealing here with the school's contribution to the general effects of social problems and not to specific services or provisions for particularly disruptive or troubled children.

So far I have argued that the excess of children's problems in cities is largely explained by greater family adversity of various kinds. This observation is a view not infrequently echoed by teachers when they complain that their problems walk in through the door. By this is meant not only that families cause the children's problems but also that attempts to deal with these at school are frustrated by lack of interest or support from the home. However, this seems to be too simple an explanation. Firstly, it is known that schools with very similar catchment areas have very different rates of problems and that this is unlikely to be due simply to differences in intake (Rutter et al 1979, Power et al 1972). Secondly, the research evidence suggests that the majority of parents, even from disadvantaged circumstances, are not alienated from the idea of school, although they may be puzzled about its practices. Thus, 74% of Wilson and Herbert's multiple-problem families in Birmingham saw the teacher as someone whose authority ought to be supported. The parents of reading-retarded children studied by Sturge (1972) were keen to help their children and among the West

Indian parents in our study 84% had taken their children to a library in the previous three months. Half had bought the children books during this time and half had checked the children's homework regularly. In general their educational aspirations were high. What they may have lacked were techniques. This means that conventional measures of parental interest in school work, such as attendance at open days or PTA meetings, are not adequate. These occasions are often intimidating even for middle-class parents, let alone disadvantaged ones. A study of a thousand disadvantaged families quoted by Midwinter showed that 59% of mothers tried to help with reading and practical work, 74% felt teachers were interested in parental views, but on the other hand 80% felt that they knew very little about the methods being used in the school (Midwinter 1978). Parents in these circumstances may lack expertise or knowledge as to how to help their children, but there is little evidence that they do not care. However, both their lack of expertise and their lack of suitable circumstances for promoting the children's educational development, means that the expectations which schools should have about the possibility of home support must be realistically based.

I would like to conclude with a suggestion that, despite the primacy of home background in determining many aspects of child development, the school nevertheless can act as an ameliorating factor in the face of disadvantage. In our comparative study it was possible to construct an index of school adversity in the same way as we had for family problems, using such indicators as high teacher or pupil turnover, the proportion of children on free school meals, etc (Rutter and Quinton 1977). When the rates of deviant school behaviour in London were examined in relation to school and family adversity, the results were illuminating. Children from low-adversity homes who were in low-adversity schools showed no behavioural problems. High school adversity appeared to have the effect of raising the rates of disturbance in children from low-adversity homes while low school adversity attenuated the effects of family problems. The ways in which the low-adversity schools had this effect could not be determined in this study but the important point here is that schools can have this effect.

I have tried to show that many social disadvantages do not always, indeed do not usually, imply that homes lack warm or supportive relationships — rather they may lack skills and resources to help facilitate cognitive and intellectual development. There is little use in

schools complaining about this. It is their job and their claim for special status that they are able to develop skills in children; it is their responsibility to find ways of doing so. This seems to me to provide the clearest way out of disadvantage for many children. The best hope of schools having positive effects is for them to concentrate on promoting those skills which increase the child's mastery of the environment and provide a path out of disadvantaged circumstances. This almost certainly involves developing basic and relevant skills rather than attempting to provide a primarily supportive but not skill-oriented atmosphere. This does not mean schools should not be warm and supportive places, but it does mean that they should not try to be substitute families.

One example in support of this argument can be given. The failure of a child to develop reading skills has well known adverse consequences. There is another side to this issue however. The development of such skills may have a protective effect. Figure 14.2 shows the relationship between poor, average and superior reading skills, family adversity and behavioural deviance in school. The upper line shows the findings for children from deprived and disadvantaged homes. The bottom line shows the results for those in more favoured circumstances. In both cases the rates of disorder are considerably less among children of above average attainment. Of course, this may simply mean that more intelligent children are more resilient to stress, but the study by Rutter et al (1979) shows that behavioural problems in the classroom are strongly influenced by the extent of reward and encouragement for scholastic attainment.

Reading is just one obvious example of the possible

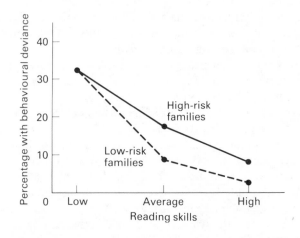

**Figure 14.2** Reading skills, family adversity and behavioural deviance.

positive effects of the development of ability. It is likely that many other skills of both an academic and non-academic sort have similarly powerful effects in ameliorating the effects of disadvantage. This leaves us in a distinctly hopeful situation. Schools can do nothing about low social status, large families, bad marriages or poor housing. It is more than encouraging to know that those abilities they can foster can have a powerful impact in ameliorating the effects of the greater amounts of disadvantage that cities unfortunately contain.

## References

Brown G W, Bhrolchain M N and Harris T 1975 Social class and psychiatric disturbance among women in the urban population *Sociology* **9** 225–54

Davie R, Butler N and Goldstein H 1972 *From Birth to Seven: a Report on the National Child Development Study* (London: Longman)

Department of the Environment 1975 *Inner Area Study, Lambeth: Poverty and Multiple Deprivation* (London: Department of the Environment)

—— 1977 *Inner Area Study, Liverpool: Social Area Analysis* (London: Department of the Environment)

Edwards A 1973 Sex and area variations in delinquency rates in an English city *British Journal of Criminology* **13** 121–37

Gilje E K 1976 Migration patterns in and around London *Greater London Council Research Memorandum* 470

Hare E H and Shaw G K 1965 *Mental Health on a New Housing Estate, Maudsley Monograph No. 12* (London: Oxford University Press)

Jephcott A P and Carter M P 1954 *The Social Background and Delinquency* (Nottingham: University of Nottingham)

Midwinter E 1978 Family functioning and educational performance *Social Work Today* **10** (12) 12–13

Milgram S 1970 The experience of living in cities *Science* **167** 1461–8

Moore N C 1974 Psychiatric illness and living in flats *British Journal of Psychiatry* **125** 500–7

Newman O 1976 *Defensible Space: People and Design in the Violent City* (London: Architectural Press)

Newson J and Newson E 1963 *Infant Care in the Urban Community* (London: Allen and Unwin)

—— 1968 *Four Years Old in an Urban Community* (London: Allen and Unwin)

Power M J, Benn R T and Morris J N 1972 Neighbourhood, school and juveniles before the courts *British Journal of Criminology* **12** 111–32

Quinton D 1980 Cultural and community influences on child development *Scientific Foundations of Developmental Psychiatry* ed. M Rutter (London: Heinemann Medical)

—— 1981 Parents with children in care. I. Childhood experiences and current circumstances *Journal of Child Psychology and Psychiatry*

Richman N 1974 The effects of housing on pre-school children and their mothers *Developmental Medicine and Child Neurology* **16** 53–8

Robinson W S 1950 Ecological correlations and the behaviour of individuals *American Sociological Review* **15** 351–7

Rutter M and Madge N 1976 *Cycles of Disadvantage* (London: Heinemann Educational)

Rutter M, Maughan B, Mortimore P and Ouston J 1979 *Fifteen Thousand Hours: Secondary Schools and their Effects on Children* (London: Open Books)

Rutter M and Quinton D 1977 Psychiatric disorder: ecological factors and concepts of causation *Ecological Factors in Human Development* ed. H McGusk (London: North Holland)

Schaffer H R and Schaffer E B 1968 *Child Care and the Family* (London: Bell)

Sturge C 1972 Reading retardation and anti-social behaviour *M. Phil. Dissertation* University of London

Taylor G and Chave G 1964 *Mental Health and Environment* (London: Longman)

Tonge W L, James D S and Hillam S M 1975 Families without hope: a controlled study of 33 problem families *British Journal of Psychiatry Publication No. 11*

Ward C 1978 *The Child in the City* (New York: Random House)

West D J 1969 *Present Conduct and Future Delinquency* (London: Heinemann Educational)

West D J and Farrington D P 1973 *Who becomes delinquent? Second Report of the Cambridge Study in Delinquent Development* (London: Heinemann Educational)

Wilson H and Herbert G W 1978 *Parents and Children in the Inner City* (London: Routledge and Kegan Paul)

Yancey W L 1971 Architecture, interaction and social control *Environment and Behaviour* **3** 3–21

# 15 Urban Inequalities, Market Processes and Local Government

*by B E Coates* et al

**Market processes and their spatial outcomes within cities**

Spatial inequalities and territorial injustice exist in every 'economic city'. A city contains human resources ranging from the most talented and aggressive to the most marginally employable, and occupations stretching down from the directors of multinational corporations through their managers and research staff, workers in the mass production units requiring a narrow range of skills, to the most menial jobs in the service trades. These differences are compounded by, as well as related to, ethnic, religious and racial characteristics, life-cycle, age and sex structure, health, education, vocational training, endeavour, perseverance and sheer opportunity.

The unequal distribution of income is one result of this plethora of conflicting pressures and interlocking attributes of city life. This, in turn, produces unequal competition in the market place. It is necessary to stress that this basic inequality is not simply a result of the economic forces alone, but emanates from a complex interplay of social and political as well as economic forces. Economic reality for certain groups is, for instance, influenced by the extent of discrimination, both social and political. Relationships within a city are of great complexity, but a common result of the unequal distribution of income is residential segregation and the creation of ghettos. Such features within a city are the spatial expression of deep seated economic and sociological realities. They bear witness to the basic fact that the extent to which benefits accrue in the market system is a function of income, wealth, educa-

tion, health, race and so on. The market mechanism is designed to discriminate through its rationing function against those who have purchasing power inferior to that of other consumers in the market (Henderson and Ledebur 1972). If this is allied to racial, ethnic, religious, class or caste discrimination, which excludes those discriminated against from access to jobs and other forms of income, then an 'apartheid situation' will exist in the city. Within the ghetto the process of cumulative circular causation may operate in a similar fashion to the one depicted in figure 15.1.

In the intra-urban situation the market mechanism results in the exclusion of some individuals from participation in market activity because of their failure to conform with social, educational or other standards established by the institutions that control market behaviour and participation. Conversely, the market mechanism works in favour of the wealthy who also tend to cluster in space, choosing desirable locations in terms of positive attributes (externalities) such as quality of physical environment, access to commuter services, proximity to suppliers of luxury goods, and security investments in property, and avoiding areas accumulating negative attributes (externalities) such as pollution, crime, poverty, disease, poor schools and inadequate health and recreational facilities. The institutions of the economic city, for example banks, estate agents, mortgage companies, lawyers, are geared to help the affluent attain their supposed 'needs' and maximize their advantages.

Clearly people compete to live in certain socio-economic environments. Once they are established in the area of their choice, that is in areas which accord with their own 'mental map' of the city, they will act to maintain its character to protect their status, family position and financial investment. Their actions pro-

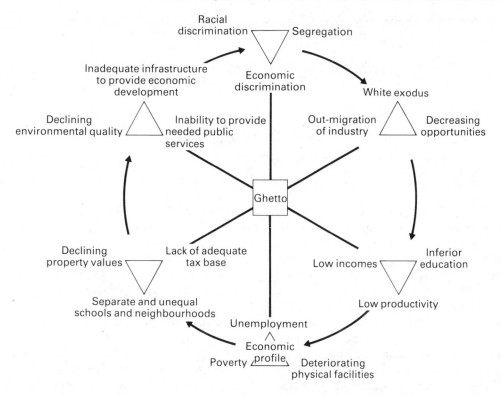

**Figure 15.1** The circular and cumulative process of socio-economic relations in the ghetto. (From Henderson and Ledebur (1972) *Urban Economics, Processes and Problems*, John Wiley and Sons, New York.)

duce conflict over 'goods' which they themselves do not produce — such 'goods' are generally termed externalities (Cox and Reynolds 1974). The actions of others produce benefits or costs, that is, positive or negative externalities. In the 'economic city' of capitalist society one of the more important positive externalities for which housholds compete concerns property values. The value of its homes and its land are indicative of the socio-economic status of a neighbourhood and thus of its inhabitants, who often view their property as a major speculative investment as well as a home, and they will act to protect their capital by fighting for positive externalities and against negative externalities — a feature which accords with the objectives of the estate agents, banks, insurance companies and building societies, as well as with those who raise local taxes on property rated by its monetary value.

These institutions are in a position to alter the status of a neighbourhood. It can be raised by channelling investment into domestic properties and by developing positive externalities such as well equipped schools, rigid controls on industrial development and pollution, access to the main commuter links and so on. On the other hand, a neighbourhood can go downhill as property investment finance is withheld and as negative externalities are allowed to accumulate in the area — the public incinerator, the demolition scars of 'slum redevelopment', the ageing schools, the serried ranks of cars left by commuters from another area, and the children playing in the streets for want of a safer place in which to burn up a portion of their abundant energy. Herbert and Johnston (1976, p 8) state:

In the conflict for these externalities, the competitors do not begin on an equal footing. Those with wealth and power are better able to manipulate the socio-economic and political systems which distribute — sometimes unintentionally — the externalities, and they dictate the form of the urban residential mosaic, allocating other groups to areas outside that which their social and economic superiors normally visit.

## Local legally bounded spaces

### Local government systems

All but the very smallest nation states have a network of lower-tier authorities for political/administrative purposes. Such authorities may be multipurpose ones, such

**Table 15.1**   Units of local government in the United States, 1942–72 (from US Bureau of the Census)

|  | 1942 | 1952 | 1957 | 1962 | 1967 | 1972 |
|---|---|---|---|---|---|---|
| Counties | 3 050 | 3 052 | 3 050 | 3 043 | 3 049 | 3 044 |
| Townships | 18 919 | 17 202 | 17 198 | 17 142 | 17 105 | 16 991 |
| Municipalities | 16 220 | 16 807 | 17 215 | 18 000 | 18 048 | 18 517 |
| School districts | 108 579 | 67 355 | 50 454 | 34 678 | 21 782 | 15 781 |
| Non-school special districts | 8 299 | 12 340 | 14 424 | 18 322 | 21 264 | 23 885 |
| Total | 155 067 | 116 756 | 102 341 | 91 185 | 81 248 | 78 218 |

as the states (50), counties (more than 3000), municipalities (c 18 500), and townships (about 17 000) of the United States or the metropolitan (6) and county (39) councils, 296 county districts and 36 metropolitan districts of England (outside Greater London), or they may be responsible for a single function, for example a school, a park, a forest reserve or a sewage district, a regional hospital board, a public transport undertaking, or a water board. Table 15.1 indicates the tremendous profusion of legally bounded spaces in the United States. Though the overall number of units was halved between 1942 and 1972 this fall was accounted for by the disappearance of 90 000 school districts. Over the same period the number of non-school special districts went up from 6299 to 23 885. Overall, the density of local legally bounded spaces is greater in the US than in any other country (Soja 1971).

In many countries, such as the United States, Canada and Australia, the power of these locally created general and special purpose districts is very considerable. The nation state is an amalgam of formerly independent states and the latter remain as the prime administrative units within the nation, having a considerable impact on the nature and quality of public goods bestowed on individuals and localities. A similar case can be made for non-federal countries which often work through a network of lower-tier authorities which already exist. Thus local legally bounded spaces in a unitary state have the power to influence the services provided, and by their spatial allocation of resources they can create, fortify or rectify inequalities. Clearly, local autonomy over how money is spent is much greater in the United States than in the United Kingdom. In the United States a far greater proportion of the money spent is raised locally, mainly from taxes levied on property (the state fixes the limits of debt held by the local units however, and this is one of the reasons for the creation of so many special districts — see below).

In the United Kingdom, on the other hand, more than 60% of the money spent by local authorities comes from the central government by way of rate-support grants. The central government is in a strong position, therefore, to intervene in order to even out what it considers to be gross inequality in patterns of expenditure by local authorities, and can put pressure on the local spenders to make changes in the priority they give to different services.

In England, the map of local legally bounded spaces has evolved over the last thousand years or more. But during the past century or so the power of the central government has increased at the expense of local governments, and in 1974 the former carried out a major reconstruction of local government throughout the country. However, the past is not 'swept under the carpet'. Even in the major reforms of local government in the 1880s, 1890s and 1970s most of the 'new' local government boundaries were not drawn on a 'clean slate' in relation to the supposed needs of the 'present', but followed ancient parish, wapentake and shire boundaries. The 'new' patterns are the result of political compromise. The Acts of 1888 and 1894 did not embody many of the most important recommendations of the Royal Commission set up to advise Parliament. In particular, the number of county boroughs created was raised from a recommended handful to more than fifty by the time the Bill had passed through Parliament. Similarly, the 1972 Act recognized fiercely held traditional loyalties to existing counties and embodied a radically different structure to that proposed by the Redcliffe–Maud commissioners (Redcliffe–Maud Report 1969). Once again, 'an attitude of sacred inviolability, on the basis of which many local councils staunchly guard their own vested interests, had created a barrier of inflexibility' (Douglas 1968) — a barrier which the central government is either unwilling or unable to crush.

Faced by these barriers to change, the central government in a unitary state, rather than beat against the weight of vested interest, often bypasses the existing forms and meets a particular demand by setting up a series of new administrative bodies, each with its own structure and spatial pattern: these new units often cover a larger territory and benefit from economies of scale but they are no longer subject to local control. Functions are thereby taken from the local authorities and instead of a clearly defined set of multifunctional authorities able to coordinate activities, the system spawns administrative complexity, a disintegrated set of authorities and a division of responsibility between local and central government which sometimes allows each to blame the other for 'output failures'. Though all concerned might acknowledge the need to govern all parts of the state as fairly and as efficiently as possible, the spatial political/administrative system adopted for the purpose emanates from a combination of historical factors, political compromise and administrative convenience.

Similar problems occur in the Tokyo, Paris, London, New York, Chicago and Los Angeles metropolises and on a smaller scale in scores of 'metropolitan' areas throughout the world, not least in the burgeoning cities of the Third World. The poorest members of the 'city region' are trapped in deteriorating environments (in terms of both physical conditions and public service provision) and the richer members of the economically interdependent region flow out into 'the green and pleasant lands' beyond the administrative reach of the central city of great need and financial burden. In all such cases the political fragmentation produces a markedly inefficient situation in which the scale of the *de jure* territory and the scale of the movement network are incongruent — the more incongruent they are the greater will be the conflict between localized populations and between the administrative units and the network space (Cox and Reynolds 1974). Incongruency is most marked in the great urban centres of the world, the very places where 'conflict' can be most harmful to millions of individuals and families.

### Local autonomy and spatial inequalities

Local legally bounded spaces have been created in recognition of the fact that despite a high degree of interdependency between groups and areas in a modern state much of human activity is localized. Spatial interaction of the individual and group is largely determined by locational attributes, especially distance and its major correlate, accessibility. Political and administrative structures have therefore been developed to organize local space in response to demands from local communities whose social and economic activities require the political partitioning of space for their needs to be met. We have seen that a state is not divided to assist in the creation of territorial justice. A government (local or central) may favour not only certain groups (by the way it taxes income, property, and wealth, for instance) but also particular areas. A spatially biased system may indeed be a deliberate aim of government. Many examples of spatial bias may be cited; for instance, overt discrimination against the South after the American Civil War, action against selected nationalities in the Soviet Union in the 1920s and 1930s, the acceptance of discriminatory policies towards African areas in South Africa and Rhodesia, exploitation of the rural peasantry for the benefit of a small, privileged urban class in Latin American dictatorships, and a wide range of policies adopted in many plural societies throughout the world (including Northern Ireland). Such policies operate at different scales: there are Bantustans and African townships, and there are black city ghettos and black rural slums across the South.

Locally, a dual housing market may operate to the advantage of certain groups and the disadvantage of others. Or urban landowners (including public authorities) may place restrictive covenants on land use to protect some areas from unwanted invasions of certain groups. Similarly, mortgage companies, building societies, banks, and both public and private developers may, within the law, operate a blatantly unequal allocation of resources to different parts of the urban area, thus initiating or reinforcing intra-urban inequalities (Harvey 1973). Such inequalities arise from local government policies on land-use zoning, grading of council tenants for allocation to certain properties and estates, relocation of slum dwellers, provision of recreational space, location of schools and health clinics in relation to the public transport system and so on. Even local government action in allocating house improvement grants in the United Kingdom may, in practice, further disadvantage rather than positively assist, as was intended by the central government, many of those most in need, as planners in the public sector and builders in the private sector concentrate on those areas where their objectives seem most assured rather than on the lower-income families and housing-stress areas (Duncan 1974). These and many other examples could be analyzed in detail.

In the next section we discuss just two examples of the way in which local autonomy affects inequalities. Both examples illustrate the way in which inequality is accepted by the operators of local legally bounded spaces, and, supported by the full weight of a nation's law, the system of local autonomy over the way in which money is spent is thus, by its actions, a partner in the production of patterns of territorial injustice. We look first at the practice of zoning in the United States and second at the provision of public education in the United Kingdom.

## Case studies

### Zoning to maintain inequalities

There are tens of thousands of local legally bounded spaces in the United States. Each of these territorial units instils a sense of community and separateness. Many pursue narrow local interests at the expense of the larger functional community. 'Moulded by nineteenth-century notions of the moral and ethical values of local autonomy and self-determination, much of the US has become shattered like a pane of glass into a system of often competitive and mutually suspicious fragments' (Soja 1971). Most of the major northern cities find themselves surrounded (strangled?) by a resistant white collar of wealthy, autonomous suburbs determined to maintain their high-quality, low-cost schools, to keep out noxious facilities, to minimize negative externalities of all kinds and to sustain their 'right' to provide their own rich community with services paid for by their own relatively low taxes.

Zoning is the favourite device employed to maintain the 'exclusive' and 'separate' character of political space. Toll (1969) has traced the evolution of zoning (a system of laws regulating and restricting the use of land in particular areas) in the United States. He rejects the view that the struggle for zoning was championed by reformers seeking the general welfare of the community against the powerful resistance of the rich, propertied groups. On the contrary, Toll shows that it was the ruling economic élite itself which introduced and diffused zoning laws to reinforce their power and prevent the 'invasion' of undesirable groups into their territory by controlling land use within it. Soja (1971) concludes that zoning was, and is, used to: (1) protect suburban property owners; (2) discriminate against particular racial and economic groups by artificially keeping

house prices high; and (3) protect local interests, thus accentuating problems of urban fragmentation with its attendant social and economic inequalities.

By manipulation of zoning ordinances, suburban municipalities can prevent free movement of certain groups in the population into suburban communities. Typically these groups are comprised of the

low-income families, the elderly, the large families, the welfare families (particularly those with a female head of household) and, above all, the racial minorities. Thus much suburban zoning is now viewed as discriminatory against low-income groups, racial minorities, and the like, and there can be no doubt that in many cases suburban zoning is accomplished primarily for discriminatory reasons. It is thus extremely difficult to break down, even though the cities are anxious to achieve a better balance within the metropolitan areas as a whole. The inner-city territories are in fact bearing a disproportionate amount of the nation's cost of providing welfare, education, special services, and the like. The burdens and advantages associated with a form of spatial organization set up to generate the social surplus are thus unequally distributed among the political jurisdictions. There is, in short, a pattern of allocation which is socially unjust (Harvey 1972).

The inevitable consequence of such local government action is that within the girdle of rich, white suburban legally bounded spaces one finds central city deterioration, black ghettos, zoning and planning based on greed or fear and 'visual pollution' of the urban and rural landscape (Bunge 1975).

Can the system producing such spatial inequalities be changed by a restructuring of politically partitioned space? Harvey (1972) observes that the forced integration of the suburbs has so far been rejected and, as the suburbs now hold the balance of political power in the country, he sees too forceful action amounting to political suicide. Moreover, he argues that the pricing system of many public (and private) utilities (such as water, electricity, sewage and sanitation) systematically discriminates against the poor, giving an implicit transfer of income from the poor to the rich. Reorganization of government would mean the rich losing economic and social as well as political advantages. Noxious facilities which at present are put in the least politically powerful areas might well be distributed more evenly. Areas with poor resources and thus poor public provision of services and facilities might, given a redistribution of political power, get better services at the cost of increasing taxes in the hitherto protected suburban areas. In so far as interest groups in a city meet their objectives by exercise of social influence, wealth, and political power, a changed geographical organization of governmental and political structures would thus undermine the power base of those groups which have arranged geographical space to serve their particular interests.

The problems involved in restructuring such an unequal system are immense, therefore. Furthermore, Stetzer (1975), after a painstaking analysis of the patchwork quilt of 150 school and 196 special districts in Cook County, Illinois (main city, Chicago), concludes that the advantages of these districts outweigh the disadvantages. Many officials considered that their district was about the right size! Economies of scale were rarely considered, and the important factors in creating new units were deemed to be anticipated growth, expected support and other governmental forms in the area. Against this background it is extremely difficult to reconcile effective local control and appropriate scale of organization. In the smaller local districts it is also difficult to internalize the benefits and costs within the same political jurisdiction, and the unequal distribution of resources within Cook County is the result of differences in location of taxable enterprizes, socio-economic status, degree of development, degree of obsolescence and so on. The crucial fact is that most of the income of the SDs comes from taxation and therefore the amount of taxable property within the district is the main determining factor in the provision of public goods.

*Educational provision and inequalities*
In England and Wales the provision of education for all children has been compulsory for more than a century. The central government currently provides the bulk of the money spent on educational provision but the local authorities have considerable power over the way in which the money is spent. Expenditure patterns establish that local authorities are worth treating as political systems in their own right. As table 15.2 shows, in the school year 1973–4 the pattern of expenditure was far from being a uniform one: the Inner London Education Authority (ILEA) spent, on average, £344·56 on each secondary school child, Bradford disbursed £323·41 and Southampton £319·27, whereas cities such as Bootle, Halifax and Wigan spent less than £221 on each pupil; that is, less than two-thirds as much as the ILEA. Taking another measure of educational provision, that is the percentage of the relevant age group given a grant by the local authority to attend university, the range was from more than 15% in Caernarvonshire, Richmond, Solihull and Cardiganshire to less than 3·5% in several working-class cities of rapid

nineteenth-century growth.

Boaden's analysis (1971) of the factors affecting the provision of local government services in the county boroughs (large towns and cities) of England and Wales elucidates the way in which the system operates. Boaden first tests whether the local authorities are simply the agents of central government as providers of

**Table 15.2** Spending per pupil 1973–4 (after Chartered Institute of Public Finance and Accountancy 1975, and Department of Education and Science 1975, Crown Copyright)

| Highest spenders | £ | Lowest spenders | £ |
|---|---|---|---|
| | | PRIMARY | |
| Cardiganshire | 206·84 | Cornwall | 131·64 |
| ILEA | 201·22 | Wakefield | 131·52 |
| Oxford | 181·89 | Bradford | 130·46 |
| Kingston upon | | Burnley | 129·83 |
| Thames | 178·63 | Southport | 126·63 |
| Hertfordshire | 176·61 | Blackpool | 125·30 |
| Newham | 176·59 | | |
| | | SECONDARY | |
| ILEA | 344·56 | Eastbourne | 226·17 |
| Bradford | 323·41 | Worcester | 226·03 |
| Southampton | 319·27 | Wakefield | 224·42 |
| Stoke-on-Trent | 310·40 | Bootle | 220·82 |
| Grimsby | 308·94 | Halifax | 217·97 |
| Norwich | 307·64 | Wigan | 211·37 |

Percentage of age group receiving full and lesser value of awards at university

| Highest | | Lowest | |
|---|---|---|---|
| Caernarvonshire | 17·3 | Gateshead | 3·2 |
| Richmond | 15·6 | Warley | 3·2 |
| Solihull | 15·3 | Preston | 3·0 |
| Cardiganshire | 15·1 | West Bromwich | 3·0 |
| Barnet | 13·7 | Newham | 2·9 |
| Harrow | 13·6 | Barking | 2·1 |

Pupil/teacher ratio

| Most favourable | | Least favourable | |
|---|---|---|---|
| | | SECONDARY | |
| Bradford | 12·7 | Isle of Wight | 19·1 |
| Southampton | 14·6 | South Shields | 19·1 |
| Chester | 14·6 | Worcester | 19·2 |
| Richmond | 15·3 | Blackpool | 19·2 |
| Waltham Forest | 15·4 | Southend on Sea | 19·3 |
| Stoke-on-Trent | 15·5 | Wigan | 20·2 |

education (and of other services such as police, libraries, fire protection and so on) or whether by their own actions they are capable of producing inequalities. He tests the following hypotheses relating to central control and rejects them. (Here we give selected items on education. It should be emphasized that Boaden is testing his hypotheses in relation to eight services, not just education.)

(1) *County boroughs (CBs) will display broadly similar levels of activity within any service area because of the operation of central pressures and controls* — In terms of educational expenditure (rate of spending per 1000 population), Boaden finds the range is from £28 093 to £17 263, with a mean of £23 970 and a standard deviation of £2100.

(2) *Smaller CBs will be more likely to conform to central wishes and submit to central control* — Among the twenty smallest CBs ten spent above the overall average on education, six were in the top quartile and six fell in the bottom quartile.

(3) *The poorest authorities will be more likely to conform to central wishes and submit to central control* — Five of the sixteen poorest CBs were in the highest quartile and twelve were above the overall average in per capita expenditure.

Having established that CBs are not simply agents of the central government but active actors in their own right (a view supported by the work of Nicholson and Topham 1972), Boaden formulates a model relating expenditure (a measure of output, or policy) to three sets of independent variables: needs, disposition and resources (see figure 15.2). For each service (education, children, health, welfare, libraries, police, fire and housing) operational indicators are devised for

need, disposition and resources. The principal measure of output used (except for housing) is the annual expenditure per 1000 population met from local rates and central rate-support grant sources. (In housing, the main dependent variable taken is the proportion of new local houses provided by the local authority.)

In education, per capita spending is most obviously related to (1) the size of the school-age population (need), (2) Labour-controlled councils (disposition), and (3) the ability of the poor authorities to spend most (resources; that is, other services were being given a lower priority in favour of education). Labour-controlled councils spent more on education whatever the context of need in which they found themselves, and confirmed 'their general orientation to government activity and to education in particular' (Boaden 1971). Thus the political party in power locally does affect the level of spending and priorities, with the Labour-controlled councils most active in services with a significant impact on the overall role of government (that is, they are bigger spenders on the bigger services), and in areas of the community supporting them (especially in the field of education). On the other hand, Nicholson and Topham's investigations (1971, 1972) of the determinants of investment in housing by local authorities and the effect on investment of size of authority found no support for the view that a local Labour majority will affect spending. They suggest, however, that the influence of the local bureaucracy is not unimportant, especially with regard to the expenditure decisions of large authorities.

Though 'the central piper does not call the tune as much as is supposed' (Boaden 1971), no system of local finance seems likely to provide sufficient resources for needs to be met satisfactorily. Central equalization

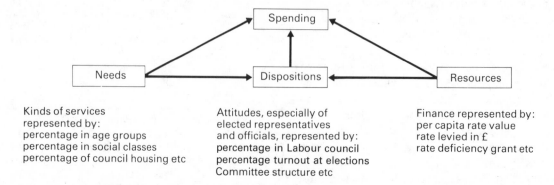

**Figure 15.2** A model of local government spending (from Boaden (1971) *Urban Policy-making*, London, Cambridge University Press).

schemes are thus absolutely essential if severe territorial inequalities are to be avoided. Despite the activity of Labour councils, the map of educational opportunity in England and Wales (including the non-CB areas which Boaden does not consider) bears a striking resemblance to the map of socio-economic status, indicating that education is being provided, and encouragement being given, in a biased manner towards certain groups. In addition, of course, the higher socio-economic status groups are better able to 'opt out' of the state system and buy educational privileges in the private sector.

Education — the provision of training so that individuals may realize their potential and the nation benefit from the developed skills of its citizens — is a crucial public good. If equality of opportunity is considered to be a child's birthright, a facilitative environment must be produced which is spatially biased in favour of those areas where socio-economic factors operate to impair a child's educational progress. In this regard, provision of local services in the United Kingdom is more redistributive than in the United States where governmental districts divide up the city into neighbourhoods. Thus inter-governmental conflicts in the United States tend to become intra-governmental conflicts in the more extensive and multipurpose local authorities in the United Kingdom. Populations in local government areas in the latter are rarely homogeneous and, though demands for education may be articulated by particular sections of the population, the resources are often allocated more or less evenly across the whole area.

Inequalities in educational provision between local government areas are considerable, however, and only 'positive discrimination' (Halsey 1972) in favour of the less fortunate areas can rectify the imbalance (the favoured areas being unwilling to be dragged down to an 'average' level). Such 'positive discrimination' can only be organized by the central and local government if they recognize the strength of the following: (1) the marked social gradient in the degree of interest in education and understanding of the educational system; (2) the vicious circle of limited life chances, apathy and bad economic and social conditions leading to limited life chances and so on for the next generation; (3) the need for positive discrimination in favour of those whose educational progress is restrained by material and cultural handicaps; (4) that variations in social conditions are more important in accounting for children staying in full-time education beyond the compulsory leaving age than are variations in standards of provision of secondary education — this means governments could make a hefty redistribution of educational resources in favour of deprived areas without lowering the proportion staying on in the more favoured areas/regions; and (5) only a fifth of school children stay on at school or enter higher education — the other four-fifths leave school at the first opportunity, though many may return as part-time students (Coates and Rawstron 1971, Boaden 1971).

Attention has been drawn to a few of the variations from place to place in England and Wales, and some of the reasons for these massive variations have been tentatively suggested. There can be little doubt that spatial inequality exists and that its removal requires much effort in the spatial levelling of educational opportunities. Territorial justice is of course a necessary prerequisite for social justice, but do we know how to distribute resources to achieve agreed goals and how are the goals themselves to be agreed? How should we measure educational output and how do we combine inputs to achieve the desired outputs? These and other questions are not dealt with here but answers to them are vital in any programme devised to spread equality of opportunity more evenly across England and Wales. If British society adheres to the equality norms of the 'welfare state', then local and central governments must ensure a more equitable spatial distribution of basic 'public goods' such as primary and secondary education. There is a long way to go, especially if the Regional Affairs Correspondent of *The Guardian* is correct in stating (Ardill 1975): 'For all its vast bureaucratic machinery of regulation and control, local government continues to blunder along in the dark with no clear notion of whether its policies and programmes are meeting real needs and solving real problems.' Turning attention to local government, Ardill concludes that 'the structure of governments prevents, or hinders, the sensitive adaptation of national policies to local circumstances.'

## References

Ardill J 1975 Regional policies 'doing more harm than good' *The Guardian* 21 July

Boaden N 1971 *Urban Policy-making: Influences on County Boroughs in England and Wales* (London: Cambridge University Press)

Bunge W 1975 Detroit humanly viewed: the American urban present in R F Abler *et al Human Geography in a Shrinking World* (North Scituate, MA: Duxbury Press) pp 147–82

Coates B F and Rawstron E M 1971 *Regional Variations in Britain: Selected Essays in Economic and Social Geography* (London: Batsford)

Cox K R and Reynolds D R 1974 Locational approaches to power and conflict *Locational Approaches to Power and Conflict* eds. K R Cox, D R Reynolds and S Rokkan (New York: Halsted)

Department of Education and Science 1975 *Statistics of Education* 5

Douglas J N H 1968 Political geography and administrative areas: a method of assessing the effectiveness of local government areas *Essays in Political Geography* ed. C A Fisher (London: Methuen) pp 13–26

Duncan S S 1974 Cosmetic planning or social engineering? Improvement grants and improvement areas in Huddersfield *Area* **6** 259–71

Halsey A H (ed.) Educational priority areas — problems and policies *Educational Priority* Vol 1 (London: HMSO)

Harvey D W 1972 Society, the city and the space economy of urbanism *Resource Paper No.* 18 (Washington, DC: Commission on College Geography, Association of American Geographers)

—— 1973 *Social Justice and the City* (London: Edward Arnold)

Henderson W L and Ledebur L C 1972 *Urban Economics: Processes and Problems* (New York: Wiley)

Herbert D T and Johnston R J (eds) *Social Areas in Cities* Vol 1 (London: Wiley)

Nicholson R J and Topham N 1971 The determinants of investment in housing by local authorities: an econometric approach *Journal of the Royal Statistical Society A* **134** 273–320

—— 1972 Investment decisions and the size of local authorities *Policy and Politics* **1** 23–44

Redcliffe-Maud Report 1969 *Royal Commission on Local Government in England 1966-69* Cmnd 4040 (London: HMSO)

Soja E W 1971 The political organization of space *Resource Paper No.* 8 (Washington, DC: Commission on College Geography, Association of American Geographers)

Stetzer D F 1975 Special districts in Cook County: toward a geography of local government *Department of Geography, University of Chicago, Research Paper No.* 169

Toll S I 1969 *Zoned American* (New York: Grossman)

US Bureau of the Census *Census of Government 1967 and 1972*

# Section IV
# The Limitations of Planning

## Introduction

Although planning has become firmly established as a method of State intervention in urban development its powers are limited, its objectives inconsistent and its theoretical basis unclear. It claims to control the physical environment and to influence the spatial distribution of economic activity but its relationship to those areas of policy-making which are responsible for development remains ambiguous. Planners tend to defend their intervention in the land and property market on grounds of public interest. Regional policy has been supported by a variety of social, economic and political criteria such as economic balance, or equality of opportunity. But the case for planning remains insecure whether looked at in terms of outcomes or in terms of theoretical justification. Unlike those forms of State intervention such as housing or transportation which control investment (considered in Section V), the outcomes of planning are difficult to measure. Competing theories of planning have, in the main, failed to relate the concept to practice or to provide an adequate justification for planning as a method of State intervention with clear purposes.

A theoretical justification for planning must rest on two related premises. One is that the market left to itself would either fail to provide certain goods or services or distribute them in such a way as to lead to gross inequalities or loss of amenity for certain groups. The other is that a private market in land would result in benefits being secured for some landowners at the expense of others and against the public interest. Planners have tended to show little interest in the distributional issues which lie at the heart of planning. Financial interests linked to property development are regarded as beyond their competence and the costs and benefits involved in land-use decisions are lost sight of when individual applications to develop are decided on their individual merits. The sense of weakness and confusion evident in planning lies partly in its lack of a coherent theory which explains the relationship of planning to the social context in which it operates.

The theoretical confusion is the question tackled by *Kirk* in the first extract in this section, taken from her book *Urban Planning in a Capitalist Society*. She suggests that there are as many as four theoretical approaches which might be used to explain planning, each concerned with the distribution of power. The *pluralist* approach assumes a stable, consensual political system in which power is diffused among interest groups which compete on an equal footing. Government is open and accessible and responds to these pressures by seeking to maintain a balance among them or by acting as a referee. Such an approach, she argues, ignores the inherent ability of certain interest groups (such as property developers or the environmental lobby) to exert influence while others are powerless to defend their interests. Pluralism tends to assume that the only important issues are those which are visible and give rise to political activity. What this approach does not allow for is the fact that people may be inactive because they perceive themselves to be powerless and without influence. This idea of latent conflict which does not manifest itself in activity is a major debate in political science and is a theme raised by Dunleavy in the final section of this reader.

The *bureaucratic* approach recognizes the power exerted by complex and impenetrable bureaucracies which are relatively insulated from and insensitive to pressures from localized interest groups. Both this and the pluralist approach neglect the social and economic context in which decisions are taken and they regard the background as neutral and unproblematic. This is not true of the *reformist* perspective which takes inequality as its theme and recognizes that planning operates to benefit some interests against others. At a political level this approach argues for changes within the existing political framework which will redress the social and economic problems it identifies. The fourth approach, termed by Kirk *political economy*, relates the process of urban development to that of capital accumulation. Planning, as part of the State apparatus, becomes an instrument of domination, integration and conflict resolution advancing the interests of the powerful while seducing the powerless with various reformist concessions which reduce pressure and curb the worst excesses of capitalism. It is clear that Kirk favours the latter approach both as explanation and as the basis for defining action to bring about fundamental change.

It is not necessary to choose any one perspective as revealing the ultimate truth, for all offer insights. All stress the significance of power and identify planning as a political process in which choices are made. The very concept of State intervention is political. The idea that planning can or should control the market and the fact that planning decisions can have a distributive effect are political issues. More explicit and coherent theory can provide planning with a clear identity of its social role. Even if planners were more concerned with theory it must be said that the powers available to them provide little scope for successful implementation. The British planning system is a classic compromise in which planners possess power to prevent but not to promote development.

This is particularly true of *land-use planning*, the system of controls which is operated by bureaucrats and politicians in central and local government to whom the term 'planner' is most commonly applied. The main features of this system are *development plans* covering each area of the country (county and district councils) which set the policy framework. These plans are backed by powers of *development control* whereby proposals to develop are granted or refused. Development plans tend to be couched in vague, ambiguous terms in which objectives are inconsistent leaving them open to varying interpretations. Long-term plans are unable to adjust to the speed of economic and social change and this encourages a flexibility which renders them all but meaningless. Lacking powers of investment, plans become dependent on those activities which make development possible such as developers with the financial power to make things happen. On a different scale and with different objectives and powers is *regional policy* which aims to overcome spatial inequalities in economic development. In this field, too, powers of control are important and take the form of industrial development certificates (IDCs) and office development permits (ODPs) designed to limit industrial and commercial development in the more prosperous regions. But regional policy does possess complementary financial incentives aimed to encourage industry to invest in the assisted areas with characteristically higher rates of unemployment and other problems. Even here the ability of regional policy to prove effective in the face of market forces quite beyond its control is doubtful as the extracts from Keeble and Hudson demonstrate.

The limitations of planning both in theory and practice may be assessed by recourse to empirical evidence. What evidence there is suggests that planning is not, as many planners seem to suppose, a neutral activity making olympian judgements of what is in the best interests of all but that it necessarily and inevitably responds to sources of power. At the local level certain interests can command the resources necessary to achieve their objectives. They may be developers anxious to secure profitable portfolios or the relatively affluent wanting to retain high standards of amenity, two groups able to make their demands felt where it counts. At the regional level the response of industry to international market processes far outweighs attempts by policymakers to restore imbalances in the economy. The remaining extracts in this section present some of the available empirical evidence and focus on three major themes — the inherent weakness of planning, its political character, and its relationship (or lack of it) to those other processes and policies responsible for the distribution of investment necessary for urban development.

The first of these by *Blowers* is taken from his book *The Limits of Power* which reflects a bureaucratic theoretical approach in its analysis of the distribution of power in the determination of local planning policies. The extract presents a series of case studies of local land development on the fringe of two provincial towns in southeast England, Luton and Bedford. It describes the interaction of local politicians, planners, developers and pressure groups at the point when specific decisions about land use come to be taken. Considerable conflict is evident requiring political solution. In principle the solutions could be available from the various long-term plans intended as a basis for decisions on land development. In practice such plans are too flexible and vague and can be called upon to support conflicting interests. The examples underline the power of developers whether in the private sector or local authorities to achieve their ends, and the decisive influence that can be exerted by well orchestrated pressure groups. They suggest that planning policy does not control nor even guide development decisions, it merely represents the balance achieved at any one time between various sectional interests.

Regional policies aimed at providing a balanced spatial economic development have been pursued with varying degrees of vigour by different governments in post-war Britain. The impact of such policies has proved difficult to evaluate and has led to questions about the appropriate means, effectiveness and even the relevance of such policies. This is at the core of the debate between *Keeble* and *Hudson* which was conducted in the 1970s in the columns of *Area*. Although at one level the debate is about the interpretation of statistical evidence, at another it is concerned with alternative theoretical perspectives on planning. Keeble adopts a reformist position arguing that regional policies have been successful in achieving convergence between regions but have become misdirected as the problems of the inner cities have attained greater significance. What is required in his view is a shift in the emphasis of policy to support the inner cities and to foster efficient industries outside the assisted areas. Hudson's perspective is more radical. He questions the efficacy of regional policies and finds the debate of marginal relevance in the wake of mounting unemployment and the loss of manufacturing jobs brought about by mergers and the restructuring of industries. These structural changes can only be tackled by changes or a reversal of current policies with the aim of creating full employment. This debate at the broad level is illuminated by the empirical findings for the electrical engineering and electronics industries presented in the article by Massey in Section II of this book. She lends some support to Keeble's view of convergence in unemployment between regions but underlines the changes in levels of skill and the overall decline in jobs brought about by restructuring and largely unaffected by regional policies which is the basis of Hudson's argument. It is clear from this debate that regional policy cannot be viewed in isolation from other policies which bear upon industry and employment.

The issue of the relationship of planning to other policies of government is made explicit in the final extract from *The Containment of Urban England* by *Hall et al*. This is part of a concluding chapter of a major and wide-ranging study of the role and consequences of the British planning system. The authors make an ambitious attempt to demonstrate the potential outcomes in urban form of alternative policies for inflation, housing, transport, land, industrial location and local government reform and other policies. These relationships are expressed diagramatically and reveal that a common outcome for a variety of policies is the peripheral expansion of cities and larger towns and the areas around the great conurbations — precisely what has happened in Britain. Moreover, the policies favoured by successive governments — Labour and Conservative — have contributed to this end and have each contained contradictions. The most fundamental contradiction has been between those policies encouraging development especially of private ownership in housing and those aimed at conservation of the environment through green belts, agricultural priority and other policies. The result has been 'urban containment' in which the planning system has conspired to concentrate development within the major cities and towns while protecting the countryside. In the process the losers have been the majority confined to high-density living, or poor standards of amenity, deprived of facilities and experiencing high costs for transport and housing. The gainers are the affluent living in rural areas protected by policies designed above all to conserve their inherent environmental advantages. 'The story of urban development in post-war England essentially is the story of their triumph. It has been a triumph achieved through power and influence' (Hall *et al* 1973, pp 431–2).

These four extracts make uncomfortable reading even in the context of the 1960s and 1970s, a time when the planning system was widely perceived as necessary and useful. Now that planning is under serious challenge from those who see it as an impediment to market processes and irrelevant to the social, economic or even physical problems faced in a period of decline, the conclusions they offer are even more disturbing. Although the extracts draw only on British experience they underline the difficulties which planning encounters in any advanced economy. British experience is apt since it has a well developed planning system against which to set theories of planning and draw upon evidence of its outcomes. The implications do not provide for sanguine expectations for developments in other countries. As the basic components of the British planning system — new towns, community land ownership, structure plans and regional policy — are diluted or even abandoned, planning, unsupported by a coherent theory or even a consistent ideology, offers little by way of self-defence even from its advocates. It is no longer a question of whether planning can achieve such goals as social justice or equality but whether planning should exist at all.


## Reference

Hall P *et al* (1973) *The Containment of Urban England* (London: George Allen and Unwin) Vol. 2


**Andrew Blowers**

# 16 Theoretical Approaches to Urban Planning

## by G Kirk

Four theoretical approaches which purport to provide some explanation of aspects of the land-use planning system are considered here. They are the pluralist, bureaucratic, reformist and Marxist approaches. These four headings are used for the sake of convenience. Such labelling may give the impression that they apply to distinct clearly articulated bodies of theory, though in fact this is not the case in any strict sense. The purpose of the headings is merely to impose some structure within which different contributions can be discussed. It should be pointed out that these theoretical perspectives are not directly comparable. They have been employed for different purposes and have not always been formulated specifically in the context of urbanism, though applications to this context can be developed. Between the various approaches there are areas of overlap and areas of divergence, such that there are both competing and complementary aspects. The different perspectives may be distinguished in terms of a number of criteria: the extent to which they take account of the wide range of factors involved in urban situations and land-use planning, and the degree to which the interplay of spatial, social, political and economic aspects is acknowledged and taken into account.

### The pluralist approach

This perspective has been particularly pervasive and influential, and merits consideration in some detail. To some extent several of the ideas associated with it have been passed into conventional wisdom in what Dennis (1972) called 'the modern theory of democracy',

Source: Kirk G 1980 *Urban Planning in a Capitalist Society* Ch 2 (London: Croom Helm) pp 55–94.

popularized by Schumpeter (1943) and subsequently such writers as Lipset (1960), Dahl (1961, 1967) and Polsby (1963), with its stress on the passivity, even apathy, of the individual, whose political activity is generally restricted to voting in elections. From this viewpoint, the overall aim of the political system is seen as the maintenance of the stability of society as a whole.

From the pluralist view, society consists of a collection of interest groups competing for control over government action through the electoral process. Pluralists 'see society as fractured into a congeries of hundreds of small special interest groups, with incompletely overlapping memberships, widely differing power bases, and a multitude of techniques for exercising influence on decisions salient to them' (Polsby 1963, p118). Power is said to be distributed in a diffuse way so as to guarantee that no one group can dominate any particular segment of society. Different groupings of individuals align themselves in various combinations according to the issues. If a particular interest threatens to gain the upper hand, opposition groups will emerge to challenge the powerful group, and thus the equilibrium will be maintained. Playford (1971) distinguished two slightly different roles for government. In the 'balance of power' variant, government was forced to accommodate itself to a number of conflicting interests, among which a rough balance was maintained. This was differentiated from the 'referee theory' of pluralism where government supervised and regulated the competition of interests so that none could abuse their power to gain mastery of some section of social life. The political process is characterized as 'open' and democratic, with easy access to decision-makers and numerous channels of communication for individuals and groups. In practice only a minority may feel strongly enough about a particular issue to be politically active, but in theory there is nothing to stop anyone who wants to do so from becoming active. All it needs is a little effort.

There are three central themes associated with this perspective, as summarized by Dunleavy (1977b). First, the polity was viewed as 'a weak unit lacking in any developed ideology or particularly separate identity, operating in an environment of strong external influences and controlled by politicians who concentrate overwhelmingly on building and maintaining an electoral majority' (p 194). Such a polity, it was argued, would be very responsive to pressures from the community. Secondly, there was the assumption that influence is unidirectional: that public opinion plays a decisive role in securing changes in public policy, though the possibility of indirect influence being exerted by elite groups on elected politicians, for example, is rejected. Thirdly, political activity was determined by people's interests. Since it is those people with an interest in the outcome of a political decision who become politically active, and since the weak polity is responsive to external pressures, the outcome will faithfully reflect the balance of interests in a community. Broadbent (1977, p 205) saw pluralism as a market theory, and claimed that it was 'the dominant accepted social theory in the UK.'

In support of this view of the distribution of power in contemporary Britain, a number of points can be made. First, there has been a remarkable increase in pressure-group activity in recent years, roughly since the mid-1960s, at both national and local levels, and this is nowhere more noticeable than in the sphere of land-use planning. There has been opposition to the proposed location of motorways, international airports, large-scale power stations and tanker terminals; campaigns for safer roads, more playing space, additional housing, rehabilitation of existing housing rather than demolition and redevelopment, and so on. Much of this local pressure group activity has had some measure of success, and several guides to pressure-group organization and accounts of campaigns have been published.

Then there has been an increase in personal and geographical mobility. With the spread of suburban development, the job mobility associated with an expanding economy of the late 1950s and early 1960s, and the more general availability of personal transport with increasing car ownership, it has been argued that communities based on particular locations are tending to be replaced by communities of interest where members are not joined by their common area of residence, place of employment or social class, but come together due to shared interests in particular issues or activities. This kind of community of interest may be taken as evidence in support of the pluralist formulation, with its stress on changing alignments and interest groups, for it involves the forming and reforming of social groups and organizations based on the shared interests of a relatively mobile population.

The pluralist perspective also has an inherent appeal, for it contains an implicit egalitarianism, with its stress on the right and ability of people to organize around issues that concern them. If individuals are thought to have equal status then it can be argued that they have equal opportunities for organizing themselves to press their particular interests, and theoretically equal chances that such activity will achieve its objectives. Thus the pluralist perspective legitimizes existing social and political arrangements, as fundamentally just and fair, involving the idea that inactivity is a sign of satisfaction with the situation as it stands or, at worst, disinterest. The emphasis is on activity, and it seems probable that it will appeal to people who are themselves active in local or national politics and who, perhaps as a result of this involvement, tend to underestimate the problems it may present for others, who may be less articulate, or lacking information, expertise and resources.

Despite the attractiveness and apparent plausibility of this pluralist approach, its validity must be seriously questioned. There are three main criticisms: the question of the equality of the various competing interest groups; the assumption of the one-way nature of political influence from the bottom upwards; and the stress on activities and associated methodological emphasis on 'key issues' for study.

At a national level pressure groups form to promote particular interests and points of view and to persuade central government to introduce or amend legislation or government practice. There are wide variations among both national and local pressure groups. They differ in the scope and type of issues with which they are concerned: their funds and facilities available, the organizational skill of their members, and their ability to articulate demands and mobilize support.

In the realm of land-use planning, for example, there are several national organizations which act as pressure groups on central government including the professional institutes: the Royal Town Planning Institute, the Royal Institute of British Architects, the Royal

Institute of Chartered Surveyors, the Institute of Auctioneers and Valuers. The professional bodies are often consulted by government and make representations about existing or prospective legislation. They are in the habit of making press statements, submitting evidence or memoranda giving the official view of a particular institute or association, as for example submissions to the Skeffington Committee's inquiry into public participation in planning and their comments on the Committee's report (Royal Institute of British Architects 1968, 1970, Royal Town Planning Institute 1968, 1971) and opinion concerning the Community Land Act (Royal Institute of Chartered Surveyors 1974). Then there is the National Association of Property Owners, the Country Landowners' Association, the road transport lobby, large contractors and construction firms which make up what Colenutt (1975) called 'the property lobby'. Colenutt discussed their influence on government policy. The developer's philosophy is that government intervention in the property industry inhibits development, holds up economic growth, and is restrictive and unfair. Of the regulations that cause concern, four are outstandingly important: rent control, taxation, land nationalization and local authority planning restrictions. These organizations and individuals can be expected to use whatever opportunities are available then to press their interests. This will include informal discussions as well as formal lobbying, for these interests are well represented in both Houses of Parliament, and on both sides, though more closely connected to the Conservative Party. Other national groups with an interest in land-use planning matters include the Town and Country Planning Association, the Civic Trust, the Council for the Protection of Rural England, the National Farmers' Union and environmentalist groups like Friends of the Earth, Intermediate Technology Development Group, Earth Resources Research and so on.

Much more numerous in land-use planning are the many local organizations which are concerned to try to influence local authorities. Here it is necessary to distinguish conservation and amenity societies on the one hand, and tenants' and community action groups on the other, and to note differences in personnel, resources, objectives and strategies. Amenity societies are interested in furthering the cause of conservation and preservation of existing buildings, street patterns, open space and so on, the self-appointed caretakers of the local and national heritage and environment. They

are likely to have middle-class people as members, people who are articulate, and who have (potentially) useful professional contacts both at work and socially for expert advice and information. They usually organize themselves in a way that is acceptable to local councillors and officers, often with a formal constitution and committee structure, and make their demands through acceptable channels, such as lobbying. Further, there is a formal mechanism for involving representatives of conservationist groups in local planning authority policy-making through Conservation Area Advisory Committees, established under the Civic Amenities Act, 1967. The aims of conservation societies may not always be in accordance with those of the local planning authorities they are trying to influence. The objective of conservation may conflict with business and some farming interests, as for example with the conservation and rehabilitation of old buildings, rather than demolition and redevelopment, and there is also the possibility that conservation societies may conflict with local authorities wishing to collaborate with business over redevelopment proposals. Moreover, some local councillors, particularly those from traditional working-class backgrounds, may oppose conservation proposals as not being in working people's interests.

In contrast to amenity organizations, community action groups are usually formed with the aim of bringing about a redistribution of scarce resources in favour of working-class people. In addition they may have a looser organizational structure, less professional expertise, fewer financial resources, and be forced to resort to more 'militant' tactics of direct action to draw attention to their views.

One additional point concerns political affiliation. Conservation groups often claim to be 'non-political' and may have members from various shades of political opinion. Community action groups are very often explicitly political, and may have links with the Labour Party or some members who are involved in smaller parties on the left — the Communist Party, Workers' Revolutionary Party or Socialist Workers' Party. These differences between organizations are particularly important in the context of a discussion of public participation in planning. They are not on an equal footing as regards resources, organizational skills and objectives.

A second criticism of the pluralist perspective concerns the stress on activity, and the consequent lack of attention paid to inactivity. Pluralists offer two interpretations of inactivity: it is either attributed to satisfaction with the system as it currently operates, or to disinterested apathy. If no complaints are heard, it is assumed that people are either satisfied with the *status quo*, and have no complaints, or else they are not sufficiently interested to register complaints if they have them. What this approach does not allow for is the fact that people may be inactive because they perceive themselves to be powerless and without influence. There may be dissatisfaction which they do not bother to voice, either because they do not know how to do so or because they feel that government, or whatever organization would be the likely target, would take no notice.

Crenson (1971) was interested to look at the 'non-issue' of air pollution, which did not surface as a political question for some years, despite its prevalence in urban America. This is in contrast to such issues as poverty and racial discrimination, which clearly had achieved the status of political 'issues'. He compared the experience of two areas: Gary, Indiana, and East Chicago. In Gary, a town dominated by US Steel, industrialists never openly involved themselves in the clean-air issue, which emerged in the political arena only very gradually. In East Chicago, by contrast, industrialists took an active interest in this question, and legislation to combat it was passed relatively quickly. Crenson suggested that a pluralist interpretation would attribute more political power to industrialists in East Chicago as compared to those in Gary, since they were more politically active in this issue. His own analysis ran directly counter to this, however. Clean air was slow to be taken up as a political issue in Gary, a one-industry town, exactly because of the economic dominance of US Steel. The firm did nothing to push the issue, as clean-air legislation could be expected to increase their operating costs, and the politicians also assumed that the company would have little interest in clean-air legislation. Thus Crenson explained a lack of activity on the part of Gary industrialists as a sign of their power, rather than the other way around.

Dunleavy (1977a, b) looked at the issue of high-rise housing, a common form of public housing in Britain in the 1960s, despite its unpopularity with residents and people generally. He chose this question specifically because it did not emerge as an open political issue, and explored the various power relationships in operation, involving local authorities, professional architects, large-scale building and construction firms, and working-class tenants and prospective tenants.[1]

This is what Bachrach and Baratz (1970) called 'the second face of power' and their critique of community power studies rested here. For they argued that power may be exercised by restricting the scope of decision-making to relatively 'safe' issues, creating or reinforcing social and political values and institutional practices that limit the scope of the political process to public consideration of only those questions which are comparatively innocuous. 'Innocuous, that is, to privileged interests', added Miliband (1969, p 156). Bachrach and Baratz talked of the power of 'non-decision making', of business preventing certain questions from being considered. 'Prevention' carries the suggestion of activity on the part of business, and implies conscious effort to ensure that policies are not questioned or challenged. Westergaard and Resler (1975) took this critique further, stressing that the pluralist approach could not take account of those issues which did not come into dispute at all. Still more important than the power of 'non-decision making' was:

the power to exclude which involves no manipulation; no activity on or off stage by any individual or group; nothing more tangible than assumptions. They predetermine the range of issues in dispute, and limit it to those in which negotiation, competition or conflict has some practical chance of shifting the balance between the contending parties.

This is power that is 'anonymous, institutional and routine' (p 247).

Finally, against the pluralist view of the unidirectionality of influence from constituents to politicians and officials, there is the question of the resources which can be deployed by dominant groups against lesser groups. As Dunleavy (1977b, p 198) put it, this includes the prevention of the 'accurate perception of their interests by the powerless' and the inhibiting of 'mobilization, organizational development and protest success'. The writings of Bachrach and Baratz, and Westergaard and Resler already cited speak to this point, as does Dearlove's (1973) study of local government. Elite theorists like Crenson (1971) and class theorists such as Aaronovitch (1961), Gutsman (1969), Miliband (1969) and Poulantzas (1973) all pointed to the power of dominant political groups.

To sum up this discussion, the strength of the pluralist perspective lies in the attention that is paid to the activity of local and national organizations and interest groups — their aims and aspirations, membership and support, activities and tactics, in influencing particular policies. Its several weaknesses, however, outweigh this strength. There is over-emphasis on activity at the expense of any consideration of apparent inactivity.

The various organizations do not compete on more or less equal terms. Also there is the erroneous assumption that influence is only one-way, from the mass of ordinary electors to official policy-makers who are constantly anticipating possible loss of popularity and electoral defeat. This approach can be contrasted with a view of the distribution of power which emphasizes the positions of politicians and public administrators, and it is to this perspective that I shall turn next.

## The power of public bureaucracy

A stress on the power of public bureaucracy is implicit in studies of local authority decision-making, particularly in the spheres of housing and land-use planning, as in the work of Dennis (1972), Davies (1972) and Rex and Moore (1967). Both Dennis and Davies gave accounts of local authority decision-making where residents' groups were in conflict with the local councils. The councils in question, Sunderland and Newcastle, were not the weak institutions of some pluralist studies, however, receptive and responsive to pressure from below. In this work it is assumed that local authorities are complex structures, difficult to penetrate and largely impervious to influence from local residents. Moreover, within the local authorities the relationship between councillors and officers does not appear to be in practice what it is conventionally conceived to be in constitutional theory. The councillors themselves had little detailed knowledge about the planning and housing issues in dispute and were reliant on the expertise of the full-time officials of the planning and housing departments. It is not only members of the public who have difficulty obtaining information from the bureaucracy, but also local councillors.

Rex and Moore (1967) and Pahl (1975) viewed the city as a relatively discrete social system and looked at the constraints operating within it and at the distribution and use of scarce resources. Rex and Moore focused on the immigrant communities of Sparkbrook in Birmingham, particularly the housing which was available to newcomers. These authors were concerned with the bureaucratic rules of resource allocation, for the immigrant workers did not qualify for mortgages, having no capital, nor for council housing, since they could not meet the five-year residence qualification. Hence they had to rely on the privately rented sector and found themselves in lodging-houses, often in poor condition, with makeshift facilities. The city council was concerned with the deterioration of a neighbour-hood which had once been the home of white people, and wanted to stop the spread of lodging-houses in other areas. Ultimately, a private Act of Parliament allowed the council to regulate the condition of houses in multiple occupation, and defaulting landlords were prosecuted. Yet these landlords, many of them of Asian or West Indian origin themselves, were providing a service for immigrant workers, which neither the local authority nor the building societies were prepared to do. Furthermore, Rex and Moore argued, as some immigrants who had been living in Birmingham for the qualifying five-year period became eligible for local authority tenancies, they were discriminated against, perhaps unconsciously, and were offered accommodation from amongst the poorest of the council's housing stock. These authors raised important questions, for they made problematic the actions and interests of bureaucrats in local housing and planning departments, and urged that more housing be made available to immigrants by government, and that non-discriminatory allocation policies be pursued.

This problem was amplified by Pahl (1975 Ch 12) who was concerned not only with housing but with the allocation of resources generally. He emphasized the importance of researching into the operations of the managers of the urban system, the local technocrats and 'social gate-keepers' who mediate in the allocative processes and who have the capacity to shape the socio-spatial system. These were identified as housing department officials, estate agents, planners, private landlords, social workers and so on.

This general perspective has several strengths. First, there is acknowledgement of the power and complexity of central and local government bureaucracies, and the gradual expansion and increasing control of government activity in everyday life. It is a reflection of this increasing pervasiveness of government activity that there has been such a growth of pressure-group activity at local level. Focusing attention on the power of public bureaucracy points to the difficulty that such groups may experience in their attempts to penetrate these expanding bureaucracies, though local groups are not equally placed in this, as I argued above in connection with pluralism. Some groups are better able to cope with the bureaucracy than others, are less intimidated by the complexity of the organizational structure, have some idea who to see and what to ask for.

There are additional points to be made about the apparent plausibility of this bureaucratic approach, regarding the researcher's working environment, which concern ease of access to information and sources of research funds. A researcher interested in a local organization — especially if this represents some disadvantaged section of the community — will perhaps tend to see the situation under investigation through their eyes, and with their definitions, aspirations and expectations. Hence, if they find the local authority impermeable, or bureaucrats insensitive, this will be reflected in the researcher's conceptualization. It might be argued that it is precisely this ability to share the perceptions of others which makes for valid research. However, this sympathy may lead to a tendency to attribute considerable authority and control to local government officials, for example, at the neglect of the roles of central government or employers, this attribution flowing from a local organization's view of local government officials as their chief targets.

A further related point is the question of access to adequate information. If the emphasis is placed on local organizations at the neglect of how local bureaucracies operate, and how local government policy is formulated, this may be partly because researchers do not always have easy access to local authority decision-making. Then there is the possibility that some academics may be susceptible to a view which stresses the power of public bureaucracies partly as a result of their own personal experience of large university and college bureaucracies.

If one is interested in local allocation and decision-making procedures, then the main focus of attention will be on local authorities. The research interest narrows down the field of the relevant and pin-points certain issues, giving a secondary status to others which are not of immediate concern, or taken as given. Dennis and Davies, for example, were interested in local housing issues and concentrated on the particular relationship between local organizations and the respective local authorities. This interest restricted their viewpoint, and did not incorporate more general questions concerning the relationship between land-use planning and the land market, for example, or central government policies on housing standards or housing finance.

To recap, the strength of this approach lies in the attention it pays to the ways in which government bureaucracies operate, their power, the complexity of their structure, and the fact that they are not necessarily responsive to pressure from below. Because of this it is considerably stronger than the pluralist approach. As has been noted, the scope of Dennis's and Davies's work was restricted by their interest and was limited to the relationship between the local organizations and the respective local council, resulting in an anti-bureaucracy-cum-profession diagnosis, in which the only structure recognized was an economically empty structure of authority. Land-use planning decisions were taken by officers and councillors according to their prejudices, convictions and technical expertise, with little regard for the interests of 'the planned'. From this view, presumably, cities just happen by accretion of such decisions.

Both the pluralistic and bureaucratic approaches treat conflicts of interest as taking place within an economic vacuum, without any reference to the fact that in a capitalist society governments operate and make decisions within the constraints of a capitalist economic system. The pluralist perspective, for example, does not suggest that owners of capital have more power than non-owners. Nor does the approach which stresses the power of public bureaucracy question the relationship between local and central government and private business interests, as, for example, with the redevelopment of inner-city areas. It is as if the economic context of political decision-making is taken as a non-problematic 'given', a neutral background against which political and cultural differences are worked out, an influence which can be taken completely for granted. Basic facts of life in Britain and America, such as substantial inequalities in wealth and power, the fundamental role of private ownership of property and market forces, the power of business interests and assumptions about their validity and continuance, are glossed over or ignored altogether, presumably because they are not seen to be relevant. This leads me to consider a third theoretical perspective, reformism, which does not make this omission, and which does attempt to relate political decision-making, both in land-use planning and more generally, to an analysis of the economic structure of capitalist society.

## Reformism

This is a label which should be applied with some caution to the various writings and activities so designated, yet such an approach seems to me to be sufficiently clear for it to be differentiated from the others I am considering, though there is an obvious overlap with Marxist approaches discussed below. It is the perspective of some academics, some activists in the trade union movement, Labour Party and left-wing political groups, and some aspects of social administration — at least the declared objectives, if not always the actual outcomes of social policy. It embraces a mixed collection of work and writings, is associated particularly with poverty, homelessness and poor environment, and characterized by humanitarian, egalitarian aims, involving positive discrimination in the redistribution of resources.

The strength of a reformist perspective is the recognition given to basic structural inequalities of power, influence, income and wealth in modern Britain in a way that the pluralist approach does not, and indeed perversely ignores. The theoretical underpinnings of this approach are derived from two main sources: a humanistic liberal egalitarianism, and some variant of Marxism, and this dual legacy gives rise to varying degrees of recognition and importance being attached to the problem of reform. This perspective entails critiques of persisting inequalities in Britain and America, explanations of how these have come about, and how they are reinforced by prevailing institutions, and thus perpetuated.

Harvey's work (1973), which has been particularly influential, was concerned with the issue of inequality, specifically in the context of urban areas, and mechanisms which governed the distribution of income within the urban system. These included the price of accessibility to facilities and the cost of proximity to sources of noise and pollution, the ability of groups with financial resources and education to adapt to changes in the urban system more rapidly than other groups, and the uneven distribution of externality effects. This raises the question of the scope for redistributive measures within the land-use planning system: who is intended to benefit from planning schemes, who actually benefits, and whether one person's or group's benefit always involves another's loss.

Within the constraints of the existing economic system satisfactory solutions to the problems of deprived areas are not easily found. Indeed, considerable change is necessary in the present priorities for resource allocation, though this point does not always seem to be appreciated. Donnison (1973), for example, proposed that local service centres be established in deprived areas both to improve the services available to people and to act as a focus for local political activity. The centres would combine the functions of a citizens' advice bureau and offices of local government departments, such as housing, social services and education. They might also have library facilities, a planning team, housing aid services, rent officer, probation officer and so on, depending on local circumstances and needs. On the political side, local people would be encouraged to take responsibility 'and to nominate their spokesmen to consultative groups attached to statutory services' (p 398). They must be given information, and offered space in the centres to hold meetings and for other activities. A management committee might consist of local councillors, aldermen, perhaps the local MP, and a few co-opted members. Central government should give grants to local authorities to establish such service centres, and financial support to local groups. Donnison recognized that political activity over the distribution of scarce resources necessarily involves conflict, and suggested that the 'aim is to promote more productive conflict and furnish procedures for successive, temporary arbitrations and agreements' (p 400).

A reformist perspective contains an inherent contradiction. The analysis of urban or social problems these writers presented, though not a uniform one, tended to stress the inevitability of poverty and disadvantaged groups under capitalism. The logical prescription for action from this analysis would be radical change, admittedly a severe problem for implementation, whereas many of the actual recommendations and activities tend to be reformist, allowing that there will probably be no major changes in the economic and political system, at least in the foreseeable future. The contrast between this position and the fourth perspective to be considered here partly concerns this issue of reformist solutions. Marxist approaches give an account of urban issues which tie them firmly into the capital accumulation process, and offer an implicit or explicit critique of the exploitation of working people under capitalism. There is a clear integration of political and economic aspects of urbanism, and an associated interest in political activity at national and local level which attempts to radically alter the current distribution of power in society in favour of the working class.

## The political economy of urbanism

Much of this work is very recent, explorative in nature and to some extent tentative, often at a very high level of generality, and acknowledged by the authors to be inadequately formulated or as yet incomplete. However, this approach appears to have several inherent strengths not found in the perspectives which I have already discussed. Crucially important, it does not treat economic factors as non-issues. It can take account of power 'exercised' through day-to-day assumptions about how a market economy operates, and it is addressed to the characteristics and purposes of the city development and redevelopment process. One of the fundamental weaknesses of the pluralist and bureaucratic approaches to the study of urban issues and land-use planning is their neglect of economic factors, accepting that economic questions constitute a separate area of study from sociological and political questions and that it is unnecessary to relate these two areas of concern.[2]

Fundamental to writers in this Marxist tradition is the relationship between the development of cities and the capital accumulation process. Lamarche (1976, p 86), for example, viewed the city and urban problems as the 'local consequences of capitalist accumulation'. He delineated a specialist capital, property capital, whose primary purpose was to plan and equip space in order to reduce the indirect circulation costs of production. Property capital had a planning role in the way it selected sites, and an equipping role in the types of buildings developed. Its commodity was floor space, let by the square foot, and its profits depended on the difference between construction costs and rent extracted. Property capital mainly catered for commercial, administrative and financial users, rather than providing housing, for example.

Since he is not a financier, that is to say he seeks something other than the average rate of interest on the capital he advances, the developer will not operate in the field of housing, unless he can convert the advantages provided by the environment into profits (p 96).

The developer will be able to demand a higher price if these environmental advantages are not equally distributed in space, but will only be interested in luxury residential accommodation as a sideline, ancillary to larger developments. In order to maximize rental income, property capital will concentrate its developments in areas with good situational advantages, favouring high-rise buildings in such areas, and large developments, where tenants have complementary functions.

Lamarche noted the importance of public investment to developers, who extracted higher rents for sites which are close to railway stations, government offices, schools, parks and so forth, though this additional value was dependent upon the activities of public authorities rather than private developers themselves.

Further, he claimed that the development plans of the authorities could only be realized if they were subordinated to the interests of developers, and that urban renewal was governed by the priorities of property capital. Lamarche drew on examples from Canadian cities. Despite some differences in scope between the urban planning systems of Britain and Canada, his characterization shared common features with some accounts of land-use planning in Britain, as for example in Counter Information Services' anti-report on housing in London (1973, p 51):

The history of planning shows that private ownership of land and private initiative in, and profit from, its development, set forces in play which will always break through the obstacles of any planning measures, which start from an acceptance of their legitimacy and permanence.

Lojkine (1976, pp 119–46), like Castells whom I shall consider next, based his work on a structuralist reading of Marx following that of Althusser (1969) and Althusser and Balibar (1970). The Althusserian concept of the social formation contains four distinct practices: economic, political, ideological and theoretical, although the economy is determinant 'in the last instance'. Lojkine argued that the capitalist mode of production requires a spatial organization which facilitates the circulation of capital, commodities, information and so on, and that the capitalist city can be seen as a spatial form which, by reducing the indirect costs of production, and costs of circulation and consumption, speeds up the rotation of capital. Lojkine stressed the tensions and contradictions inherent in capitalism, and hence the capitalist city and land-use planning. Capitalist relations of production, together with modern industry, bring about a growing tendency towards urban concentration and agglomeration. 'Rational, socialized planning of urban development' is necessary to continued capitalist accumulation, but may not be forthcoming due to the existence of three limiting factors (p 128).

The first problem is that of profitability. Many collective means of consumption — housing, schools, parks, medical facilities, for example — are by their very nature 'opposed to the imperatives of profit'. Secondly, competition between firms tends to lead to concentration of investment in some areas, and a widening gap between well equipped and poorly equipped regions, towns or neighbourhoods. Thirdly, there is the problem of the private ownership of land.

It is through State intervention that collective means of consumption exist at all, one of the roles of the State being to maintain the cohesion of the social formation as a whole, which may involve economic as well as legal and ideological measures. Lojkine noted three common characteristics of the urban policies of the developed capitalist States: that State intervention enables the capitalist system to resolve the immediate contradictions that no individual capitalist agent either can or wants to resolve; that it has permitted public financing of unprofitable means of communication and collective means of consumption enabling capitalism to promote simultaneously the development of all the general conditions of production; and that urban planning, while it has had very uneven results, has enabled immediate difficulties to be resolved, as with public health legislation and working-class housing. Attempts at land collectivization have enabled limited planning experiments to be undertaken, such as the New Towns. According to Lojkine, this intervention does not function solely as a safety valve, but is the reflection of class struggles and worker pressure, obliging the State to limit the spontaneous tendencies of capital accumulation.

Castells' work, like that of Lojkine, was based on a structuralist reading of Marx, and for Castells the economic, political and ideological practices of the social formulation were the keys to understanding urban society. 'Urban' refers to 'reproducing labour power', meaning that all aspects of life outside the workplace are bound up in this process. This would include the basic provision of housing, food, heating, clothing, and more broadly the provision of education, health and social services, recreation and leisure-time activities, which train the workforce and maintain its capacity for work. Thus, 'urban units' (the city and the city region) are to the process of reproducing labour power what enterprises are to the productive process, and the 'urban system' is 'the particular way in which the elements of the economic system are articulated within a unit of collective consumption' (1976, p 153). Analysis of this urban system leads directly to the study of urban politics which can be divided into two constituent elements: urban planning and urban social movements. Urban planning is the intervention of the political system in the economic system at the level of the urban unit, in order to regulate the process of the reproduction of labour power, and the reproduction of the means of production. Thus the study of urban planning is the study of the structures governing the nature of urban society, and of the attempts to resolve or minimize the contradictions or dislocations between aspects of the system. The study of urban movements is the study of the practices of class agents. The first leads to an understanding of the configuration of the urban system, and hence of the social formation; the second to understanding the processes by which it is maintained or transformed. Although this distinction is crucial at the theoretical level, in practice they are 'two sides of the same coin', since structures are articulated practices of class agents, and practices are the interactions between and within social classes as determined by the particular structural configurations of the society.

Having thus set out the conceptual framework, Castells proposed his 'experimental hypothesis'. Since the State expresses at the political level the combined economic and political interests of the dominant classes, then planning cannot be an instrument of social change, but only one of domination, integration and conflict regulation. On the other hand, a process of social change (as distinct from 'reform' which can be achieved by the planning system or within the field of conventional politics) does emerge from the new field of social conflict to which the social movements give expression, when popular mobilization occurs, when social interests become political wishes, and when new forms of organization of the reproduction of labour power (e.g. welfare state services) clash with the dominant capitalist ethos of accumulation, competition and profit. Thus, Castells argues, it is the urban social movements and not the planning institutions which are the source of change and innovation within the city. Hence he is interested in organizations which mobilize around particular issues, such as urban renewal in Paris, a proposed motorway in Quebec and squatter settlements in Allende's Chile (1977, Ch 14). His methodology involves identifying the contradictions, the stakes for each social group intervening, the characteristics of these organizations and an analysis of their activities and effects.

The fourth contribution to be considered here is that of Olives (1976), who utilized Castells' approach in his analysis of local resistance to a proposed urban renewal project in Cité d'Aliarte, a suburb of Paris. Olives concluded that every protest movement springs from the perception of a *stake*, by a *social force*, in this case the attempt to evict immigrant workers, mainly Africans, from their hotels or hostels. The size of the stake must be large enough for urban or political effects to result, but such effects are only possible where there is a minimum degree of organization of the *social base*. It is neither sufficient for the stake to be large and defined on a social base, nor does the presence of an organization lacking a social base result in urban or political effects. Legal action undertaken by a purely protest organization — petitions, letters, requests and so on — can never lead to effective results, because it fails to mobilize people. Given the presence of an organization, the appropriate type of action and social base, it appears that it is the size of the stake which is decisive in determining whether urban or political effects are obtained.

I suggested that this Marxist work appears to have an inherent strength for gaining an understanding of urban issues and land-use planning in its emphasis on the inter-relationship between city development and the workings of a market economy. However, it has been subjected to much criticism, varying in scope and intensity. It is important to discuss the criticisms in some detail, to assess how damaging they are. There appear to be six points to be considered: the lack of attention given to the characteristics of local organizations; the methodology employed; the claim of universal applicability; a theory of social change; the question of uniqueness of capitalism; and the nature of the role of the State.

### Characteristics of local organizations

Castells focuses on the effects achieved by social movements, and is barely concerned with any other aspects. He is mainly interested in the results of an organization's activities, not with details of the kind of people involved, their degree of politicization, the issues which have predisposed them to be active, nor the practical details of their campaigns. In criticism of this emphasis, Pickvance (1976, Ch 8) stresses the importance of organizational resources in contributing to a group's survival and the achievement of its aims. Of course, any organization needs resources, both personnel and finance, above some minimum level, to operate at all. Different organizations vary enormously in their resources — money, people's time and energy, information, professional expertise and influential contacts. However, successful groups often have greater resources than unsuccessful groups. If this is so, then Pickvance's criticism of Castells is less telling, for a stress on effects would implicitly incorporate the point about resources. If one focuses on the effects of local action, one is interested to ask whether or not it achieved its objectives, and if not, why not? In practice there is often an evaluation of any particular campaign, perhaps trying to identify if it was wrongly directed;

how more people could be contacted, interested and mobilized in support; what further activities might be planned, publicity attracted, and so on.

I would accept that it is necessary to know much more about the characteristics of local organizations than Castells suggests, especially concerning mobilization — the type of people likely to be active in particular organizations, and the issues and activities likely to attract support. Pickvance, however, tends to overstress the importance of resources, and to overlook the importance of a group's aims which are related to success and resources. Groups with demands which threaten the *status quo* tend to have fewer resources than those which do not, though they all, with the exception of big business, have fewer resources than the State. Groups with radical demands also tend to be less successful, but I am arguing that this is at least as much to do with the nature of their aims as their relative lack of resources.

### Methodology

A criticism of structuralist work put forward by Dunleavy concerned the criteria employed for the selection of issues to study, which he also levelled at pluralist studies as mentioned above:

So far as I can see structuralists study virtually any 'struggle' which is accorded significance in general Althusserian Marxism. Some British studies on structuralist lines go further and analyze cause groups whether they use protest tactics or not. Thus Pickvance uses evidence about *a local branch of the United Nations Association* to derive hypotheses about the role of organization in urban social movements, while Lambert *et al* pass directly from a study of a residents' association involved in the redevelopment process in Birmingham to a typically structuralist account of urban conflict, even though the association functioned as no more than a normal, well-integrated interest group. Such indiscriminate studies of the micropolitics of the city cannot substantiate structuralist hypotheses.

A study of protest which is to be more than theoretically suggestive and which is to transcend the intrinsic interest of a particular instance of protest must illuminate a fundamental power relationship within the urban system, particularly one usually tending to keep an issue latent (1977b p 200).

His own study of local community action over the issue of mass high-rise housing was formulated in the light of these criticisms (Dunleavy, 1977a,b).[1] First, he selected a topic which has not surfaced in conventional politics, namely the building of high-rise housing for local authorities, despite its unpopularity with tenants and prospective tenants. He went on to explore the relationship between specific local authorities, central government policy for financing council housing and the role of large construction firms, which were influential in getting comprehensive high-rise housing schemes accepted, allowing them to use systems building technology.

The activity of local organizations needs to be put in the context of more general inactivity, in both the spheres of production and consumption: in the workplace and in the community. I argued above that routine assumptions about power operate constantly to keep issues latent, and to restrict the scope of debate and decision-making to relatively 'safe' questions. This happens in several ways: the issues are obscured, either consciously or unconsciously; people have inadequate and fragmented scraps of information which have to be pieced together for an understanding of the full picture; there often seems no point in contesting issues which appear not to be negotiable.

## The claim of universal applicability

Structuralist writers claim universal applicability for their formulation, criticized by Pahl (1975, pp 280–4), who argued for the uniqueness of particular capitalist countries, based on their different historical development and local variations in current practice. Pahl stressed the need to look at British capitalism specifically, without unquestioningly assuming that models which derive from French experience will be appropriate.

I am concerned to understand land-use planning in a capitalist economy, and have argued above that work in the Marxist tradition appears to have several strengths for a study of urban issues. The fact that some of this work has been developed in France should not render it immediately valueless for a study of land-use planning in Britain. There is a sense in which every settlement is unique, just as every person is unique. Settlements, like individuals, have their own specific histories, though having said this, it must be acknowledged that individuals are part of wider societies and settlements are part of the nation-state and are affected by general economic conditions, the law of the land and the general political situation. It is the focus on settlements that is not helpful, and this discussion requires the introduction of the concept of class, for deprivation is not random, neither socially nor spatially.

## A theory of social change

Pahl widened this critique of universal applicability and criticized the structuralist perspective as a theory of social change, claiming that there is little evidence from British experience to support a view of urban social movements as leading to radical change in British society. To illustrate this position he took two issues which one might expect to have elicited a general reaction from local groups and consequently coordinated and cohesive activity across local authority areas. These were the 1970–4 Conservative government's Housing Finance Act 1972, and the more general issue of property speculation which had been particularly prevalent from the mid-1960s to the early 1970s in many British towns, though most notably in London. Neither of these issues provoked collective organization on a national basis, however. Many attempts on the part of tenants' associations to organize against the Housing Finance Act were not successful (Sklair 1975, pp 250–92). At parliamentary level the Bill was contested in committee, and subsequently some individual local authorities attempted to salvage a limited advantage by negotiating rent increases below the level specified in the Act. Sklair showed that the protesting authorities, and the local authority tenants who engaged in rent strikes, were characterized by their isolation and lack of support. Similarly, the activities of property speculators did not stimulate working-class collective action against the private ownership of urban land although they did stimulate many working-class organizations to campaign at local level, but it was industrialists, according to Pahl (1975, p 274), who apparently urged the government to take action and hence to adjudicate the conflict of interests between finance capital and industrial capital.

## The question of the uniqueness of capitalism

A final criticism of Castells and others by Pahl (1975, 1977) concerned the stress laid on the uniqueness of capitalism and their view of urban issues and problems in capitalist societies being a product of *capitalism* specifically. He maintained that useful insights can be gained from comparative studies, taking into account urban problems in eastern Europe, for example. I have acknowledged above that problems of resource allocation are not unique to the capitalist mode of production, but I am not prepared to accept that these problems have exactly the same causes in centrally planned and market economies. I am arguing that it is the contradictions of British capitalism which account for the contradictions of land-use planning. Thus it is the existence of private ownership of land and capital and private initiative in development which are at the root of the problems of resource allocation in Britain. An understanding of the land-use planning process in Britain requires a theoretical perspective which can take account of the economic, political, spatial and social questions involved, and can go beyond the one-sidedness or restricted focus of the pluralist and bureaucratic approaches. I consider the general view of the city as the outcome of the capital accumulation process to be a useful one. Similarly, the concepts of stake, social base, social force, protest effects and so forth developed by Castells and Olives in their analyses or urban social movements appear to be a useful starting-point for conceptualizing the activities and structure of protest groups. In the foregoing discussion of the criticisms of Castells' and others' work I am not convinced by the arguments concerning the claim of universal applicability and the uniqueness of capitalism, as put forward by Pahl. I do accept, however, that it is necessary to know more about the characteristics of local protest organizations and the role played by the local authority, to locate activity within the context of more general inactivity, and to consider critically Castells' and Olives' formulation as a theory of social change.

## Notes

[1] See paper 21 in the final section of this reader (Eds).

[2] Urban economics sees city organization in terms of competitive bidding for sites, where land and values reflect demand. At equilibrium, rent levels are said to be equal to the marginal productivity or utility of the land. This view is exemplified by Alonso (1964) or Muth (1969). (Alonso's theory is presented in paper 6 of this reader (Eds).)

## References

Aaronovitch S 1961 *The Ruling Class* (London: Lawrence and Wishart)

Alonso W 1964 *Location and Land Use* (Cambridge, MA: Harvard University Press)

Althusser L 1969 *For Marx* (London: Allen Lane)

Althusser L and Balibar E 1970 *Reading Capital* (London: New Left Books)

Bachrach P and Baratz M S 1970 *Power and Poverty* (London: Oxford University Press)

Broadbent T A 1977 *Planning and Profit in the Urban Economy* (London: Methuen)

Castells M 1976 Theoretical propositions for an experimental study of urban social movements. *Urban Sociology: Critical Essays* ed. C G Pickvance (London: Tavistock) pp 147–73

—— 1977 *The Urban Question* English edn (London: Edward Arnold)

Colenutt B 1975 Behind the property lobby *Conference for Socialist Economists*, 1975 pp 123–33

Counter Information Services 1973 *The Recurrent Crisis of London* (London: CIS)

Crenson M A 1971 *The Unpolitics of Air Pollution: a Study of Non-Decision Making in Cities* (Baltimore and London: Johns Hopkins)

Dahl R A 1961 *Who Governs? Democracy and Power in an American City* (New Haven: Yale University Press)

—— 1967 *Pluralist Democracy in the United States: Conflict and Consensus* (Chicago: Rand McNally)

Davies J G 1972 *The Evangelistic Bureaucrat* (London: Tavistock)

Dearlove J 1973 *The Politics of Policy in Local Government. The Making and Maintenance of Public Policy in the Royal Borough of Kensington* (London: Cambridge University Press)

Dennis N 1972 *Public Participation and Planners' Blight* (London: Faber)

Donnison D V 1973 Micro-politics of the city *London, Patterns and Problems*, ed. D V Donnison and D E C Eversley (London: Heinemann) pp 383–404

Dunleavy P J 1977a The politics of high-rise housing in Britain: local communities tackle mass housing *D Phil Thesis*, Nuffield College, Oxford

—— 1977b Protest and quiescence in urban politics: a critique of some pluralist and structuralist myths *International Journal of Urban and Regional Research* 1 (2) 335–45

Gutsman W L (ed.) 1969 *The English Ruling Class* (London: Weidenfeld and Nicolson)

Harvey D 1973 *Social Justice and the City* (London: Edward Arnold)

Lamarche F 1976 Property development and the economic foundations of the urban question *Urban Sociology: Critical Essays* ed. C G Pickvance (London: Tavistock) pp 85–118

Lipset S M 1960 *Political Man: the Social Bases of Politics* (New York: Garden City Press)

Lojkine J 1976 Contribution to a Marxist theory of capitalist urbanization *Urban Sociology: Critical Essays* ed. C G Pickvance (London: Tavistock) pp 119–46

Miliband R 1969 *The State in Capitalist Society* (London: Weidenfeld and Nicolson)

Muth R 1969 *Cities and Housing* (Chicago: University of Chicago Press)

Olives J 1976 The struggle against urban renewal in the Cité d'Aliarte (Paris) *Urban Sociology: Critical Essays* ed. C G Pickvance (London: Tavistock) pp 174–97

Pahl R E (ed.) 1975 *Whose City?* 2nd edn. (Harmondsworth: Penguin Books)

—— 1977 'Collective consumption' and the state in capitalist and state socialist societies *Industrial Society: Class, Cleavage and Control* ed. R Scase (London: British Sociological Association and Allen and Unwin) pp 153–71

Pickvance C G (ed.) 1976 *Urban Sociology: Critical Essays* (London: Tavistock)

Playford J 1971 The myth of pluralism *Decisions, Organizations and Society* ed. F G Castles, D J Murray and D C Potter (Harmondsworth: Penguin Books/Open University Press)

Polsby N W 1963 *Community Power and Political Theory* (New Haven: Yale University Press)

Poulantzas N 1973 *Political Power and Social Classes* (London: New Left Books)

Rex J and Moore R 1967 *Race, Community and Conflict: a Study of Sparkbrook* (London: Oxford University Press)

Royal Institute of British Architects 1968 Evidence to Skeffington Committee *Journal of the Royal Institute of British Architects* **75** (10) 402

—— 1970 People and planning *Journal of the Royal Institute of British Architects* **77** (1) 36–7

Royal Institute of Chartered Surveyors 1974 What next for property? *Investors' Review* **82** (15) 8

Royal Town Planning Institute 1968 Public participation in planning. Memorandum of evidence submitted by the Institute to the Skeffington Committee *Journal of the Town Planning Institute* **54** (8) 343–4

—— 1971 Forum on public participation in planning, Sessional meeting at the Town Planning Institute *Journal of the Town Planning Institute* **57** (4) 171–5

Scase R (ed.) 1977 *Industrial Society, Class, Cleavage and Control* (London: British Sociological Association and Allen and Unwin)

Schumpeter J A 1943 *Capitalism, Socialism and Democracy* (London: Allen and Unwin)

Sklair L 1975 The struggle against the Housing Finance Act *Socialist Register* ed. R Miliband and J Saville (London: Merlin Press) pp 250–92

Westergaard J and Resler H 1975 *Class in a Capitalist Society* (London: Heinemann Educational)

# 17     The Politics of Land Development

*by A Blowers*

## Land planning at the local level

The importance of changes in the planning, governmental and financial framework lies not so much in terms of long-term planning policies and the general distribution of wealth and real incomes as on their effect on the balance between various interests in land at the local level. Decisions on land tend to involve ideological, financial, and planning considerations. Decision-makers are susceptible to pressures from the various interests involved. Those interests which are most likely to be successful are those most able to articulate their case and to gain access to the decision-makers. Thus land policy is likely to represent the balance achieved at any one time between various sectional interests. It is an intensely political matter although planners profess to operate here as elsewhere as disinterested protectors of 'public interest'. 'The belief that planning is non-political seems pathologically well-entrenched even ironically among those planners who accept the validity of the suggestion that "planning is political" (Reade 1977). In other words, professional planners tend to equate 'political' with politicians, and fail to recognize the significance of their own values in presenting choices. That their values and hence their choices differ among themselves and between them and politicians has already been evident.[1]

As well as treating politics with circumspection, some planners may also be hostile to the private sector, regarding it as 'motivated by greed, steeped in speculation, staffed by furtive entrepreneurs and marshalled by a landed profession sunk in abject cynicism' (Ratcliffe 1976). The distinction between interests, especially between private and public interests, may be far from clear. The public interest is not a unitary unequivocal conception. Instead of one common public interest there are many sectional interests — public and private. These interests may be in conflict or in concert. The distinction becomes even more problematic when decision-makers acting on behalf of the public are themselves involved in or influenced by private interests.

Broad generalizations about the role of interests in land and their relationship to the organization of the State and a capitalist economy are unlikely to provide insight into the process of urban development. Such insight must be sought by the use of detailed empirical evidence gathered at the local level. 'However, the use of case studies raises the obvious question of how "representative" the cases chosen are' (Elkin 1974). It would be impossible to claim representativeness for the case studies which follow. They are concerned solely with urban fringe development and are taken from one area in the United Kingdom. These examples do illustrate different aspects of the politics of land development and indicate the nature of the interests involved, the types of conflict to which they give rise and the relative importance of planning considerations in arriving at decisions. They also help us to interpret the effect of national legislation upon local land-planning policy. These examples cover the period 1973–7, a period of administrative and political changes in Bedfordshire. It was also the time during which the Community Land Act was introduced and the country structure plan developed. The major aims of the structure plan, which was submitted in 1977, were to meet the county's housing and employment needs in a context of low population growth. The countryside and agriculture would be conserved by concentrating future development around the major towns.

### Bedford

The first case study concerns a land transaction that

Source: Blowers A 1980 *The Limits of Power: the Politics of Local Planning Policy* (London: Pergamon) pp 119–35.

took place in 1973–4 just prior to local government reorganization and before the Community Land Act. Post-war urban development in Bedford has been largely on the north, east and southwest fringes of the town. Much of this development was undertaken by the former borough council buying up land well in advance of need, granting planning permissions, preparing the sites, designing the layout and allocating the land to private builders and developers or using it for the council's own developments. In this role of estate developer the council was able to secure betterment on its transactions and to control the nature of the developments, a process not unlike that envisaged by the Community Land Act. Purchases were made as opportunities arose, and therefore urban development tended to be influenced by land availability and financial considerations.

The issue concerns three plots of land to the northwest of the town[2] (see figure 17.1a). The developer was the Harpur Trust[3] which had sold off a prime town-centre site occupied by one of its direct grant schools for redevelopment as a shopping centre with an associated car park. An integral part of this deal was the relocation of the school on a site on the periphery of the town. The trust proposed to purchase 32 acres of land adjoining its new school (and owned by the council) for playing fields (site A on figure 17.1b) enabling it to release its existing playing fields which occupied two sites of 9 and 19 acres (sites B and C), separated from the new school by a road. Part of the smaller of these two sites (site B) would be dedicated to the council for open space (4 acres), an option for housing development would be granted to the council on the rest of that site (5 acres), while the developer would sell his larger site for private housing (site C).

In planning terms the trust would clearly benefit by having playing fields as an integral part of its new school campus. The council, too, would gain the immediate benefit of open space and a site for its housing but would forfeit the future potential use of its own large site. The transfer of the 19-acre site (C) from private playing fields to private housing development would mean the loss of open land but the gain of housing. The financial aspects of the deal proved controversial. The council's land would be sold at existing use (agricultural) value plus a premium of £2000 per acre, thus realizing £64 000, but the developer's land would be sold at residential value (at that time possibly as high as £50 000 per acre) realizing potentially £1.2 million for the 24 acres it was proposing to sell. There

was little doubt that the developer's own land should be valued at residential prices since it had originally been bought at such values, neighbouring sites had been sold at such values, and there was every indication that housing was an appropriate use for land in that location. What was at issue was the potential use and therefore the potential value of the council's land.

In favour of the price offered it could be argued that it was a good offer for hilly and unproductive land which was unlikely to be used for residential purposes. The site (site A) would be landscaped and would form an attractive wedge between Bedford and a large village to the north. Opponents of the deal (the Labour oppos-

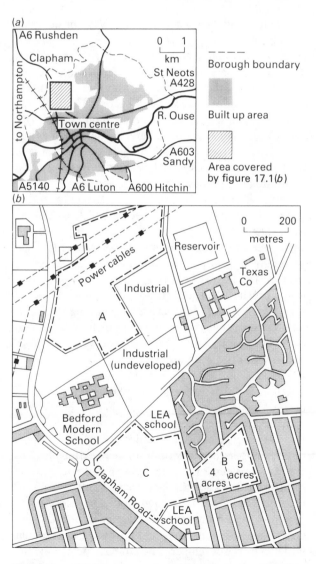

**Figure 17.1** Bedford: (a) location of study area, (b) location of individual sites.

ition on the council) were concerned to establish its appropriate value, not to determine its use. They argued that it could be valued at a higher price since adjacent land had been sold for industrial purposes and land elsewhere on the urban fringe in similar locations had been sold at residential values. In any case the council, as planning authority, was able to determine its use and, as owner, to fix its price. In the general interest of the community the council should either retain the land for future public benefit or sell it at the highest price possible.

The interesting feature of this case is that a developer was able to initiate a transaction on conspicuously favourable terms to itself. In this instance the developer was helped by the fact that there was considerable overlap between its representatives and members of the council.[4] Any potential conflict of interest on the part of these members could be rationalized by identifying the interests of the community with those of the trust. Early negotiations were kept secret. Even in the absence of a close relationship between seller and buyer and in a less secretive atmosphere, the developer would have held a considerable advantage as the initiator of the transaction. The council was put into the position of reacting to a proposal rather than considering all the possible alternatives both in the present and future.

Planning considerations played very little part in the issue. Details of the particular scheme were discussed only after the financial terms of the transaction had been agreed in principle. Objections from the public were restricted to the specific planning application which precluded any debate about the potential use of the council's land. Financial and planning decisions were kept separate, and public comment on the financial aspects — the central issue — was disregarded as irrelevant to the planning issues.

*Conclusions* In its residential developments, Bedford Council has already nurtured the kind of relationship with developers envisaged by the Community Land Act. Developments had proceeded according to opportunities rather than according to any master plan. In the case considered here the borough had not made its future intentions for the land clear. Under the Community Land Act the council would be expected to designate land for relevant development. If it did not do so, there is nothing to preclude a deal similar to that initiated by the developer. The power to determine the use of the land and its price would remain, as it did before the Act, with the council.

With the development of structure and local plans the public are afforded an opportunity to debate the alternative uses of land, which was not the case with the deal considered here. However, such plans are intended to be flexible and can be revised in the light of changing circumstances. Their implementation is likely to be incremental and the original objectives of a plan may well be subverted over time.

It is difficult to evaluate the public interest in land since this is composed of several interests both now and in the future. In the case considered here the present population could benefit by the acquisition of some open space (site B) in an older housing area but a future population could be denied a potential amenity (site A). The public could be said to benefit from the sale of land (site A) at above the existing use value but lose the opportunity of future sale at a possibly higher value. The benefit to the trust was not merely a private one, since the acquisition of playing fields brought educational benefits to that part of the public who could afford or achieve access to its schools. The distribution of costs and benefits embraces all those opportunities, present and future, foregone or created both financially and in terms of planning as a consequence of the deal. Such a comprehensive evaluation of a transaction is neither practicable nor politically realistic. At any one time where one interest recognizes an opportunity from land development it is likely to achieve its objectives so long as there is no clear and generally accepted alternative in sight. In the Bedford case the Harpur Trust, quite aside from its access to decision-makers, had a clear objective. Its opponents were forced into objecting to the scheme as proposed without presenting any clear alternative. Discussions of the financial aspects of the deal were precluded from public debate until the crucial features had been settled. Where alternative proposals for land development do exist and are marshalled by competing authorities each purporting to be acting for the public interest, the outcome is less certain, as the next case demonstrates.

*Luton: the green wedge*
Luton, the largest town in Bedfordshire, is a product of growth in the car industry. The Chiltern Hills and Luton Hoo (areas of high amenity) to the south and east of the town have been largely responsible for its eccentric shape since outward growth has been confined to the north and west of the town centre (figure 17.2). Dunstable and Houghton Regis to the west have begun to coalesce with Luton. The remaining open space

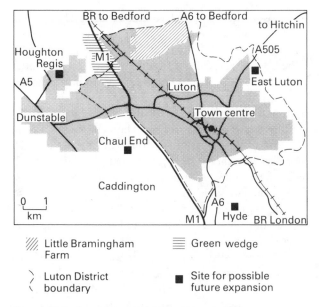

**Figure 17.2** Luton, Dunstable and Houghton Regis –
location of sites.

///// Little Bramingham   ≡ Green wedge
Farm

⟩ Luton District         ■ Site for possible
⟨ boundary                 future expansion

between these three settlements — the green wedge —
has been the subject of pressure for residential
development and other urban uses. The case studied
here reveals the problems which arose in attempting to
define the use of this land. It occurred after local government reorganization and during the preparation of
the county structure plan.

The area consists of 97.5 acres just north of the
built-up area of Luton, bounded on the west and south
by residential land, on the east by the London–Birmingham motorway M1, and to the north by the boundary
of Luton with South Bedfordshire, beyond which is an
area of open land. The whole area (475 acres) lying
between the three settlements and stretching across the
M1 is all that remains of the green wedge which the
1944 Greater London Plan (Abercrombie 1945) indicated should stretch over 2500 acres and maintain the
separate identity of Luton and Dunstable. Since World
War II the wedge has been progressively eroded as the
suburbs of Luton and Dunstable have been developed
and the village of Houghton Regis transformed by the
introduction of London 'overspill' population. The
issue was whether that part of the wedge remaining
within Luton and owned by the borough should be
developed for council housing, thus further reducing
the green wedge.

Earlier decisions had indicated the conflict between
housing pressure and the need to preserve an open area
(figures 17.3 and 17.4). After a public inquiry, which

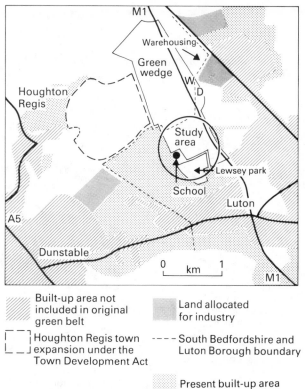

**Figure 17.3** Luton, the green wedge.

///// Built-up area not      Land allocated
included in original       for industry
green belt

⌐ ⌐ Houghton Regis town   - - - South Bedfordshire and
∟ _ ∟ expansion under the       Luton Borough boundary
Town Development Act

Present built-up area

released 850 acres in 1956, the Minister commented,
'Although the existing green wedge between Luton and
Dunstable cannot be fully preserved, some green space
should be maintained between the communities, particularly in the area flanking the proposed motorway'
(Bedfordshire County Council 1974). Part of the eastern edge was released for warehousing, though part was
later refused for similar purposes. On this occasion
(1971) the Inspector reporting the public inquiry
stated:

I accept that visually the green wedge is marred by electricity power
lines and also that the clamour of the motorway is a predominant and
ugly feature, nevertheless it is possible to find reminders that this was
once pleasant country. I do not feel, however, that the loss of visual
amenity can be regarded as a major issue in this case. What does
matter is the retention of open land on the outskirts of a large town
where there is obvious pressure (Bedfordshire County Council 1974).

To the north of the wedge the London overspill scheme
was designed to retain a substantial area of green space,
but there were pressures from a private developer for
housing land. It appeared that resistance to further
development was hardening when proposals to develop
a hypermarket and housing in various parts of the

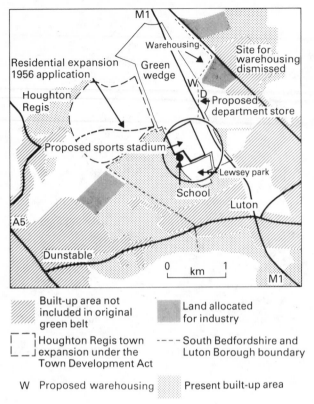

Figure 17.4 Luton, the green wedge – past planning proposals in the area.

wedge were refused. By 1974 a large part of the wedge had been absorbed for various kinds of development, but at each stage there had been reference to the desirability of maintaining some open land between the three converging settlements.

It was in these circumstances that Luton Borough Council proposed to develop the remaining land within their boundary, aptly named Pastures Way. They proposed that part should be retained as open space (32.5 acres), part for schools (41 acres), and the remainder (24 acres) should be developed for council housing (figure 17.5). It was the residential proposal that brought the borough into direct conflict with the county council. The county council argued that 'retention of the character of the area is of paramount importance' (Department of the Environment 1976). They were prepared to concede the use of the land for schools or recreation, which were not felt to be incompatible with the concept of the green wedge. But they feared that the release of land for housing would make it difficult to resist the pressure for development on the remaining part of the wedge. In sum their position was:

'The residential proposal would be completely contrary to the established policy to retain the "green wedge", and if permitted, would create a serious precedent whereby it would be difficult to resist other similar proposals in this area' (Bedfordshire County Council, Environmental Services Committee 1974). The borough council who owned the land argued that the land was essential to meet their housing needs. Luton was a housing stress area with a waiting list of 4000 families, and there was no immediately available land, apart from the site, to help meet their housing needs. They did not consider that residential development 'would represent more than marginal visual intrusion detracting from the character and quality of the area' (Bedfordshire County Council 1974).

The conflict led to divisions on political, professional and geographical lines. Consultations were held between the two councils. In June 1974, the county's Environmental Service Committee reaffirmed its view that the proposal was contrary to established policy which aimed to retain the open land on the outskirts of Luton and Dunstable and to prevent their coalescence.

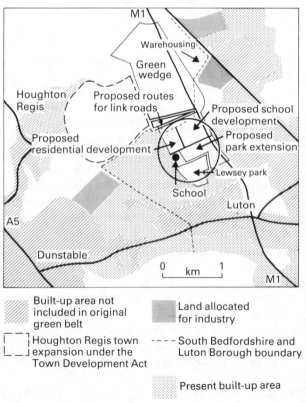

Figure 17.5 Luton, the green wedge – the Borough Council's proposed development.

Luton's pressure on the county council was maintained, and it was eventually decided to arrange a joint meeting between the three councils involved (Luton, South Bedfordshire and the county council). At this meeting the divisions between and within the councils became apparent. The officers of the county council and of Luton maintained opposing positions, whereas South Bedfordshire's officers were prepared to accept a small amount of residential development but not the whole 24 acres. However, the South Bedfordshire members at the joint meeting supported the county council's line. Luton's members were, as expected, fully in favour of the proposal. The county members were divided. At the subsequent meeting of the Environmental Services Committee to discuss the consultation, 7 members supported Luton (all Labour, 4 from Luton) while 11 opposed it (7 Conservatives and 4 Labour, 3 of whom were from South Bedfordshire). There were also 3 abstentions (1 Labour and 2 Liberals).[5] Early in 1975 the matter was referred to the Minister as a 'substantial departure' from the development plan, and he decided to call the application in and hold a public inquiry.

At the inquiry the county council maintained that the principle of the green wedge had been vindicated despite the loss of much of the original land. They also poured doubt on Luton's housing programme, claiming that the borough owned sufficient land to accommodate an increased building programme, which, though somewhat short of the borough's proposed rate, was much faster than had been achieved in the past. In addition there was further land available within the borough to accommodate 3500 dwellings within the next five years and a number of sites outside the borough with reasonable rail or road access. Luton had failed to recognize the need for overspill. 'The only merit in developing the Pastures Way site is its cheapness to the borough which has shown a parochial approach to land acquisition and a lack of foresight. There is no insuperable shortage of alternative land' (*Luton Evening Post*).

Luton countered this by suggesting that policy on the green wedge had been ambivalent over the years. They concluded that:

the original Green Belt concept had long been abandoned with tacit consent from Ministers. Now any suggestion of keeping the built-up areas of Luton and Dunstable separate is irrelevant. What is left is the green wedge which, within the Borough, has little visual importance, its only possible function being the preservation of *some* open space between residential communities and the M1 (Department of the Environment 1976).

The land was essential to help Luton achieve its housing programme. The alternative land suggested by the county council was not immediately available, was not owned by the borough, was not provided with services and would result in heavy construction costs. Failure to build at Pastures Way would leave a gap in the building programme. The county council's approach was 'the planning approach which ignores the delays and snags which a housing authority must first overcome' (Department of the Environment 1976). 'It is a theoretical approach to the problem but we have got to look wider than pure theory' (*Luton Evening Post*).

Not suprisingly the Inspector regarded Pastures Way as a 'finely balanced case in which two important issues are in direct conflict' (Department of the Environment 1976). He accepted the county council's arguments that the proposed housing programme could not be achieved and that other sites were available. He concluded: 'I think the time has come to call a halt to all further inappropriate developments within the green wedge unless there are overwhelming reasons for allowing them.' He did not consider the housing needs to be overwhelming and therefore recommended that the permission be refused. However, the Minister disagreed with this conclusion. He felt the housing needs were the prime consideration and the immediate availability of the site had been largely ignored. Moreover, he emphasized the economic arguments in favour of the site which had been ignored by the county council and not given great prominence by the borough. He said 'at a time when savings must be made in all fields of public expenditure the comparative cheapness of the application site must be counted an important consideration.' He made the obligatory reference to the green wedge concept and his sentiments were not very different from those voiced when the wedge had been breached on previous occasions. 'This decision is not to be taken however as indicating abandonment of the green wedge which is considered in general to be sound and valuable, nor is it meant to be significant in relation to any other development proposals within the green wedge' (Department of the Environment 1976).

*Conclusions* In this case the debate was over the use of a piece of land — not over its value — since it was already in public ownership and there was no question of its sale. Both protagonists could claim to be serving the public interest, the county asserting the long-term desirability of maintaining an amenity and Luton stressing immediate housing need. The long-term strategic concept of a green belt had, by a process of incremental-ism, been subverted and transformed into the more modest aim of protecting some open land as a green wedge. Planning policy in this area — and arguably in many similar areas — had responded to immediate pressures and reflected the political and economic situation existing at different points in time. The county council could maintain a strategic view regarding the development within the context of potential (and less environmentally damaging) development elsewhere whether in Luton or outside. Luton, embarking on an ambitious housing programme, were unimpressed by such arguments and concerned to occupy the site as the most readily available. The forces were evenly matched at local level, leaving the issue to be settled by means of a public inquiry.

The inquiry process is supposedly neutral, weighing up the different arguments to provide an informed judgement. In this case the Inspector was over-ruled by the Minister who was able to provide a different interpretation to the evidence presented. The introduc-tion of new arguments by the Minister to support his disagreement with the Inspector's findings was unpre-cedented. But pressure had been applied on the Minis-ter both by the MP for Luton West and by the chair-man of the Luton Housing Committee, both of whom were Labour. It may well be that such informal approaches were decisive. Certainly the County Plan-ning Officer described it as a 'political decision'.

Such a reaction misses the central point that such decisions are intensely political. Planners offer advice based on their own ideological predilections. 'There simply cannot be any objectively correct answer to the question of how resources should be allocated. It is a question for politicians, not experts' (Reade 1975). If it were possible to argue objectively, then we might assume that all planners would arrive at similar conclu-sions. This was manifestly not the case here. Evidence was selected, sifted and interpreted in arriving at recommendations. It is also important to recognize the significance of the circumstances prevailing at the time the decision was taken. A little later the county plan-ners might have been strengthened by a much greater resistance from the Ministry of Agriculture unwilling to lose valuable agricultural land. A little later, too, the downturn in the economy, widespread public expendi-ture cuts, a new (Conservative) administration in Luton less concerned to build council houses, might have weakened or removed altogether the case for hous-ing at Pastures Way. The conflict between short-range

incremental planning against long-term strategic plan-ning is always likely to be resolved in favour of the former. It is unlikely that the Community Land Act or the structure plan would have changed the situation. The land was, in any case, publicly owned, and the structure plan would only deal in broad land-use terms, leaving local land allocation to be resolved through local plans.

It is interesting to compare Luton Borough's argu-ments about their housing needs with actual perfor-mance once the permission was granted. The borough had stated that 129 houses would be built in 1976/7 and the remainder in 1977/8. By the middle of 1977 only 75 were under construction and altogether four phases were proposed, the second to be completed by 1980. The arguments advanced at the time may well have been very genuine but it does show how rapidly events (political and economic) can alter the circumstances of a case.

### Luton: Little Bramingham Farm

The largest remaining undeveloped plot of land within Luton was Little Bramingham Farm (375 acres) to the north of the town (see figure 17.2). In this case the dispute was not over the use to which the land should be put, there being general agreement that it was suit-able for housing, but whether it should be used for public or private housing. The issue arose in 1976 just at the time when the Community Land Act came into operation, when the draft structure plan for Bedford-shire was published and when political control in Luton changed from Labour to Conservative.

The site, in private ownership, was capable of accommodating a population of about 7000, an indus-trial area and the facilities necessary to create a residen-tial neighbourhood. Both the county and borough councils were committed to the principle of housing development for this area. The Ministry of Agriculture was opposed to the loss of this grade 2 land, but despite this and the structure plan's emphasis on conservation of the best land, an outline planning permission was granted to the borough in 1976. It was accepted at that stage that the land would be purchased under the Community Land Act as a mixed public and private development but with a substantial amount of council housing to meet the borough's housing needs. This would enable the development to correspond to the structure plan's aim to meet local housing needs by concentrating future development around the major urban areas.

After the change in political control from Labour to Conservative in May 1976, the new administration in Luton published an outline development plan which appeared to threaten the strategy of the structure plan. It envisaged that some of the development would be for private housing, although it favoured giving local people priority in the purchase of dwellings. Phased release of land 'to meet the needs of Luton's natural increase only is vitally important if the strategic aims outlined by the county structure plans are to be achieved' (Borough of Luton 1976). The county planners argued that there were practical difficulties in achieving this since there was evidence that half of the purchasers of private houses came from outside the county. The level of public housing proposed was too low to meet the housing needs of those unable to purchase houses. If the development attracted people from outside the area then land would have to be released elsewhere to meet Luton's housing needs. This could result in a higher rate of development than was proposed in the structure plan. Despite efforts to discuss the issue, the borough as the housing authority was able to develop the area in any way it wished, arguing that a phased release of the land would enable it to meet local demand as it arose.

It is significant that the Labour Administration had decided to purchase the land under the Community Land Act, which was designed largely to deal with private housing development, rather than through the Housing Acts which gave loan sanctions for council housing projects. This seriously weakened attempts made by the Labour MP for Luton West to persuade the Minister to intervene in the issue. He argued that loan sanction had originally been given for the purchase of the site to enable Luton to tackle its housing problems:

The officers of the County Council from the Chief Planning Officer down have asked if there is any way I can help to make the Council face up responsibly to its obligations. The new Conservative Council has added massive stupidity to ideological spite in their housing policies. . . . It may not be possible to save the Council entirely from its follies but a quiet word now could avoid a future disaster.[6]

The Minister replied that the loan sanction had been granted without conditions as to the mix of public and private housing and, in any event, the Community Land Act was intended to ensure a sufficient supply of land for private development. He adopted the conventional ministerial stance of non-interference with local authorities, saying it was up to the council to determine its programme. The MP was not satisfied: 'it is difficult to think of anything more irresponsible for a housing authority to do than to wash its hands like Pontius Pilate as Luton is doing of the problems of thousands of families. I hope this hand washing does not become infectious.'[7] He regarded the policies of Luton as 'inspired by a pathological class hatred of council tenants.'[8] However, it was clear that the Minister was not prepared to intervene.

*Conclusions* Little Bramingham Farm is of interest since it was debated at a time of considerable change in the national and local circumstances governing land transactions. It exposed a crucial weakness in the structure planning process, namely, that housing and growth policies left much to the interpretation and determination of local housing authorities. Both Labour and Conservative administrations in Luton, with vastly differing ideologies on housing, could claim they intended to meet the housing policy of the structure plan. Once the county council had accepted the principle of housing development in the area there was little more they could do to influence the type of development pursued. Likewise, whatever the postures struck by the local MP, the Minister forbore to intervene in such a sensitive local issue as housing or to risk a clash with a local authority of a different political complexion. Indeed, the Community Land Act, with its emphasis on achieving quick returns on private housing development, lent support to the policy pursued by the borough council. At least it gave them potentially more control over the development than if it had been undertaken by the private sector. In any case, if the council had wished to develop predominantly council housing it had the necessary compulsory purchase powers and loan sanctions available to it. But such a large development could have implications for the future strategic development of Luton and the problem of finding land for future expansion.

## The expansion of Luton

The question of land for the future expansion of Luton had been raised in the debates over the green wedge and Little Bramingham Farm. It became a major controversial issue in the preparation of the structure plan — here I shall deal with one of the issues it was expected to resolve. The broad aim of the structure plan was to make the best use of existing resources and to conserve agriculture and the countryside by concentrating future growth around the existing urban centres. Projections of future population for the Luton area indicated that room for about 3600 dwellings would be needed beyond land already committed by 1991. It had been accepted that such land should be found within the county rather than outside it at Milton Keynes. Four potential sites had been identified (see figure 17.2). One, to the east of Luton, was generally acceptable as a natural extension of the existing built-up area, but another, to the south at Caddington, was ruled out on grounds of aircraft noise from the flight path associated with Luton Airport. The county planning officers preferred a third site to the northwest at Houghton Regis on grounds of good transportation links and because it would not interfere with high-grade agricultural land. They also argued: 'Development at Houghton Regis could have the added positive value of helping to improve the provision of social and commercial facilities' (Bedfordshire County Council 1976). Certainly the GLC's housing development at Houghton Regis had given rise to social problems (Price 1976). A fourth site, to the southeast of Luton, at Hyde on the Hertfordshire border, had been suggested during consultations with Luton Borough Council. 'A possibility not identified in the Strategy is development in the Hyde area which is largely free from constraint, is closer to the town centre facilities than any site to the north and which could utilize road and rail without adding immediate congestion to existing patterns of travel' (Bedfordshire County Council).

Houghton Regis had originally been selected by the Environmental Services Committee on the insistence of the County Planning Officer who managed to persuade a small majority to support it at the end of a long meeting with only a few members present. This was the idea put forward for public consultation leaving little scope for public debate about the merits of the Hyde site. After the period of public participation on the structure plan, the Luton Labour members succeeded by a narrow majority in deleting Houghton Regis and substituting Hyde. The plan was twice debated by the county council. On the first occasion the Labour group with the support of the Mayor of Luton and the Leader of Luton Council (both Conservatives who were also members of the county council) maintained the preference for Hyde.

The County Planning Officer described the move as 'a political decision steam-rollered through' but he was reminded that he had employed similar tactics to secure the adoption of Houghton Regis in the first instance. Certainly the Luton Labour members had applied their political muscle. Their motive ostensibly was that the natural area for growth was to the south of Luton, though the fact that Hyde adjoined the Luton Hoo Estate, part of which had long been coveted by Labour politicians, was not unconnected with their decision. The support of leading Conservatives had been decisive, indicating that it was still possible for political differences to be submerged when the interest of the former county borough was at stake. What had clearly emerged was the antipathy of some members to the idea of development at Houghton Regis. The local member claimed that the area should be left alone to recover from the problems already experienced through rapid development in the past.

The council's decision released a barrage of criticism on the grounds that public debate on the merits of alternative sites had been avoided. There was a well orchestrated protest from Hertfordshire led by the Harpenden Society. Local authorities, councillors and organizations in Hertfordshire registered firm opposition to the proposal. Nearly a thousand individual protests were received before the council met to make its final decisions on the plan before it was submitted to the Minister.[9] The opponents were concerned that no opportunity had been given for objections to be made during the period of public participation. Their objections were based on the loss of amenity and agriculture, the transportation problem and the high construction and sewerage costs that would result from any development at Hyde. Behind the protests also was the fear that Luton's overspill would adversely affect the environment of Harpenden, a salubrious commuter dormitory on the Bedfordshire border.

The scale of the protest caused the Labour group to reappraise its position and it modified its decision. At the county council meeting (February 1977) when the plan was finally approved, the group passed a motion that should any future land for Luton's needs be required then the county council would examine other sites close to the urban area, including Caddington and Hyde. Although Hyde had not been positively excluded (as a defeated Conservative amendment would have intended) it remained open to debate. Hertfordshire County Council found the proposal 'completely unacceptable' and resolved to carry its objections to the examination in public of the structure plan.

Meanwhile the county council elections produced a Conservative council that might be more sympathetic to the arguments against any expansion at Hyde. In the 'rejoinder' (Bedfordshire County Council, Planning Advisory Group 1977) to the objections to the structure plan the new council maintained that it was probable 'that there will be a need for some additional land to be released to meet the housing needs of the urban area towards the end of the Plan period.' Such need, if it arose, would be accommodated in East Luton. 'Beyond this, the County Council would be prepared to accept an amendment to the submitted Structure Plan to reflect a greater degree of caution about whether additional sites would be required.' Thus all references to specific sites could be deleted. If, however, monitoring showed that additional land was required, then 'possible sites close to the urban area will be investigated in conjunction with the District Councils involved.'

The retreat was not enough to allay the fears of the Hertfordshire interests. This issue accounted for 136 out of a total of 241 objections to the structure plan. Of the 136 (127 from within Hertfordshire) 118 were from individuals, 12 from organizations, 5 from local authorities and 1 from an MP. The opposition was carried to the examination in public where the councils involved, two MPs and the Harpenden Society represented by a QC, explained their opposition with only the Luton Trades Council speaking in favour of development at Hyde. The opponents argued that higher-density development on committed land could accommodate much of the increase and that some diversion of growth to Milton Keynes could be accepted. If further land was to be identified, then all the statutory processes would have to be gone through before any decision was made. The Harpenden Society reiterated the objections to Hyde and suggested that the best approach was to identify the constraints on development around Luton and determine how much could be accommodated within them. They pointed out also that the county council's *volte-face* in substituting Hyde for Houghton Regis brought about conflicts with other major policies in the structure plan. Thus a manoeuvre by a few Labour members late in a meeting when several members had left had been responsible for a *cause célèbre*, unleashing the passions and encouraging the organized resistance of people from beyond (though only just) the county's borders.

*Conclusions* In this example the planners' original proposal had been overturned by the politicians who, in turn, had capitulated in the face of concerted public protest. It demonstrated that where the particular interest of an existing population is directly threatened it can, if properly focused, succeed in preventing development. The ability of well organized and articulate pressure groups to prevent or modify undesirable development has already been demonstrated.[10] It is an incidental but ironic feature of this issue that it aroused more public participation — and that from beyond the borders of Bedfordshire — than any other aspect of the structure plan.

The question of the expansion of Luton illustrates once again that any attempt at long-term strategic planning is vulnerable to attack and change. Instead of presenting a clear indication of the location of future growth, the structure plan had been forced, by local political circumstances, to defer a decision and to keep the options open. It always proves easier to avoid conflict by not making a decision. But the existence of conflict leads to future uncertainty, as was the case here. As with the other examples described in this extract it appears that the urban land development process is characterized by short-term incremental changes rather than responding to a grand design.

## The nature of land development

In the United Kingdom the process of land development is a combination of the market and attempts to influence it on the part of the public planning authorities. At the national political level, land policy has long been a controversial issue. Successive Labour governments have tried to increase the control of the planning authorities over the market and to secure the profits in land development for the community. Conservative governments used planning controls and taxation as means to regulate the market while maintaining the profit motive. Conservatives reject direct intervention, as is evident in their distaste for the Community Land Act. But, as is often the case in British politics, the ideological differences suggest greater divergences than have occurred in practice. There has been a general acceptance of the principle of development control and of the desirability to secure at least some of the betterment in land for the community. Proponents of the unrestrained free market, or of complete State control through nationalization of all land (Brocklebank *et al*) remain a relatively small minority and have failed to convert governments to their views. In broad terms the experience of national legislation on land so far has been to accept the balance of power in society rather than to attempt any radical redistribution of wealth in land.

The detailed pattern of land development remains largely a matter for local markets and local authorities. We have seen how the Community Land Act will take a long time, even if it survives, to have any marked impact on local land policy. The Bedford casestudy suggests that private initiatives will continue to influence both the use and value of development land. Decisions affecting specific parcels of land will reflect the conditions prevailing at a particular point in time. Each of the case studies has emphasized the tendency for land development to be an incremental rather than a long-term process. This is so despite the introduction of structure planning which assumes that land development will respond to long-term strategies based on participation to satisfy the preferences of the community. Such preferences are often difficult to interpret in terms of land planning and assume a consensus of viewpoint which can only exist, if at all, in the vaguest terms. Participation is hardly representative of the community as a whole and public opinion is only aroused when, as in the case of the prospective development at Hyde, a particular group feels itself threatened by a planning proposal.

Land development policy is rarely the product of consensus but usually arouses conflict and controversy. In the case studied here conflicts occurred between different interests which were reflected in competing political and professional attitudes. There are other cases where conflict is latent or suppressed. As we shall see, the structure plan assumed a widespread consensus over many issues, but such consensus was the product of agreement over general principles which satisfied the existing power structure rather than agreement over the details of development. The appearance of consensus over the principle of the green wedge or over the use of Bramingham Farm for residential development was transformed into conflict when each issue reached the agenda for political choices to be made. The apparent long-term continuity of policy becomes, when examined, the uncertainty of implementation. Land development, like other aspects of planning, is the aggregate of many different decisions taken at different times in changing local circumstances.

## Notes

1 See the two previous case studies in Blowers (1980) from which this extract is taken.

2 This issue is described in more detail in Blowers (1974).

3 The Harpur Trust was a charity administering 4 schools — 2 independent and 2 direct grant (i.e. with a mixture of private and State-assisted pupils).

4 Eleven of the thirty governors of the trust were also members of the borough council, and four were members of the council's Estates Committee which dealt with land transactions.

5 At the following meeting when the committee received a formal application from Luton the voting was 5 for (all Labour) and 10 against (5 Conservative, 3 Liberals and 2 Labour). As a result the matter was referred to the Minister.

6 Letter from the MP for Luton West to the Parliamentary Under-secretary of State, Department of the Environment, 8 November 1976.

7 Letter from the MP for Luton West to the Parliamentary Under-secretary of State, Department of the Environment, 27 December 1976.

8 Letter from the MP for Luton West to the Chief Executive, Borough of Luton, 3 January 1977.

9 The objections included 4 local authorities, 11 organizations, 134 individuals and 790 signatures to a petition.

10 See the previous chapters in Blowers (1980) from which this extract is taken.

## References

Abercrombie P 1945 *Greater London Plan* (London: HMSO)

Bedfordshire County Council (1974) *Luton/Dunstable Green Wedge Joint Appraisal* November 8

—— 1976a *County Structure Plan, Draft Written Statement* September

—— 1976b *Public Participation Phase I Report*, p 20

Bedfordshire County Council, Environmental Services Committee 1974 *Agenda Item* 23 April 26

Bedfordshire County Council, Planning Advisory Group 1977 *County Structure Plan: Response to Objections* August 4

Blowers A T 1974 Land ownership and the public interest: the case of Operation Leapfrog *Town and Country Planning* November pp 499–503

—— 1980 *The Limits of Power: the Politics of Local Planning Policy* (London: Pergamon)

Borough of Luton 1976 *Little Bramingham Farm, Outline Development Plan* September

Brocklebank J *et al The Case for Nationalising Land*, (London: Campaign for Nationalizing Land)

Department of the Environment 1976 Proposed residential development on land off Pasture's Way, Luton *Inspectors Report* paragraph 86

Elkin S J 1974 *Politics and Land Use Planning, The London Experience* (London: Cambridge University Press)

Luton Evening Post *The Grim Battle to Provide Homes*

Price M 1976 *Houghton Regis Town Development Scheme: A Study* (Cranfield: Cranfield Institute of Technology)

Ratcliffe J 1976 *Land Policy* (London: Hutchinson Educational) p 100

Reade E J 1975 An attempt to distinguish planning from other modes of decision making *University of Manchester, Department of Town and Country Planning Seminar Paper*

—— 1977 Some educational consequences of the incorporation of the planning profession into the State bureaucracy *Paper presented to the Conference of Sociologists in Polytechnics Section of the British Sociological Association, April 1977* p 13

# 18 Spatial Policy in Britain: Regional or Urban?

## by D Keeble and R Hudson

**David Keeble:**

Ever since the 1930s, spatial policy problems in Britain have been viewed by most policy-makers and academics primarily in terms of inter-*regional* differences in social well-being, economic opportunities and living standards, and economic and employment growth. True, the regions involved have sometimes been of apparently limited extent, but the population and employed workforce concentrated in these areas has always represented the bulk of that living in the key problem regions and countries of northern England, Northern Ireland, Scotland and Wales. And for the last ten years, the development areas have explicitly covered virtually the whole of these regions, together of course with Merseyside and the far southwest.

The history of inter-regional spatial policy, with its concentration on influencing the location of manufacturing industry shows very marked intensification since the early 1960s. The remarkable intensification of the last ten to fifteen years is strikingly illustrated by the increase in annual average government expenditure on regional assistance to manufacturing industry from only £6.4 million in the 1950s to some £111·5 million in the 1960s and £225·4 million in 1970–3, the last two figures at constant 1970 prices. By 1974–5, it was no less than £441 million with a 1976–7 forecast of £575 million. The latter figure will however be reduced somewhat by the July 1976 Regional Employment Premium and other changes. The current intensity of regional policy in Britain is thus almost certainly greater than in any other western industrial country, none

of which operates such a powerful combination of regionally differentiated manufacturing growth controls and incentives.

*Reduced regional disparities?*

In the past, geographers and economists have not infrequently criticized British regional policy as largely ineffective and only a political cosmetic. Early judgements, however, were of course based on a very much less powerful regional policy than has been formulated since the mid-1960s; also allowance must be made for undoubted time-lags in impact following regional policy intensification. It should not be surprising, therefore, that evidence of a substantial regional policy impact upon inter-regional economic disparities should only now, in the second half of the 1970s, become available. Some of this evidence, relating to regional shifts in manufacturing and total employment, in unemployment rates, in incomes, and in net migration, is briefly summarized for selected regions in table 18.1.

As this shows, the period since about 1965, and in fact particularly since 1970, has witnessed striking convergence of nearly all these different indices of regional economic performance towards the national average. Thus the three major assisted regions all increased their relative shares of national manufacturing employment between 1965 and 1975, whereas both southeast England and the West Midlands recorded a decline, the former by no less than 2·37 percentage points to a value of 25·54%. Total employment trends were less clear cut: but the three assisted regions taken together did at least maintain their share of UK employment, whereas that of the two traditionally prosperous regions declined by 0·5%. Unemployment rate convergence has been very striking, with a doubling of the West Midlands' rate, 1965–75, relative to the Great

Source: Keeble D 1977 *Area* **9** (1) 3–8
— 1978a *Area* **10** (2) 123–5
—— 1978b *Area* **10** (5) 363–5
Hudson R 1978a *Area* **10** (2) 121–2
—— 1978b *Area* **10** (5) 359–62

**Table 18.1** Regional economic indicators in Great Britain, 1965–75 (after Department of Employment *Gazette*, and *Abstract of Regional Statistics*)

| | Percentage share of UK manufacturing employment 1965 | Change in manufacturing employment share 1965–75, in percentage points | Percentage share of UK total employment 1965 | Change in total employment share, 1965–75, in percentage points | Unemployment rate as a percentage of Great Britain average 1965 1975 | |
|---|---|---|---|---|---|---|
| Southeast | 27·91 | −2·37 | 32·33 | −0·10 | 63 | 69 |
| West Midlands | 13·84 | −0·21 | 10·10 | −0·36 | 48 | 99 |
| Scotland | 8·47 | +0·04 | 9·17 | −0·03 | 215 | 127 |
| Wales | 3·63 | +0·60 | 4·45 | −0·05 | 187 | 138 |
| North | 5·36 | −0·70 | 5·50 | +0·08 | 185 | 146 |

| | Average weekly earnings of male manual workers relative to UK average | | Average annual net migration per 1000 of regional base year population | |
|---|---|---|---|---|
| | 1962–3 | 1974–5 | 1962–7 | 1973–5 |
| Southeast | 104·1† | 103·5 | +0·64 | −3·82 |
| West Midlands | 104·5† | 100·3 | +1·28 | −1·35 |
| Scotland | 92·7 | 100·7 | −8·09 | −1·07 |
| Wales | 101·7 | 97·7 | −0·23 | −2·75 |
| North | 95·3 | 100·7 | −3·05 | −1·81 |

† Statistics for 1967–8: earlier data for these regions are not available.

Britain average, but very substantial falls in the unemployment relativities recorded by Scotland, Wales and northern England. Earnings data reveal a closely similar picture, with a particularly marked convergence towards the United Kingdom average in the cases of the West Midlands, Scotland and northern England. Lastly, the net migration figures again suggest a very significant improvement in the economic performance of the assisted regions, including a striking *reversal* in the direction of net migration between 1962–7 and 1973–5 in the case of Wales, and a sharp fall in migration losses from Scotland. Both southeast England and the West Midlands are now losing population by net migration.

Clearly, factors additional to regional policy are also involved in these somewhat dramatic shifts. Thus, for example, recent net migration and manufacturing employment losses sustained by southeast England partly reflect a spilling-over of industry and population from this congested region to neighbouring areas of East Anglia and southwest England, reflecting in some ways a widening of the functional region oriented to London. Again, recent unemployment convergence may partly reflect differences in regional industrial structure and hence the timing of response to the worst world recession since the 1930s although the relative consistency of the trends since about 1970 argues against this as being of major importance. Industrial structure, it has been argued, may also have influenced unemployment convergence through the regional impact of unexpected national-level demand shifts for particular key products such as coal and motor cars. Certainly regional earnings trends have been affected by the marked national-level convergence in incomes in Britain over recent years, as a result of government wages policy. And in the Scottish case, North Sea oil must play some part, with over 50 000 workers out of a UK total of 80 000 employed directly or indirectly on oilfield development work.

But while acknowledging the role of some of these additional factors, there is also now substantial evidence that the major single influence at the development areas/southeast and Midlands level has been regional economic policy, working through shifts in manufacturing location and especially industrial movement. The scale of these shifts, and their attribution to intensified regional policy, is clearly brought out in a variety of recent independent statistical analyses, at both regional and subregional levels, using different data sets from employment counts, firm movement statistics and company surveys (Moore and Rhodes 1973, 1976a, b, Keeble 1976, pp 38–43, pp 89–115).

Indeed, the variety of corroborative evidence is now such as to force even erstwhile sceptics such as MacKay (1976 p 240) recently to agree that despite all the difficulties of evaluation, which he documents in detail, 'the overall picture remains surprizingly clear. Without the incentives and controls of regional policy, the level of employment in the development areas would now be considerably lower. There is sufficient evidence to indicate that there was an important and obvious response to the intensification of regional policy in the early sixties.' And a substantial shift of relatively highly paid manufacturing employment to the development areas in turn helps to explain trends in migration and earnings.

Regional convergence in employment, unemployment and other economic indices, largely as a response to policy measures, however in turn raises at least two further major questions. Is convergence, especially of manufacturing employment, in any way related to the national manufacturing decline which has also intensified since the mid-1960s? And is regional-scale policy, at the broad development areas/southeast and Midlands level, still the best framework for tackling spatial social disparities in Britain? After all, the single most powerful motive for regional policy has always been considerations of social equity and the equalization of living standards and economic opportunities. Are spatial differentials in this respect within Britain still best conceived of in broad regional terms?

*Inner-city problems*

The point here of course is that a growing body of opinion takes the view that the most significant social if not economic problems demanding government attention in Britain today are those specific to the inner-city areas of the country's major conurbations, in *whichever* region these are located, rather than in the broadly defined development area regions as a whole. That development area socio-economic problems are largely concentrated in such older conurbations as Clydeside, Tyneside and Merseyside has of course long been appreciated. The 1971 census revealed, for example, that by far the greatest single concentration in Britain of enumeration districts recording the highest levels of urban deprivation, measured by overcrowding, lack of basic amenities and male unemployment, was on Clydeside, and the unemployment situation in these development area conurbations is significantly worse than in their surrounding regions, with Tyneside, for example, recording an unemployment rate (July 1976)

of 9·7% compared with only 8·0% in the rest of the northern development area.

However, that significant socio-economic problems are also now to be found in conurbations — London, Birmingham and Manchester, all outside traditional development areas—is a relatively new viewpoint, though one now actively being propagated by the GLC and other conurbation authorities. Their argument centres on the very rapid rate of population and job loss in their areas, the latter notably of manufacturing employment, on personal and local authority income problems, and on the unemployment situation especially amongst inner-city coloured populations. Certainly the spatial pattern of current manufacturing job losses in Britain, as during the 1960s, is strikingly focused on the country's seven official conurbations, six of which are amongst the seven leading areas of decline (the exception is Tyneside). In marked contrast, two-thirds (35) of all other non-conurbation sub-regions distinguished recorded manufacturing employment *growth* over this period. On incomes, the GLC argue that real incomes of the poorest 25% of London's households have been *falling* over the last decade, while those of their counterparts in the rest of the country have been rising. In addition, inescapably high local authority costs force Londoners to pay a significantly larger proportion of their incomes as domestic rates than elsewhere in England and Wales. Lastly, relatively high unemployment rates do undoubtedly now characterize both Birmingham as a whole — 7·6% compared with the UK average of 6·3%, July 1976 — and parts of inner London, such as Southwark and Tower Hamlets.

Of course, the GLC and Birmingham have a vested interest in arguing for a rethink of regional policy. Clearly, there is an acute need for more reliable and up-to-date monitoring of social and economic trends in Britain's conurbations, especially their inner-city areas. But with the announcement in September 1976 by Peter Shore, Secretary of State for the Environment, of 'a new policy to bring back jobs to inner cities,' including a commitment 'to see what can be done to stem the tide of manufacturing jobs moving out, and the possibility of reversing it', together with his and more recent parliamentary statements, re-evaluation of current spatial policy does seem to be under way. The apparent success of intensified regional policy over the last ten years, suggesting as it does that substantial government expenditure *can* influence the geography of welfare in Britain, may thus paradoxically be one factor in refocusing spatial policy away from the traditional broadly assisted regions and towards needy conurbations wherever these may be located.

## Ray Hudson:

The core of the argument advanced above by David Keeble would seem to reduce to the proposition that regional inequalities, on various social and economic indicators, between 'traditioanlly prosperous' regions (southeast England and the West Midlands) and Scotland, Wales and northern England have been reduced over the period 1965–75 as the result of the implementation of a 'strong regional policy'. Consequently, it may therefore be appropriate to refocus spatial policy from a regional to a conurbation, intra-regional level.

This view can be contested on two grounds. First, it is not clear in what sense a 'strong regional policy' operated over the period in question. Keeble makes no reference to available evidence (such as that contained in the House of Commons Expenditure Committee Report for the Session 1972/3) which suggests that for many major manufacturing companies regional policy measures are of little importance as a factor influencing their investment decisions. Further, it is by no means clear that a narrowing of regional inequalities reflects the implementation of a 'strong regional policy'. The shrinking of regional inequalities has not necessarily been achieved by an *absolute* improvement in the position of the three development areas selected by Keeble; for example, unemployment rates there rose sharply between 1965 and 1975 and again between 1975 and 1976 (see table 18.2).

Clearly, the relative reduction in inter-regional differences in unemployment rates reflects the more rapid rate of growth of unemployment in the 'traditionally prosperous' areas as compared to the three development areas. In suggesting that attention now perhaps be focused on intra-regional differences in unemploy-

**Table 18.2** Unemployment rates (seasonally adjusted) for selected standard regions in 1965, 1975 and 1976 (second quarter) (from Central Statistical Office 1976, Crown Copyright)

| Standard region | 1965 | 1975 | 1976 |
|---|---|---|---|
| Southeast | 0·8 | 2·4 | 3·9 |
| West Midlands | 0·6 | 3·4 | 5·3 |
| North | 2·4 | 5·2 | 6·8 |
| Scotland | 2·8 | 4·5 | 6·5 |
| Wales | 2·4 | 4·8 | 6·8 |

ment, is Keeble tacitly endorsing the abandonment of 'full employment' as a national policy goal? If so, one must recognize the implications of abandoning a commitment to full employment: for example, the national costs of unemployment in the period since 1974 have recently been estimated at £20 000 million (Field 1977).

Related to this is a second point on which one can take issue with Keeble. This is a failure to consider the shrinking of regional disparities in manufacturing employment shares in the context of the aggregate loss of manufacturing employment in the British economy. While the share of national manufacturing employment rose in the three development areas over the period 1965–75, this was not because of strong manufacturing employment growth there; indeed, manufacturing employment was stagnant in the northern region and Wales and fell sharply in Scotland, particularly for men. Rather the relative improvement in the position of the three development areas reflected the considerable loss of manufacturing employment in the 'traditionally prosperous' areas (table 18.3).

This national aggregate loss of manufacturing employment is surely the key issue raised by Keeble's paper. Rather than alterations from inter- to intra-regional in the resolution level of policies intended to reduce spatial inequalities, the fundamental policy

**Table 18.3** Employees (in 000s) in manufacturing industries in selected standard regions, 1965 and 1975 (from Department of Employment 1976, Crown Copyright)

| Standard region | 1965 | | | 1975 | | |
| | Males | Females | Total | Males | Females | Total |
|---|---|---|---|---|---|---|
| Southeast | 1656 | 733 | 2389 | 1342 | 571 | 1913 |
| Midlands | 832 | 353 | 1185 | 738 | 284 | 1021 |
| North | 345 | 115 | 459 | 334 | 120 | 454 |
| Scotland | 502 | 223 | 725 | 438 | 199 | 637 |
| Wales | 234 | 77 | 311 | 234 | 83 | 317 |

issue is that of a national programme of investment in manufacturing industry that is adequate to reverse this tendency of de-industrialization and decline in manufacturing employment and provide the necessary basis for a return to full employment. This is not to deny that such a programme would necessarily require a spatial dimension, which might well involve resuscitating the conurbations. However, such a programme would of necessity involve a component designed to eliminate inter-regional unemployment differences.

**David Keeble:**

Hudson and I are, I think, agreed on at least one major issue: namely that Britain's long-term *national* manufacturing decline is an even more fundamental problem than regional or inner-city decline. The interesting questions here, from a geographical viewpoint, are first, whether some component of national decline reflects regional policy constraints on manufacturing investment or efficiency, as suggested by some evidence and secondly, whether the government's post-1975 'industrial strategy' is right still to channel the bulk of national public investment funds to the assisted areas, despite the concentration in *non*-assisted regions of many of the industries in which Britain still retains an internationally competitive edge.

On other points, I must however disagree with Hudson's comments. It would, for example, be tedious to list the many researchers who concur that the later 1960s and earlier 1970s witnessed the strongest-ever implementation of regional policy in Britain, with the possible exception of 1945–9. Again, the powerful impact of intensified policy upon manufacturing location in Britain is indicated by a host of different studies. Anecdotal evidence in the form of particular firms' public statements to the 1972/3 Expenditure Committee as cited by Hudson, is hopelessly inadequate as a basis for policy evaluation. Of course, it is inevitable that with substantial *national* manufacturing job decline since 1966, this effect has manifested itself as the maintenance rather than growth of aggregate manufacturing employment in the development areas (with the converse of losses in southeast England and the West Midlands through policy diversion of industrial growth which are well *above* the national average). But this in no way invalidates the policy effect — without this, manufacturing employment in the development areas would now be much less than it is.

A similar point applies to Hudson's comment on rising levels of unemployment in the development areas. Of course unemployment has risen in the development areas, as nationally, given the impact on Britain since 1973 of the worst international recession since the 1930s. What is significant in this context is not that development area unemployment has risen — it would be quite extraordinary if it had not — but that in direct contrast to the trend in previous recessions it has risen *less* rapidly than in the traditionally prosperous regions.

My last comment relates to Hudson's suggestion that advocacy for social reasons of a policy to ameliorate intra-regional disparities in some way implies abandonment of the goal of nationally 'full employment'. On the contrary, one could argue — though I am not sure that I would — that the former might be a better approach to achieving national full employment than recent inter-regional policy. After all, there are more workers currently unemployed (October 1977) in Britain's seven largest conurbations (633 000) than in all the scheduled development and special development areas (507 000), on which of course the great bulk of recent regional policy effort has been concentrated. The main point here, however, is that I certainly do not subscribe, implicitly or explicitly, to any abandonment of the goal of nationally full employment. Current high unemployment is a major social evil, despite the provisions of a welfare state, and national economic policy must attempt to reduce it. At the same time, it would seem that unemployment *and* other socio-economic problems are, and could become even more, disproportionately concentrated in Britain's inner-city areas, partly as a result of their drastic recent industrial decline.

**Ray Hudson:**

Having established that David Keeble and I are basically in agreement as to what constitute desirable social and economic objectives — full employment and a reversal of the decline of Britain's manufacturing sector — it would appear that our differences principally relate to the adequacy of various means to attain these ends. Thus it seems useful to attempt further clarification of the relationship between past and current regional policies and national, regional and conurbation employment change. This is particularly so since much of current policy reflects a response to the situation of the 1960s and early 1970s, which differs markedly from that of the later 1970s. New policies are urgently needed to deal effectively with these changed circumstances but a necessary — though by no means sufficient — condition for beginning to formulate these successfully is to appreciate correctly the effects of past and current policies.

Rising unemployment in British conurbations reflects in part the response of manufacturing industry to crisis, a process of restructuring that began some time before, but was accelerated by, the post-1973 recession. Many companies with capacity located within conurbations have been compelled to reduce this, as a response to rising (international) competitive pressures, in order to remain viable. At the same time as closing older or obsolescent factories in inner-conurbation areas, many such companies have been either replacing these or expanding capacity elsewhere, increasingly overseas: between 1964 and 1974, for example, the rate of direct overseas investment by British manufacturing firms more than trebled. The paradox from the viewpoint of residents of the conurbations is that strategies which are rational for companies result in rising unemployment, not only directly through loss of manufacturing jobs (often highly skilled and highly paid) but also indirectly as these job losses in turn begin to have effects on service-sector employment levels.

This restructuring has been encouraged by various State policies, both deliberately and inadvertently. For example, Massey and Meegan (1977) have shown the impact of the intervention of the Industrial Reorganization Corporation in restructuring the electrical, electronics and aerospace industries and the consequent employment changes in inner-urban areas. Such restructuring was deliberate and as such, in the context of the 1960s, comparatively rare. More commonly, however, restructuring was encouraged 'unintentionally' during the 1960s as a result of regional policy measures, in the sense that the main stated concern underlying regional policy was to eradicate, or at least narrow, regional differences in unemployment rates. In order to create employment in assisted areas, companies (including those disinvesting from conurbations) were encouraged to replace or expand factories by investing there and many did so. Note, however, that recognition of this is not meant to imply acceptance of a simplistic thesis that regional policy 'causes' inner-city problems: restructuring, from the point of view of the companies, is and was necessary, irrespective of the incentives of regional policy. Factory closures in inner-conurbation areas would have occurred in

any case, either because companies did attempt successfully to restructure or because they made no attempt to do so and hence became uncompetitive and unprofitable: again the paradox is that, regardless of company strategy, employment in the inner city would fall.

Reflecting broader national economic policy, especially during the 1970s, regional policy measures have become increasingly explicitly directed at restructuring. The emphasis within regional policy has swung towards assistance on a selective basis, towards encouraging rationalization and mergers, and towards the substitution of capital for labour (most vividly exemplified by the abolition of Regional Employment Premium in 1976) and away from a concern with employment provision, which at best is now a secondary priority.

This is not to deny that a certain amount of manufacturing employment has arisen in the development areas as a consequence of these various regional policy measures. Given the scale of inducements available, it would, if I may borrow a phrase from Keeble, be extraordinary if some new employment had not resulted. Rather I would suggest that identifying a given number of jobs as a result of regional policy measures alone is not as simple as Keeble implies, that in any case not enough jobs have been created (as evidenced by the expansion of unemployment in the development areas), and furthermore the number of jobs is not (or, more accurately, ought not to be) the sole criterion for evaluating the 'success' of regional policy in 'creating jobs'.

Related to this is the point that while regional policy measures may have acted as a considerable inducement for certain companies to invest in development areas, particularly those investing in plants employing capital-intensive production processes, for other companies other factors may have been of greater importance. This is not to deny that regional policy measures had *some* effect in such companies' decisions to invest in development areas but to suggest that these were not the *decisive* factor. For example, for many companies investing in development areas labour availability has been a crucial factor, considerable reserves of labour being available there at a time when they were not available in the 'traditionally prosperous' areas. Thus during the 1960s female activity rates tended to be lower in the development areas than in southeast England and the Midlands, although such differences have narrowed as increasingly more women have been drawn into the labour market, partly in response to investment decisions by manufacturing companies. Furthermore, primarily as a consequence of the National Coal Board's restructuring of the coal-mining industry, considerable losses of male employment in mining and quarrying occurred in Scotland, Wales and the northern region of England from the late 1950s. Miners thrown out of work provided a source of labour for manufacturing companies seeking to replace or expand capacity in the development areas but only part of the surplus mining labour force was absorbed in this way. As a result, male employment in mining and quarrying and manufacturing fell in the three development areas, as it did in the 'traditionally prosperous' regions, and male unemployment rose.

A further point is that as companies responded to the post-1973 recession, much of the new manufacturing plant opened in development areas prior to that date was closed and in that remaining levels of utilization were often reduced. The result was a permanent or temporary loss of employment. The point here is that 'new' jobs created in development areas are not permanent, the post-1973 period merely emphasizing this, albeit sharply. Finally, Keeble refers only to the numbers of jobs created in development areas; the question of the *type* of jobs may be at least as important, for many new jobs in development areas have involved semi-skilled work for women at a time when the pressing social need was for fresh male employment.

The question remains as to the character of a regional policy that would be effective in this sense, reducing unemployment and contributing to a return to full employment. More specifically, while the emphasis in national economic policy and regional policy remains firmly on restructuring to improve international competitiveness (and profitability), is such an effective policy possible? The same question can reasonably be asked of policies which remain in this framework but whose spatial focus shifts from regional to conurbation level. It may well be that restructuring manufacturing to improve international competitiveness and profitability is incompatible with a return to full employment in Britain. It may well be that if the British economy continues to be steered by the imperatives of international competitiveness and profitability and State policies accept the legitimacy of these as steering mechanisms and facilitate and reinforce their effects, then spatially-uneven economic development will remain, whether expressed as an 'inner-city' or as a 'regional' problem.

**David Keeble:**

In his further contribution, Ray Hudson expresses certain continuing doubts about the effects of regional policy, notably over methods of measurement of its impact, the importance of labour availability in explaining the marked relative shift of manufacturing employment to the development areas since the early 1960s, the permanence of the jobs created, and their bias towards semi-skilled female workers. I share his concern over the last point, given all the evidence that new jobs have been disproportionately orientated towards low-level production tasks and to female workers (Keeble 1976, pp 238–9). However, we still differ somewhat in our judgements on the other three topics.

Both questionnaire surveys and aggregate statistical analyses of course suffer from certain problems as means of assessing causal influences on industrial location change, as I have myself pointed out on various occasions. When however there are available, as there now are, a wide range of careful analyses by different workers using both methods, all by and large agreeing that regional policy has induced, not just 'a certain' but a substantial amount of manufacturing employment in the assisted regions, it seems to me inadequate to cite only anecdotal evidence against such an impact, as Hudson's original comment did.

On the second point, I agree with Ray Hudson that manufacturing firms locating themselves in the development areas in the 1960s and 1970s have in addition been influenced by greater labour availability there. My point here, however, is that differences in labour availability do not seem to explain the big *improvement* in development area manufacturing movement, job creation and retention relative to the non-assisted regions, which occurred from the early 1960s onwards along with the marked strengthening of regional policy. Labour availability differences between the assisted and non-assisted regions, as measured by differences in unemployment rates, in fact decreased, not increased, in the 1960s as compared with the 1950s (Moore and Rhodes 1976a, p 30). Firms in southeast England and the Midlands thus had somewhat less incentive to move than previously for labour supply reasons. Pressure of demand, too (a key factor in levels of industrial movement nationally), does not seem to have been markedly different in these two decades (Ashcroft and Taylor 1977, p 96). So the unemployment situation in the assisted regions in the 1960s does not seem to explain the big improvement in industrial movement to, and manufacturing employment creation in, these areas at that time, as Hudson appears to be suggesting.

The third point concerns Hudson's claim that 'the "new" jobs created in development areas are not permanent', as demonstrated by post-1973 trends. This assertion relates, of course, to a question of crucial importance for regional policy assessment. Yet it is not accompanied by any supporting evidence. Certainly there have been individual serious redundancies in particular cases such as Bradford (Rose 1978) and Merseyside (Norman 1978). But at least over the post-1973 period to 1976, the latest date for which Census of Employment figures have been published, each of the three main development area regions as a whole more or less maintained — or increased — their share of United Kingdom manufacturing employment, compared with significant declines in the shares of southeast England and the West Midlands (table 18.4). This suggests that the findings of the earlier detailed studies of assisted area branch factory closure and job loss during the late 1960s (Atkins 1973, Townroe 1975) still hold — namely, that assisted area branches, though recording a somewhat higher closure rate than their non-assisted area parent factories, in fact maintained aggregate employment significantly *better* than the latter. This apparently even applies to 'speculative' assisted area advance factories, Slowe's research (1977 p 22) on factories occupied up to the end of 1975 concluding that 'such investment in fact appears to be, if anything, more stable than in other factories.' Lastly Mackay (1978, p 43), though rightly stressing the difficulties of projecting future effects of regional policy, concludes that his various statistical analyses do suggest 'that the impetus provided by plant movement has not merely provided a temporary shift in the expansion path of the development regions, but has raised performance to new and sustained levels.' This reflects the higher rate of investment in new, more efficient and productive capital stock—machinery and buildings— which regional policy has engendered in the assisted as compared with the non-assisted regions in the last two decades. Of course, in an important sense, these findings and evidence are interim only. In the last resort, only time itself can show whether policy-induced jobs in the assisted regions are more or less permanent, relative to national trends. But as an interim judgement, they do not seem to support the view expressed by Hudson.

I would add to Hudson's concluding questions two

**Table 18.4** Manufacturing employment in selected British regions, 1973–6

| | 1973 | | 1976 | |
|---|---|---|---|---|
| | (000s) | Percentage share of UK | (000s) | Percentage share of UK |
| Southeast | 2037 | 26·02 | 1851 | 25·55 |
| West Midlands | 1074 | 13·72 | 979 | 13·51 |
| Scotland | 657 | 8·39 | 608 | 8·39 |
| Wales | 329 | 4·20 | 303 | 4·18 |
| North | 462 | 5·90 | 438 | 6·04 |

others. Is it possible to achieve greater manufacturing investment and international competitiveness, on which expansion of both public and private sector *service* industry jobs and hence reduction in national unemployment could perhaps be based, without a more positive policy of fostering the many efficient firms in the non-assisted regions of Britain? And secondly, if special aid is to be given to certain areas to reduce socio-economic inequalities, do not unemployment, racial problems and deprivation in the conurbations demand that they, as a group, be accorded the first priority rather than the broadly assisted regions which have been favoured since the 1960s?

**References**

Ashcroft B and Taylor J 1977 The movement of manufacturing industry and the effect of regional policy *Oxford Economic Papers New Series* **29** 84–101

Atkins D H W 1973 Employment change in branch and parent manufacturing plants in the UK: 1966–71 *Trade and Industry* 30 August pp 84–101

Central Statistical Office 1976 *Economic Trends 2, Annual Supplement*

Department of Employment 1976 *Gazette* **84** (8) 39–50

Field F 1977 *The Conscript Army: a Study of Britain's Unemployed* (London: Routledge and Kegan Paul)

House of Commons Expenditure Committee (1973) Regional development incentives, 2. *Minutes of evidence taken before the Trade and Industry Sub-committee, 2nd Report, Session 1972/73*

Keeble D 1976 *Industrial Location and Planning in the United Kingdom* (London: Methuen)

MacKay R R 1976 The impact of the Regional Employment Premium *The Economics of Industrial Subsidies* ed. A Whiting (London: HMSO)

—— 1978 The death of regional policy — or resurrection squared? *Centre for Urban and Regional Development Studies, University of Newcastle-upon-Tyne, Discussion Paper* 10

Massey D and Meegan R 1977 Industrial restructuring versus the cities *Centre for Environmental Studies, Working Note* 473

Moore B and Rhodes J 1973 Evaluating the effects of British regional economic policy *Economic Journal* **83** 87–110

—— 1976a Regional economic policy and the movement of manufacturing firms to Development Areas *Economica* **43** 17–31

—— 1976b A quantitative analysis of the effect of the Regional Employment Premium and other regional policy instruments *The Economics of Industrial Subsidies* ed. A Whiting (London: HMSO)

Norman P 1978 Empty days on the dole *Sunday Times* **8089** July 23 pp 17–18

Rose H 1978 A de-skilled town *New Society* **45** July 20 pp 133–4

Slowe P M 1977 *Advance Factories in British Regional Policy* (London: Regional Studies Association)

Townroe P M 1975 Branch plants and regional development *Town Planning Review* **46** 47–62

# 19 Policy Alternatives — Past and Future

*by P Hall with additional material by R Thomas and R Drewett*

## Introduction

Physical planning controls are clearly not the only factor affecting urban growth. Controls over the location of different sorts of employment — or the lack of such controls — are basic in regulating the numbers of workers, and their families, seeking homes in any local area. The rate of inflation, which influences people's attitudes to investment, and taxation policies as applied to different types of housing, critically affect the progress of the market for owner-occupied housing. Transport policies, both for road and for rail travel, affect the cost of transport in relation to wages and salaries and may introduce a differential according to whether the trip is by road or rail, over short distance or long. Though presumably subsidiary to land-planning control policies, agricultural subsidy policies affect the prosperity of the farmer and thus the firmness of his resolve to stay on the land. Rural electrification and water supply policies, not to mention the provision of television relay stations, affect the attractiveness of rural life in relation to town life. These policies have been controlled or affected by many different agencies in Britain since World War II, and we should not expect that they would prove fully consistent with each other in their effects on the urban growth process.

## Alternative policies for the future

In considering what alternative policies might have been pursued in the past and might still be pursued in the future, it is best to be realistic. The consequences of some policies might be fascinating to trace as a theoreti-

Source: Hall P *et al* 1973 *The Containment of Urban England* (London: George Allen and Unwin) Vol. 2, pp 410–455.

cal exercise, but no one should assume that they are politically feasible. If governments had consistently pursued a deflationary or at least disinflationary policy after 1945, a situation such as that in the 1930s would have been obtained: people would have tended to speculate less in land and property, while landowners would have been more inclined to sell if they thought there were dangers of actual falls in value. Again, if agriculture were left without protection or without the generous subsidization which it has enjoyed since 1945, farmers would have been positively encouraged to sell their land for possible development, just as they did in the 1930s. But to posit policies like these, whether for Britain or any other European country in the post-1945 period, is not exactly realistic.

### Alternative housing policies

Variations in housing policy, on the other hand, might have been feasible and might still be so. Subsidies to owner-occupiers could have been reduced, to encourage more people to rent; a unified housing policy could have been introduced, giving private landlords some of the same subsidy incentives as owner-occupiers have enjoyed. On the other hand the owner-occupier subsidy element could have been increased, by extending its advantages to new groups not previously able to enter: this was the object of Labour's Option Mortgage Scheme of 1967, which in effect extended the benefits of the income tax concession to poorer families not paying enough tax to enjoy it, and it could be further increased by devices such as a 'balloon mortgage', part of which is repaid only on sale of the house. Most fundamentally of all, if there were ever a completely unified housing policy in which subsidies were attached to the family and its need, rather than to the particular type of housing occupied (Nevitt 1966) the result might well be a very large extension of demand for owner-occupied housing (for in a period of inflation, house ownership is advantageous to all) and this, in turn, would have given a strong impetus to further urban decentralization.

*Alternative land and credit policies*
There is no lack of possible alternative land and land finance policies, either. At one extreme, as the Uthwatt Committee recommended in 1942, there could be outright nationalization of development land outside existing urban areas, with compensation for lost development rights, and purchase at the point of development by a central planning authority at existing use-value (Expert Committee on Compensation and Betterment 1942). This was a logically consistent policy in which the market in land would not be required to work at all. At the other extreme, it would be possible to return to a virtual non-plan situation, at least over wide tracts of the country. This would open up wide tracts of land for potential development and allow development values to float freely — as they did in the 1930s — thus inhibiting the speculative withdrawal of development land from the market, and reducing development values generally.

Credit policies represent a closely related area where alternative policies merit close examination. The building industry and the home buyer are very susceptible to the vagaries of the economic climate which manifest themselves in the cost and availability of credit. Of particular importance is the *cost* of credit which affects the building industry and inflates development costs. However, it is not the cost of credit but the *availability* of credit which critically affects consumer demand. This is particularly important to first-home buyers. It seems curious that many basic commodities and industries are protected by government policies (such as farm subsidies) but no such policy exists to protect the home buyer from the worse effects of liquidity which are beyond his control. During such periods of economic stress it would be a great contribution if the government acted as guarantor to building societies or increased finance for housing through local authorities. This would even out the availability of credit and introduce *stability* in levels of construction and satisfy consumer needs in a basic commodity. Many families would have directly benefited during recent years from such alternative sources of financing.

*Alternative transport policies*
Plenty of alternatives also exist in transport policies. Instead of building radial motorways into the centres of the cities — as in the case of the Aston Expressway in Birmingham, the Parkway in Bristol, or the M53 Mid-Wirral motorway and second Mersey Tunnel in Liverpool — or instead of the costly inner-city ring motorways proposed for London and other major cities, emphasis could be shifted to orbital motorways around the cities, giving good communication between different parts of the suburban rings. This is a policy that has been followed in the United States, with remarkable effects on land use and activity patterns in certain cities — Washington, Baltimore, Boston. They would cost a good deal less to construct than inner-city motorways and they would give good economic benefits; they would provide a relatively economic way of allowing people a wider range of home and job choices, and of encouraging people to find cheaper homes at some distance from the city centres. There might even be justification for encouraging subsidized production of a cheap car for suburbanites, rather than supporting an increasing uneconomic public transport system in such areas (Norton 1970). The opposite extreme to this policy would be to encourage better high-speed public transport systems along radial lines into the city centres, connecting suburban homes to city jobs. They might be express bus systems, using reserved facilities on motorways, or new or rehabilitated rail-based systems.

*Alternative industrial location policies*
Industrial location policies, too, could be varied. Indeed, perhaps one of the most remarkable features of the post-war period in Britain has been the continued reliance on the control method of approach despite the evidence that in many respects it was ineffective. During the 1960s and early 1970s there was increased enthusiasm for the ideal of a differential payroll tax on congested areas — however these might be defined — coupled with a negative tax, or subsidy, in uncongested or development areas (Labour Party 1970). This system might at first supplement, and then in time replace, the present system of controls. It would have the broad effect of raising the cost of immobile city-centre activities. Those activities which had least need of a city-centre location, were most mobile and were most sensitive to costs, would move most rapidly to uncongested locations. At that point, the precise effect would greatly depend on the relationship between the

**Table 19.1** Alternative policy options and urban forms

**A  Policy options**

1  Alternative ways of managing the economy
 (a)  Faster inflation.
 (b)  About the same rate of inflation as in 1945–70.
 (c)  Slower inflation, or disinflation, or deflation.

2  Alternative agricultural policies
 (a)  More generous agricultural subsidies.
 (b)  About the same rate of agricultural subsidies as in 1945–70.
 (c)  Less generous agricultural subsidies, or none at all.

3  Alternative housing policies
 (a)  More generous/easier entry mortgages.
 (b)  About the same mortgage rates/ease of entry/housing subsidy as in 1945–70; subsidies of the same type as in 1945–70, i.e. attached to houses rather than people.
 (c)  Less generous/harder entry mortgages.
 (d)  A national home finance agency to provide easy mortgages underwritten by the State (in practice equals 3(a)).
 (e)  Rationalization of the subsidy system to concentrate it more on those in hardship (low incomes, large families). Eliminate specific subsidies tied to types of housing, e.g. subsidies for high-density and/or expensive land.
 (f)  Negative income tax to replace subsidies (in practice, effects similar to 3(e) ).
 (g)  Increase specific subsidies for high-density and/or expensive land.

4  Alternative land policies
 (a)  Nationalize development rights; development through compulsory State purchase by central authority (the Uthwatt solution).
 (b)  Nationalize development rights; development partly through private agencies paying development charge or betterment levy, partly through public agencies using compulsory purchase at a price below full market value (the 1947–53 and 1967–70 system).
 (c)  Nationalize development rights; development partly through private agencies, but paying no charge or levy, partly through public agencies paying full market (i.e. development value) prices (the 1959–67 system; from 1953 to 1959 there was a mixture of 5(b) and 5(c) ).
 (d)  Do not nationalize (or denationalize) development rights, i.e. administrative non-plan (the pre-1939 system).

5  Alternative transport policies
 (a)  Make public transport more attractive/cheaper through additional subsidies; leave private transport as it is, or make it more expensive (raise licence fees/fuel tax/purchase tax on cars).
 (b)  Proceed roughly as during 1945–70, i.e. some subsidy to public transport (hidden, though losses on public transport met by Treasury, up to 1968; thence social service subsidies to specific public transport operations).
 (c)  Do not aid public transport but encourage spread of private vehicle ownership through lower purchase tax/licence fee/fuel tax, faster road construction, etc.
 (d)  Devolve road construction from the Ministry of Transport to a roads board with the status of a nationalized industry; charges on road users to go to road construction (in effect similar to 5(c)).
 (e)  Introduce a road pricing system which makes travel on congested roads dearer, travel on uncongested roads cheaper, while keeping present level of charges on road users unchanged (the road pricing solution).

6  Alternative technological policies
 (a)  Develop further industrialized, low-cost housing techniques, involving perhaps the modification of building bylaws.
 (b)  Develop new methods of high-speed public transport, perhaps more 'individualized' than presently to make them more competitive with private transport, with the help of State subsidies.
 (c)  Develop cheaper cars which could be bought by a wider market, thus extending car ownership more rapidly than now forecast.

7  Alternative administrative policies
 (a)  Reform local government along the lines of the Redcliffe-Maud recommendations of 1969: unitary authorities for most of England; metropolitan authorities for conurbations and their surrounding rural areas.
 (b)  No reform; continue as in 1945–70.
 (c)  Two-tier reform with some powers reserved to district authorities (as in the Senior memorandum of dissent to the Redcliffe-Maud report); metropolitan authorities restricted to the continuously built-up conurbations (cf the GLC boundaries).
 (d)  Reform including a powerful tier of regional or provincial authorities, responsible for structure planning (effect similar to 7(b)).

8  Alternative employment location policies
 (a)  Introduce a graded system of financial incentives and disincentives for location in different areas: a payroll tax in congested areas, subsidies to employers in uncongested or development areas.
 (b)  Continue 1945–70 controls (IDCs and ODPs).
 (c)  Remove all controls; no incentives or disincentives.

**B  Design alternatives**

1  Peripheral growth of conurbations and major freestanding cities (the *non-plan* situation of the 1930s) with:

| | |
|---|---|
| (a) Jobs concentrated at the centre and longer radial journeys. | 1.1 Increasing dispersion of the peripheral development (scatter). |
| (b) Jobs dispersed at the periphery and local criss-cross journeys. | 1.2 About same degree of dispersion as in the 1930s. |
| (c) Any combination of 1(a) and 1(b). | 1.3 Much closer concentration with high-density housing developments and very limited peripheral growth. |

2 Development of new towns or comprehensively-planned town expansions (separately as in the Abercrombie prescription, or in groups as in the Howard prescription: the *garden city* ideal of the 1940s) with:

(a) Jobs concentrated at the centre with short radial journeys.

(b) Jobs dispersed at the periphery with short local journeys.

(c) Any combination of 2(a) and 2(b).

2.1 Lower densities (less than 12 per acre).

2.2 Densities of 12–14 per acre.

2.3 High densities (15+ per acre).

3 Peripheral growth of medium-sized or small freestanding towns

(a) Jobs concentrated at the centre with medium-distance radial journeys.

(b) Jobs dispersed at the periphery with short criss-cross journeys.

(c) Any combination of 3(a) and 3(b).

3.1 Low-density peripheral additions.

3.2 Medium-density peripheral additions (approximately 12 per acre).

3.3 High-density peripheral additions with much high-density renewal including high-rise structures.

rate of tax (or subsidy) in overspill locations around the conurbations, *vis-à-vis* more distant development area locations. But one effect would be a more rapid decentralization of jobs from the conurbations to their suburban rings. The tax might not win much approval from the big city authorities such as Birmingham or the GLC — unless they got a share of the proceeds. For it would lead to a weakening of their rateable-value base, which they very much fear. But if congestion is regarded as a problem — albeit a difficult one to define operationally — the tax would be a way of meeting it. So would a system of charging for the use of scarce roadspace in congested cities at congested hours, the effects of which would be slightly more complex.

*Alternative policies in other areas*

There are a host of policies in other fields which might have an effect on urban growth patterns. The resources allocated to inner-urban slum schools, as against suburban schools coping with rapid population growth; clean air legislation and its enforcement; action to reduce the crime rate in certain inner-city areas; the possible introduction of a negative income tax, giving poor people greater freedom to spend discretionary income, and facing them with the choice of what sort of housing they wanted to pay for, in what area and at what price (The Open Group 1969); all these, and

many others, would have their impact. The argument here is that physical planners have all too seldom considered their own narrow range of policy objectives and policy instruments in relation to the wider framework of these programmes.

**Policy alternatives and urban forms**

Some of these policy choices are set out in table 19.1. Part A consists of alternative policy choices in a number of defined areas which are closely related to physical planning, such as general management of the economy, housing, transport, agriculture and local government reform. Under several of the headings, one of the alternatives involves no change in the general policies pursued between 1945 and 1970, which are assumed to be reasonably consistent throughout that period despite changes of government and political philosophy. (In fact the shifts have been less substantial than might at first be thought.) Part B shows a number of alternative urban designs for accommodating urban growth, broken down into the most important elements: the distribution of jobs, the distribution of people and their homes and the resulting urban form (which is very closely related to the distribution of people, because of the fact that residential and associated land uses are such a dominant part of the total urban fabric). Only the most likely combinations of people, jobs and urban forms are specified in the table and in the analysis that follows.

Against this background, we can consider how the policy options in Part A of table 19.1 relate to the feasible urban-form combinations set out in Part B of that table. Figure 19.1 is an attempt to analyze these connections systematically. Essentially, it is an attempt to combine deductive logic about the effects of policies with such empirical evidence as exists about their actual effects in the period 1945–70. Though it must be speculative and tentative, some perhaps significant conclusions emerge. One is obvious: that the policies of the recent past have had a specific intent, and on the whole have had specific results. Therefore, the alternative in most cases is to intensify the policy and the trends it produces, or to reverse them. But in a number of cases, it appears that whatever is done to alter policies, the same physical outcomes will result; the options are closed. In particular, two urban-form options, peripheral expansion of conurbations and large freestanding cities, and peripheral expansion of smaller freestanding towns close to conurbations or freestanding cities, emerge as by far the commonest outcomes from a very wide variety of choices in different fields. This is only confirmed by the history of urban development since World War II, which has been dominated of course by growth in these two types of area. Only by means of a sharp and determined shift towards comprehensive centralized planning would there be a greater emphasis on new towns, or subsequently social cities which would be their natural outgrowth. And only by means of a virtual abandonment of controls would dispersal seem at all likely. Though some policy options would encourage higher-density redevelopment in cities, rental rather than owner-occupation policy, and public rather than private transport, it is not physically possible to turn the main emphasis towards high-density renewal; the main population growth, and the main urban investment, will occur through peripheral extensions somewhere or other. It is these constraints which limit the range of physical outcomes to a very few, and make it likely that barring a major shift of policy, the bulk of the growth will continue to be housed in extensions to conurbations and major cities and towns.

Figure 19.1 emphasizes one other important fact: that there are a number of policy routes to the same physical outcome. They may well be used in concert to reinforce each other, but if they are used in conflict, this will result in confusion. The chart suggests for instance that the types of urban growth which have been strongest since World War II — peripheral expansion of conurbations/major cities and of smaller towns — have been fostered by the following different policy measures: inflation in the economy, making owner-occupation attractive; a mortgage policy which also encouraged owner-occupation through tax relief; a policy of nationalizing development rights but not providing strong powers to take land for planned developments such as new towns; a policy which encouraged private car ownership by allowing the cost of motoring to rise more slowly than the general level of prices and incomes; and a failure to carry through fundamental local government reform. An alternative form of urban growth on a large scale — new towns, planned town expansions or planned city regions, for example — would probably have required shifts in all or most of these policies. A slower rate of inflation, and less generous mortgage policies, especially cheap credit for public housing through the Public Works Loan Board; the nationalization of development rights, with all new development carried through on the basis of compul-

sory purchase by a central planning authority; licensing of new building, used to limit the amount of speculative private building; fundamental reform of local government based on strong regional planning authorities; a determined effort to steer public policy towards the relief of housing conditions among the least fortunate members of society: all these would have aided a policy of new town (or planned town-expansion) building. Dispersal or scatter of growth could only have resulted from the simultaneous introduction of non-plan policies in a variety of spheres — financial, housing, transport and employment.

By and large, different governments with different philosophies have each pursued fairly consistent policy bundles since World War II, but each one has been plagued by certain inconsistencies. The Labour Government of 1945–51 tried to put all the emphasis on fully planned developments of the new town (or town expansion) type; in this, it was perhaps more consistent — and more ideologically committed — than any subsequent government. It tried hard to restrict inflation, it limited the spread of owner-occupation by a rigidly applied scheme of building licensing, it supported public housing through cheap credit, and it carried through financial provisions which in effect abolished the role of the market by making it unprofitable. But even here there were inconsistencies. The financial provisions in the 1947 Act, radical and unwelcome to many people as they were, lacked the pure logic of the Uthwatt Committee solution. They did not bypass the market, but made it function so badly that it almost stopped, and the nettle of local government reform was not grasped. The division between the cities and the counties was left as a source of future trouble.

In any case, as we have seen, the bundle of policies proved unacceptable to too many groups — landowners, potential owner-occupiers, big cities and counties bent on maintaining their freedom of action. The result was that after the Conservative return to power in 1951, the pure policy of building new and expanded towns was allowed to continue, but coupled with many reversals of policy which pointed in the opposite direction. Very quickly, the Conservatives abolished building licensing and welcomed the private builder back into business; they tried to restrict public house-building to slum-clearance schemes; they allowed inflation to proceed, and approved the continuation and extension of subsidies to owner-occupiers through income tax relief. These policies made consistent sense; they encouraged speculative building on the pattern of the

**Figure 19.1**  Alternative policy options and urban forms.

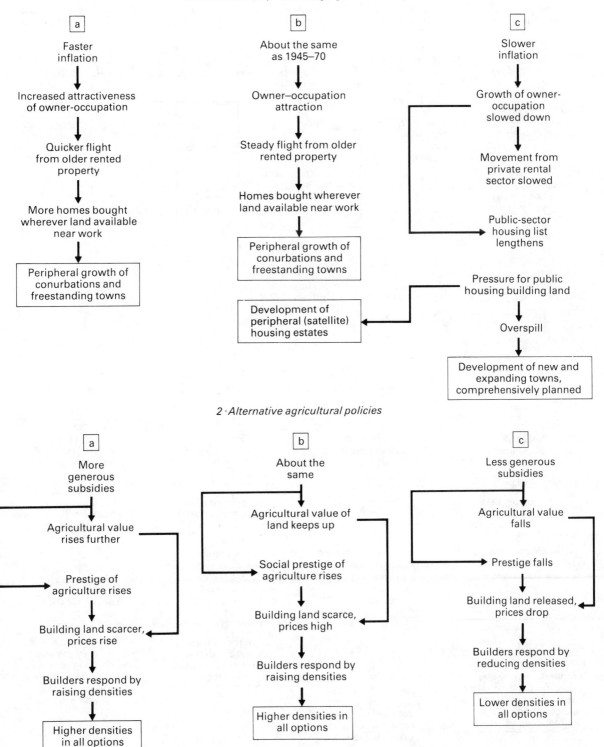

### 3 Alternative housing policies

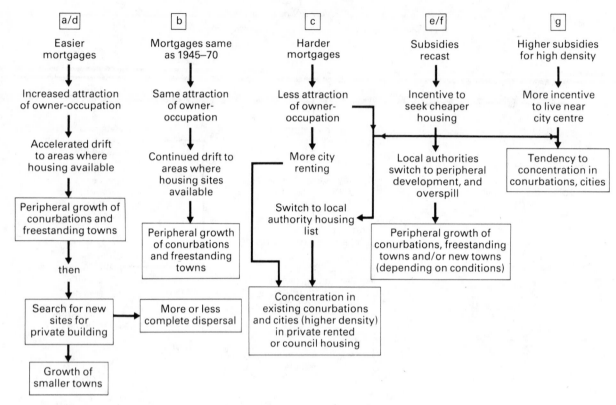

### 4 Alternative land policies

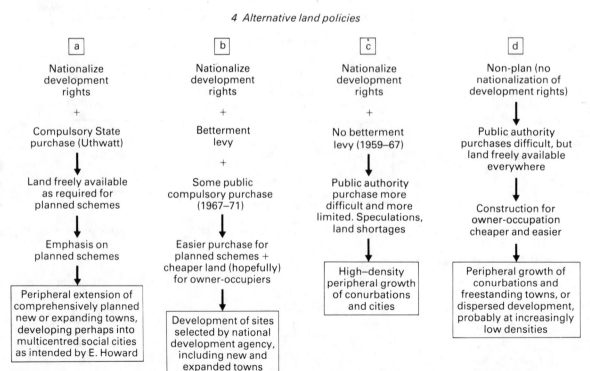

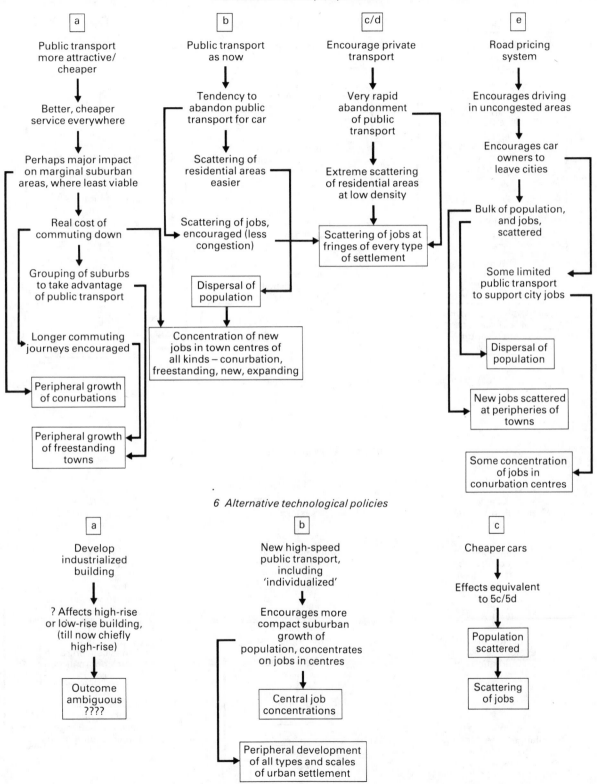

5  *Alternative transport policies*

6  *Alternative technological policies*

7 *Alternative administrative policies*

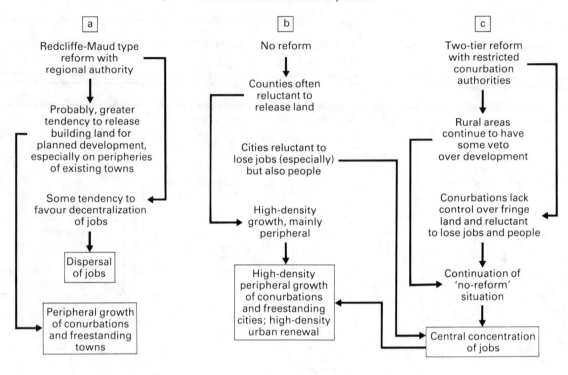

8 *Alternative employment location policies*

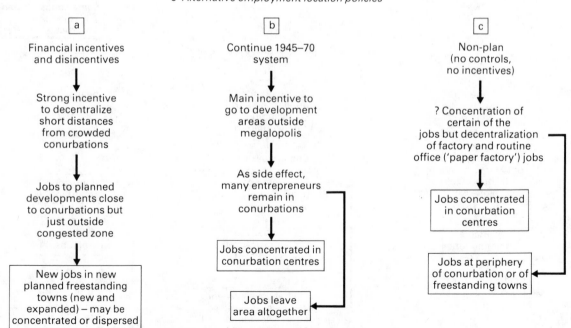

1930s. But not quite; because the Conservative Government also encouraged urban containment through green belts, and showed no enthusiasm for a far-reaching local government reform that would have brought cities and their rural rings under the same planning machine. Development in this period, from 1951 to the early 1960s, therefore tended to take two forms: high-density, high-rise building for slum clearance and renewal in the cities, and speculative building at increasingly tighter densities in the suburbs. Ostensibly, the policy benefited the aspirant home owner. But more essentially, the free system thus created was bound to benefit most those who were in possession of the right cards at the beginning of the game. It benefited landowners who could make speculative gains, ruralites who resisted development, and county councils fighting urban encroachment. It did not benefit young couples seeking cheap owner-occupied housing, nor builders wishing to build such housing, nor cities looking for extra land to rehouse their slum dwellers. In a very true sense, it was conservative of existing rights. It mainly benefited those in possession.

It might have been expected that between 1964 and 1970 a Labour Government would have introduced a very different set of policy measures. But what is most significant, perhaps, is that once in office Labour made no attempt to return to the principles of 1945. It continued the subsidies to owner-occupiers through tax relief, though confronted with evidence that this was regressive. It continued to give differential subsidies for high-rise building and for using expensive city land, though in its 1969 Housing Act it modified the operation of the former. Despite some efforts, it allowed inflation to continue, further strengthening the case for owner-occupation. It made motoring a little more expensive, but built new motorways at an accelerating rate. It designated more new towns, thereby following a trend already set by the Conservatives when they reversed their policies after 1961. It operated industrial location controls more toughly again after a period of laxity in the 1950s and introduced office location controls to accompany them. In two respects it made really fundamental innovations. It introduced the Land Commission, which might have fundamentally altered the balance of rural–urban power by acting as a strong development agency, able and willing to fight counties which were reluctant to allow development. And to the same end, it embarked on a fundamental local government reform with the avowed aim of uniting city and country. Any reform that did this was bound to represent a fundamental shift of power towards the urban interests that constitute 80% of the population. But the particular form suggested by the Redcliffe-Maud Commission in 1969, and accepted by the Labour Government in 1970, was the most radical version: unitary authorities extending over the countryside, with no lower tier to represent more local interests, would shift the weight decidedly to the cities. What is finally interesting about these reforms, though, is that they were meant to achieve quite different objectives from those of the 1945 system. The mixed economy, and the market in land, were accepted. The real intention of the Labour reforms was to make that development run faster and more smoothly.

By the return of the Conservative Government in 1970, the fate of the Land Commission was sealed; and the different formula for local government reform, unveiled in February 1971, naturally gave a greater weight to the counties and the rural areas. Thus, a quarter of a century after World War II, the pattern of the urban growth in Britain was still governed fundamentally by the uneasy compromise secured in the late 1940s and early 1950s. The pure vision of comprehensively planned communities, represented by men like Abercrombie and Osborn, was allowed to survive as a minority movement. But the great bulk of development would still continue to be carried through as a result of interaction between developers, buyers and planners; and in this complex process, the negative element of conservation would continue to dominate. The essential contradiction — whereby a whole bundle of policies worked to encourage development, while another worked to inhibit it — would be allowed to remain, unresolved.

These contradictions have been evident in almost all places and at almost all times in post-war Britain. But there has been one significant set of exceptions to the prevailing pattern. The contradictions are not, for a variety of reasons, evident in Britain's new towns. There was no contradiction between declining space standards and a desire for owner-occupation because owner-occupation is only at minimal levels in the new towns. There may have been a conflict between the city and rural interests before designation of the new town, but after such designation such conflict was reduced to a minimum by placing effective power firmly in the hands of a development corporation with plenty of money to spend, controlled by non-elected appointees. The new towns may have their traffic problems, but they have already shown that they are much more capable of accommodating the private motor vehicle than historic towns and cities. In matters of industrial location the new towns have achieved a signal success in achieving a balance between the levels of employment and population.

The contrast between the trends in the new towns and the pattern of growth elsewhere is mostly the product of the differences in the role of the 'planner'. In most parts of Britain the planner has played a rather negative role, merely responding to the initiative of others. The planner may have envisaged castles and cottages on his drawing board, but when it came to the crunch the outcome was determined largely by what private developers were able to sell, giving the fact of land shortage and high land prices. Only in the new towns was the planner free of this constraint. He himself was able to design and build neighbourhoods, shopping centres, factory estates, blocks of offices, parks and playgrounds. He has been able to experiment with new forms of housing and road layouts, and to use generous landscaping. Though he has been faced with cost constraints too, in the form of housing-cost yardsticks, they are not of the same order as those which have faced the private suburban builder. And this explains much of the difference in quality between the average new-town layout and the average suburban layout. Had land been more freely available, had its price not been bid up by artificial scarcity, the private builder too would have had freedom to experiment.

## Epilogue: future problems and prospects

There are, then, important doubts about the machinery of planning at every main level: regional, subregional, and local implementation. But even if the machinery were improved, it would not guarantee the result. From our study, two important conclusions have emerged for the future about the working methods of planning.

The first concerns the need for planning to be much more flexible and responsive to changes in the external environment it seeks to control. Nothing emerged more clearly from our study than the fact that too often between 1945 and 1970, the whole planning machinery failed to cope with changing events at anything like the necessary speed. Here a more sensitive monitoring and information service is needed. But it should not merely record and act on events quickly; it should seek more actively to predict them, even over substantial periods into the future. Long-range technological and social forecasting, which has been adopted by a number of leading industrial corporations during the 1960s, is still very little employed in urban and regional planning. As a result most local authority planners have only a very dim idea of the world they may be planning for twenty-five years hence — only five years beyond the typical time horizon of their development plans.

If this is remedied, an important part of the predictive process will be prediction of the social environment: the world of those who are planned. The planner will need to know not merely about the numbers of people to be planned for in the future and their level of material affluence, but about their views of the world. Indeed, without this knowledge he will not even be able to make the mechanical projections successfully. Thus, basic population projections — one of the most important items for the planner — have been notoriously unreliable in the past because they failed to take account of changing popular attitudes to family-building. And, to give another example, it may prove very misleading to assume that as people become richer in the future they will own more cars, if the rising number of cars happen to produce a mass aversion to them and a new emphasis on public transport.

The second point proves to be closely related. It is the lack of, and the urgent need for, an explicit element in planning procedures for measuring the welfare impacts on different sections of the population. Some groups in our study proved to gain a great deal from planning policies and associated policies in related fields such as housing and transport. Other groups have lost; but the balance of gains and losses has often been surprisingly inequitable. There is an urgent need for procedures at all levels to incorporate an evaluation, not merely of aggregate benefit against cost to society as a whole, but of specific gains and losses to different groups. This incidentally will demand that physical plans be increasingly integrated with general plans for social development at both central and local government level.

In suggesting this, there is not intended to be any implication that there is any easy way of quantifying costs and benefits to different sections of the community. Real differences of opinion must occur on such matters as the weight to be put on the conservation of land versus the need to provide adequate space standards; or the price to be paid in resources for the preservation and enhancement of amenity; or the value to be given to the interests of the present generation against the interests of future generations. These will remain, in part, subjective qualities, which can be resolved only through political processes. But insofar as these processes are bound to be imperfect because of apathy and ignorance on one hand, strong and well organized pressure on the other, there is a strong case for not relying simply on public participation to secure equity; the result, paradoxically, might be more inequitable than paternalistic planning, even if it were nominally more democratic. Instead, or additionally, planners should try to improve survey methods so as to elucidate preference patterns even among those groups which do not prove vocal or well organized. This is particularly important, for many of the environmental qualities which planning tries to achieve are by definition not available now and so are difficult to conceive of, especially for disadvantaged sections of the population.

## References

Expert Committee on Compensation and Betterment 1942 *Final Report Cmnd 6386* (London: HMSO)

Labour Party 1970 *Regional Planning Policy* (London: The Labour Party) pp 58–63

Nevitt A A 1966 *Housing Taxation and Subsidies* (London: Nelson) pp 54–169

Norton N 1970 Hop in a car, not on a bus *The Guardian* 17 August

The Open Group 1969 *Social Reform in A Centrifugal Society* (London: New Science Publications) pp 24–6

# Section V
# Public Policy and State Intervention

**Introduction**

The focus of much recent research in urban studies on State interventions affecting consumption processes and the built environment is only partially captured by the debate about urban planning. While the planning system in advanced industrial societies has always been presented as the central mechanism regulating urban problems, State intervention in fact occurs across a much wider field of public policy-making. In housing, transport, education and countless other policy areas both central and decentralized State agencies and institutions affect the space-economy and socio-cultural systems guiding urban development in multiple ways. Relatively few of these policy areas are explicitly seen as 'urban' or integrated into the structure of urban planning in any effective way. Instead policies change fairly independently in response to socio-economic and political pressures specific to each policy area. As a result the impact of largely unplanned, unpredicted or implicit 'urban' policies on the urban system often dwarfs or seems to contradict the achievements of 'planned' interventions.

The origins of such policy conflicts lie partly in the complex internal divisions within the extended State apparatus in advanced industrial societies, and partly in wider variations in the distribution of social and political power. Thus in most countries the direct provision of public services or the management of State investment in the built environment are the responsibility of agencies quite separate from planning staffs. Frequently these 'line' agencies have large staffs and budgets and jealously guard their autonomy from control by planning departments, which are typically small 'staff' departments. At the same time, the level of public support for planning forms of intervention is typically greater in market societies than for direct State intervention to produce services or investments.

Thus in Britain, France or West Germany, for example, planning is seen as legitimate regulation of market processes in the 'public interest', whereas State service provision or investment activities in the 'urban' area are far more controversial, appearing as attempts to displace market processes which are regarded as illegitimate by groups and parties politically aligned to the right. Where State agencies do embark on such directly interventionist policies they often become the focus of party political conflicts and require constant inputs of political support from favourably inclined social and political groupings if they are to survive. Equally, problems or inefficiencies can accumulate in interventionist programmes, often because the programmes are set up in ways which make their failure almost inevitable. Where this occurs then these policies have suffered from a combination of falling public support and withdrawals of political confidence and of funding.

Clearly no systematic survey of the whole panoply of State policies with significant 'urban' implications can be undertaken here. Instead the extracts in this section are concentrated on two policy areas which would generally be seen as directly affecting urban development in all approaches to contemporary urban studies, namely the provision of housing by State agencies and public investment in highways. Both of these have key implications for patterns of consumption, forming important elements of 'collective consumption' processes in most industrialized societies. Equally, both public housing and highway investment involve substantial State investment in the built environment. This investment can be used as an economic regulator increasing or decreasing demand levels in the private sector, or it can be seen as complementing investment in private production processes, or it can be used as part of an attempt to raise profit levels in the privately run industries producing these infrastructures on the ground. Within these multiple roles of housing and transport interventions, the extracts given below explore some central themes and lines of debate which have a much more general relevance across the whole field of urban policy-making.

The first reading by *Dickens* is a Marxist attempt to reinterpret a key reformist intervention in triggering the growth of public housing in Britain, namely the introduction of rent controls during World War I. Dickens' target in this extract is the liberal view of this policy change as indicating the openness of the British political system to political pressure from working-class people, and of Parliament's readiness to legislate for the remedy of legitimate grievances even when this involved acting against the interests of politically powerful groups such as private landlords and property interests. In contrast to this view Dickens argues that rent controls were never designed as a major policy change; they were an *ad hoc* expedient picked up by a somewhat weakly placed wartime coalition government as a means of damping down serious industrial unrest threatening to the war effort. Far from being a famous working-class victory, he argues that the detailed framing of the policy change was intended to minimize the damage done to powerful property interests. In this 'revisionist' account, Dickens seeks to reduce the plausibility of the liberal view of the State as a neutral apparatus standing apart from control by any one social group and capable of arbitrating social conflicts in an unbiased, open or democratic way. From his perspective, the State in a capitalist society will only intervene in market processes either where such intervention is urgently necessary to stave off social disorder and legitimate the social system, or where a policy change is broadly in line with the interests of owners of capital anyway.

The dispute implicit in Dickens' article between a liberal pluralist view of public policy-making and a Marxist perspective is taken up in an explicit form in *Dunleavy*'s paper on urban protest movements. Both views are set out in summary form, and their different approaches to analyzing instances of protest movements springing up around urban issues are examined. Dunleavy sees some striking and incongruous parallels between these apparently widely separated views, pointing out that both approaches seem to concentrate on explaining protest movements' success or failure in terms of the characteristics of the movements themselves. In contrast, he argues that the responses which government agencies make to social unrest can be a critical influence on the effectiveness of citizens' political mobilization and activity. Thus the most important focus of analysis should be the *interaction* between protestors and State agencies, with government policy changes potentially having a major impact on, and being directed towards affecting, the survival and success of citizens' movements. Dunleavy also argues that a properly selected case study of protest activity can be used to explore empirically the existence of power relations between State agencies and citizens which have the routine or normal effect of inducing apathy or quiescence amongst relatively powerless social groups. Both these points are then illustrated empirically by a case study of a particular protest movement in London where people being moved out of slums into public housing tried to acquire a say in the kind of accommodation into which they would be rehoused. The study sees this as a strategic example to focus on because it was so very unusual in Britain at this period (the late 1960s). Although public housing standards in terms of the kinds of building being provided had been worsening for years and were unpopular, the structure of the public housing programme meant that local housing authorities, their professional staffs and the industrial firms building the housing effectively monopolized decisions over public housing tenants' future accommodation. The case study explores what happened when this monopoly of decision-making power broke down in one local authority as a result of the physical collapse of a type of tower block of flats which formed the bulk of this housing authority's programme.

Case studies of short-term policy changes of this kind, however, never provide a complete picture of the policy process. Both Dickens and Dunleavy essentially concentrate on single decisions. An alternative approach followed in the last two readings in this section is to explore the overall development of public policy on an issue over a reasonably long time period encompassing a large number of separate decisions. By standing back and endeavouring to see the picture as a whole, it is often possible to discern large-scale or macro-issues which could otherwise pass unnoticed.

The paper by *Meehan* is an ambitious attempt to describe the development of public housing policies in the USA from their inception in the 1930s to the present day. Meehan concentrates on the national-level legislation enacted by Congress and its implications for local housing authorities and their tenants. His fundamental argument is that public housing in the USA is an experiment that has never been tried. Rather the terms on which policy programmes were set up were so restrictive and irrational that the policies were 'forseeably doomed', or bound to fail. Meehan thus rejects the predominant North American view that public housing programmes have demonstrated the inherent inefficiency or lack of responsiveness to consumers' needs built into any attempts by the State to substitute its activities for those of private market processes. Instead he argues that policies of direct State provision of housing have been hamstrung from the beginning by the 'programmed failure' built into public housing legislation by private property and other interests, and developed to its fullest extent in the 'privatized' public housing programmes which have predominated since the mid-1960s.

Finally, the paper by *Painter* explores another kind of distortion which liberal political scientists have seen as of key importance in influencing policy outputs, the structuring of substantive policy change by the internal organization of government. Painter's focus is on the only part of British post-war transport policy under the direct control of a central government department, namely the provision of a national system of major highways connecting the major conurbations and cities. He analyzes policy change in this area in terms of the creation of a largely self-contained 'roads policy sector' within central government, single mindedly focused on roads policy, controlling all aspects of programme development, in a favourable priority position in terms of funding and political support, and capable of insulating itself from external threats by technical criteria of programme evaluation and long-run, incremental commitments to progressively more highway provision. Painter argues that once policy was defined in a particular way, a whole range of institutional and organizational procedures were built up around it, locking the development of State activity into a path which could only be changed with difficulty. In the case of highways policy, many different groups arguing for a reconceptualization of transport policy on more integrated lines had to press for nearly ten years to achieve any change of direction. Even then their influence would have remained problematic had not this pressure on government coincided with a period of worsening public expenditure constraints and with the accumulation of 'slippage' problems in highway programmes themselves.

It will be apparent from these readings that there is, as yet, no integrated field of urban policy studies in social science. Some key general lines of interpretation and debate have emerged, especially between liberal political science or pluralist approaches and Marxist approaches. An inventory of the sources of policy influences on urban issues has begun to be drawn up, ranging from State organizational arrangements; through different modes of policy definition, justification and analysis; including overall patterns of socioeconomic and political power; up to the structural imperatives governing the development of State policies discerned by some neo-Marxist writers. These readings give quite a fair picture of the state of the art in urban policy analysis at the start of the 1980s. More significant analytic gains must await more developed work in the field. But with such gains may come a potential for improved policy formulation via improved technical advice to policy-makers or via more successful political practice by citizens, groups or classes presently excluded from political influence.

**Patrick Dunleavy**

# 20

# A Case Study of Reformist State Intervention: the Introduction of Rent Control Policies in Britain, 1915

*by P Dickens*

Some impression of the political and economic complexities surrounding State housing policy and the problems of its over-simplification can be gained by examining the circumstances surrounding a critical piece of British legislation, which is often promoted as one of the first principal victories for working-class pressure: the introduction of rent control with the Rent Restriction Act of 1915. In this war period certain urban areas came to possess a particular economic significance as centres for the production of arms. Birmingham and the Midlands comprised one such area, but what emerged as the most significant, in terms of its influence on government policy, was Glasgow — this being one of the principal centres for the manufacturing of shells and ships. As armaments production was stepped up, large numbers of workers flooded into the city and, with its long-standing traditions and inheritance of industrial militancy and bad housing, Glasgow became a focus for government concern. Landlords, finding themselves in a monopoly situation with little prospect of any significant additions to the housing stock, were able in certain areas (particularly Govan and Fairfields) to raise their rents by up to 23%.[1] These rent rises in turn caused certain key workers to threaten strike action, producing a rare combination of industrial and housing militancy, which influenced the government's legislation limiting rents to those existing before the war. On the face of it, therefore, this represented a significant advance and since from this time onwards rent control in varying forms became a per-

manent feature of British State housing policy, this also appears to have been a long-term gain.

But on closer inspection, and in relation to ongoing political events in the industrial sphere, this gain appears much less significant for a number of reasons. To return to 1915, some months before the above events, Lloyd George during a public speech pilloried a group of striking Glasgow workers. This strike largely involved skilled engineers fearing the 'dilution' of labour arising from the introduction of unskilled workers and the erosion of wage rates in comparison with those of other sections of the labour movement. It caused the Prime Minister to accuse the militants of sabotaging the war effort. He further suggested, more insultingly, that even when fully employed, these workers were so consistently drunk that they were incapable of working at the intensity that the rest of the national community deserved. In the event, the strike rapidly collapsed, partly because of whipped-up popular pressure but largely because of a sequence of legislation aimed at ensuring a regular flow of production throughout the war. The Defence of the Realm Act and the 'Treasury Agreements' secured unprecedented powers of State intervention in industry, the renunciation by the unions of the strike weapon throughout the war (this, despite intense suspicion by the militant Left), and vague (and short-lived) promises of worker participation in industry. The important point, however, from our viewpoint is that Lloyd George's attack on the strikers continued to rankle with the militant union leaders in Glasgow, and there still existed the threat that they would disrupt production. In these circumstances it becomes clear that Lloyd George's support for rent control should in large part be inter-

Source: Dickens P 1978 Social change, housing and the State: some aspects of class fragmentation and incorporation *Centre for Environmental Studies Conference on Urban Change and Conflict 1977* ed. M Harloe (London: Centre for Environmental Studies) pp 341–51

preted as appeasement of these workers, the central objective being to ensure a continued flow of industrial output.

Well before the industrial strike associated with the rents issues in Glasgow (which lasted incidentally for only one day), Lloyd George had recognized the potential threat: Chamberlain had corresponded with him over the matter in Birmingham several months earlier. But the importance of rent control, its potential for damping down industrial unrest, was brought to his attention by Lord St. David's on 9 October 1915, the principal message being that the Government should try as far as possible to ensure a wide spread for its support:

You have had lately to pitch into the Trade Unionists a good deal and to point out ways in which they and their rulers have been damaging the National Cause. They have put up with it from you in a way they would not have done from anybody else because of you championing them on so many previous occasions. It is most important, however, at the present time not only that you should be ostentatiously fair and impartial. Rents have been greatly raised for cheap houses and workmen's dwellings in different parts of the country and there have been strikes of rent payers here and there. Could you make an occasion to speak against this or write a letter against it? You understand that it is your own position and power for good that I am anxious about. Now and again you have accepted a little hint from me and I throw out this one which is, in my judgement, well worth thinking about from the position of your highest efficiency in your present office.

The Prime Minister replied two days later:

You are quite right. It is an admirable suggestion. I am collecting all the material, and shall say something about it.

Nevertheless, if the Government was alive to the threat involved some time *before* the Glasgow rent strikes had begun to be allied to industrial action and if rent control can be largely interpreted as a sop to industrial militancy, there seems little doubt that the events in Glasgow did at least accelerate Cabinet discussions and ensure, despite intense disapproval from certain Conservative politicians, a swift passage through Parliament. Rent strikes involving 20 000 households were occurring while Cabinet discussions were taking place, and at the same time certain landlords were attempting to acquire their rents through the legal stopping of wages — a procedure that led up to 15 000 workers to threaten the withdrawal of their labour. It is instructive also to note that the idea of rent control had the support of industrial capital in Glasgow. The managing director of Harland and Wolf's Govan shipyard and the management of Fairfields were fully behind the strikers' cause, having every interest in the stability of rents and

therefore of wage demands. As Engels had pointed out some thirty years earlier in *The Housing Question* (1887) any gains made in the form of reduced rents would be short-term, resulting eventually in lower wages. It is understandable that the Glasgow employers should have backed a freeze in rent levels if this meant both a freeze in wage levels and continued production. This period, incidentally, was one of profiteering on an heroic scale for industries associated with war work. Shipbuilders' profits in 1916, for example, were three times the average of the five pre-war years.

When T McKinnon Wood (Secretary for Scotland) raised the rent control issue in a memorandum to the Cabinet, it was certainly the threat to production and social order caused by the rents issue that he stressed:

I understand that the Minister of Munitions attaches considerable importance to the agitation as contributing to the unrest which exists through the Clyde districts where work of great magnitude and vital importance in shipbuilding and the production of munitions is being carried on, and where labour difficulties have caused him much anxiety. As my colleagues are aware, there is also a considerable agitation in London and other parts of England.

Wood finished his memorandum on a somewhat sinister note, stressing the urgency of the situation:

The agitation is growing, and I think it is necessary that a prompt decision should be taken by the Government, otherwise there are signs that the demands for interference will become more clamant and will expand in scope and character.[2]

In attempting to understand some of the central government concerns at this time, it is instructive to trace the permutations through which this rents legislation went as it was rushed through Cabinet discussion and the process of Parliamentary approval: only five weeks passing between Wood's memorandum and the Bill receiving Royal Assent. The overall objective was, as far as possible, to avoid causing further disruptions from interests other than the Glasgow workers. Within six days the legislative proposals before the Cabinet appear to have changed from applying to certain well defined burghs and congested areas in Scotland to applying to all towns with a population of over 100 000 throughout Great Britain. And by the time it had passed through amendments in the House of Commons another fifteen days later, the legislation had again been modified to cover all housing below certain rent levels (varying for towns of different sizes) throughout the whole country.

And it is also instructive to note how different financial interests were dealt with as this legislation (which was intended to restrict mortgage interest rates as well as rents) evolved over this short time. Initially, in response to consultations with a group representing the landlords' interests (The Owners' Association) it was proposed by Walter Long (President of the Local Government Board) to the Cabinet that *all* forms of lending institution should have their mortgage rates restricted, in order that the landlords should not unduly suffer. It is clear that he expected trouble from this proposal, as he pointed out to his Cabinet colleagues:

It cannot be pretended that the Bill is other than a purely emergency measure. It is drastic in its proposals and will, I fear, arouse a good deal of opposition in various quarters, particularly from those who are interested in banks, insurance societies, friendly societies, building societies, etc, and it is impossible to make it altogether equitable in its application. It is very difficult to make a forecast now: all I desire to do is warn my colleagues that there may be trouble from those who consider that legislation of this kind is too drastic, and is calculated to interfere with national credit.[3]

And indeed, while the legislation was under preparation, Long had received a deputation from the Building Societies' Associaton. At this stage he told the Association that the Bill would only apply to the areas where 'the agitation' was occurring. He assured them that the Government 'would be careful to avoid anything which would affect great interests like yours, or above all react on the general credit of the country.' In the event Long did not stick to his assurances, giving way it seems to pressure from Cabinet colleagues. This resulted in what the building societies saw as 'the perpetration of the gravest act of injustice that had ever been inflicted by the British Parliament' (Cleary 1965), an act that was to lead to further political contests with parliamentary representatives of the building society movement.

By the time the Bill was brought to the Commons, the proposed legislation had again been modified: a distinction was made between equitable mortgages raised primarily from banks and mortgages from other sources, particularly building societies. In the Bill before Parliament the first type of mortgage was not to be covered, since the Government's fear appears to have been that it could interfere with the banks' liquidity. The processes leading to this change are not clear, but one could perhaps surmise (at the risk of becoming over-conspiratorial) that pressure from the City had been brought to bear on the politicians before the Bill reached Parliament. At all events, by the time Long came to present his proposals to the Commons he was expressing intense concern that banking capital should on no account be threatened:

the whole of our trade and commerce in this country is carried on, not on cash principles but on credit principles — a great deal of it advanced by banking institutions, and if we do anything that strikes at the security of the great banking institutions we might bring the whole edifice down with a run (*Hansard* 1915).

This was an assessment which (judging from the absence of Parliamentary discussion on the topic) seems to have been accepted with equanimity by politicians of all political shades. The dominant needs of banking capital in the formulation of this legislation were highly significant. With the collapse of export markets during the war, the City possessed a crucial role in preserving the balance of payments, and began to maintain a considerable hold over many aspects of government policy. Such policy became based upon the over-riding concern that on no account should free flows of capital be interfered with. As a result, the needs of finance capital became highly interlocked with government policy — even, if necessary, at the expense of other fractions of capital such as industry. This domination of City interests was to be maintained in the immediate post-war period.

To return to the 1915 legislation, the building societies presented themselves, as predicted by Long, unhappy with the legislation when it was presented to the Commons. Aneurin Williams (a Liberal MP representing the building societies' interests in Parliament) foresaw in the debate the possibility of heavy withdrawals, if, for example, a new War Loan were to offer 5% interest compared with the building societies' $3\frac{1}{2}$%. Williams, in asking for an amendment allowing building societies some freedom in raising their rates, foresaw that: 'if the Bill passes in its present form many building societies will be compelled to go into liquidation' (*Hansard* 1915). Yet the mood of the House seems, on this occasion at least, to have been against the building societies. Sir Tudor Walters (a Liberal MP from Sheffield with considerable knowledge of housing finance, who was shortly to become Chairman of the influential Housing Committee) was able to attack the logic behind the building societies' case and at the same time suggest that it was in the building societies' own long-term interests to accept the legislation as it stood. He showed that of the £62 million lent to these organizations, £46 million was invested by shareholders who during this wartime period had no power to retrieve their capital.[4] And, under building society rules, the remaining £16 million need only be repaid to depositors if and when the building societies indicated that it was available. The spectacle, he argued, of building societies raising their interest rates while other less trusted individuals, such as solicitors, kept their rates steady could only damage the philanthropic image of building societies generally held by the public. The solicitors themselves, Tudor Walters suggested (whether in jest is not clear) must have been behind this amendment, in an attempt to ruin the societies' public image. The building societies were therefore not at this stage accorded by Parliament the highly favourable treatment accorded to the banks. Nevertheless, in the event the Building Societies' Association Parliamentary Committee did later secure some amendments to the legislation: the most significant of these being that mortgages repayable by monthly instalments over a period of more than ten years could be excluded. In addition the Association received the reassurance from the Minister that 'It will be our duty, if building societies find that special difficulties are created to do what we can to assist them'.

We can also gain a greater impression of the true significance of the 1915 legislation if it is placed in some historical perspective. The Act, in the event, turned out to be a decisive blow for the landlord, partly because of wartime inflation, which raised the costs of house building to very high levels. But it is important to point out that the legislation was only confirming a long-term historical trend: the State became involved in controlling rents only at a time when it was already becoming evident that the private enterprise renting form of tenure was not providing housing of a standard required by public health legislation. The 1915 Act, though important in symbolizing and accelerating the demise of the landlord, was preceded by a less celebrated piece of State intervention (less celebrated, that is, in relation to the housing rather than the land question), the Finance Act of 1909, which had already severely hit investment in working-class housing. Thus we are involved here in a familiar sequence of events in which each piece of State intervention generates new contradictions and new demands.

This kind of sequence becomes clear if we look forward beyond the 1915 legislation. The increasing unprofitability of working-class housing led many investors to cut their losses and sell to occupying tenants, a tendency which began to generate increasing business for the building societies from World War I onwards. In addition, one can assume (although this is difficult to quantify) that the small-scale investor who in the nineteenth and early twentieth centuries would have owned one or two houses became, as this kind of investment became less worthwhile, the investor in the up-and-coming building societies. And the rise of the building societies was shortly to have considerable influence on the evolution of State housing policy.

In addition, the control of rents generated a new set of opposing pressures on government policy. It is clear from contemporary reports and parliamentary debates around the time of the World War I that the original long-term aim was to return to some kind of free-market solution to the housing problem and that rent control and State-provided housing were initially seen as short-term measures for overcoming the difficulties and lack of building associated with the wartime emergency. On the other hand rent control, once introduced, was politically impossible to abandon and this meant that the Government was faced with an insoluble conflict of economic and political concerns: a conflict which eventually resulted in long-term desires for a return to a completely free market being dropped and the State assuming at least some degree of commitment to providing housing.

To summarize, this 1915 rent legislation (which is in serious danger of achieving heroic status in the annals of housing struggles) acquires, when seen in its wider historical and social context, somewhat less significance. Devised in large part as a sop to industrial militants and as a means of damping down industrial agitation and pressing on with the war effort, it was designed in such a way as to cause minimal disturbances or harm to existing social and economic interests (particularly the interests of banking and industrial capital) while at the same time appearing to be a significant victory for those objecting to high rents. It was introduced (as was State housing provision) only when it was becoming clear that private enterprise in its current form (largely because of previous State intervention) was not capable of providing decent housing. The legislation generated a series of new contradictions, new demands for State intervention and new requests for private enterprise (mediated by the State) to tackle the problem.

### Notes

[1] This paragraph and later accounts of Lloyd George's early appreciation of the significance of the rents issue draw upon McClean (1972).

[2] See Proceedings of the Cabinet PRO CAB 37/137/29.

[3] See Proceedings of the Cabinet PRO CAB 37/183/3.

[4] This information on the state of building society funds is drawn from Tudor Walter's speech in the House of Commons. Presumably the restrictions on withdrawals were a temporary measure to stabilize building society funds during the war.

# References

Cleary E 1965 *The Building Society Movement* (London: Elek) p 172

Engels F 1887 *The Housing Question* (Moscow: Progress Publishers Edition, 1975)

*Hansard* 1915 **LXXVI** House of Commons Debates, Increase of Rent (war restrictions) Bill 1915

McClean 1972 The labour movement in Clydeside politics, 1914–22 *PhD Thesis*, Oxford University

# 21

# Protest and Quiescence in Urban Politics: a Critique of Pluralist and Structuralist Marxist Views

*by P Dunleavy*

This paper begins by examining two theoretical views of urban protest movements — a liberal political science view, pluralism, and a more recent structuralist Marxist view. It then sets out to explore how well these approaches operate when applied to the analysis of a particular instance of urban protest in the late 1960s in London. The paper's central theme is that, despite the wide theoretical differences between pluralist and structuralist analyses, their views of the origins of protest failures are incongruously similar in failing to allow for the effect of State agencies' responses to protest on protest movements.

## Pluralists on protest

Pluralist accounts of urban politics have three central themes. The first is a view of the urban polity as a weak unit lacking in any developed ideology or particularly separate identity, operating in an environment of strong external influences and controlled by politicians who concentrate overwhelmingly on building and maintaining an electoral majority. Such a polity, they argue, will be very responsive to pressures from the community. Secondly the key concept in understanding politics is that of interests, by which pluralists mean individuals' subjectively defined policy preferences. The third element in this schema is a behavioural hypothesis, that political activity is determined by people's interests:

Because of differences in objective situations, few decisions of government affect citizens generally and uniformly. Most decisions have strong and immediate consequences for only a relatively small part of the population and at best small or delayed consequences for the rest. By and large only those citizens who expect the decision to have important and immediate consequences for themselves, or for those with whom they feel strongly identified, try to influence the outcome (Dahl 1961, p 927).

Since the people who become politically active are those with an interest in the decision, and since the weak polity is responsive to external pressures, the outcome will faithfully reflect the balance of interests in a community.

For pluralists protest is primarily important as a means by which minorities can demonstrate differential preference intensities, the salience of particular issues and outcomes for them. According to Dahl 'the normal political process' is

one in which there is a high probability that an *active* and *legitimate* group in the population can make itself heard effectively at some crucial stage in the process of decision by which I mean more than the simple fact that it makes a noise; I mean that one or more officials are not only ready to listen to the noise but expect to suffer in some significant way if they do not placate the group, its leaders or its most vociferous members (Dahl 1956, p 145). (emphasis added)

Dahl's description of the mechanism by which the policy process becomes 'the steady appeasement of relatively small groups' has been widely endorsed. Studies of participation have normally accorded considerable importance to membership of voluntary associations, which play a key role in pluralist theory as safeguards against the coming of mass society and as training grounds for political activists. According to Almond and Verba (1963) the availability of group-forming strategies and participatory modes of action contributes to the creation of a democratic political

Source: Dunleavy P 1977 *International Journal of Urban and Regional Research* **1** 193–218.

culture and to individuals' feelings of self-efficacy, which in turn enhances the resources of the ordinary citizen by making group-forming more likely. The mainstream of American post-pluralist literature is best represented by Lipsky, who views protest as pre-eminently a power resource of 'relatively powerless groups.' He defines protest behaviourally as a strategy 'characterized by showmanship or display of an unconventional nature' designed to activate the 'reference publics' of 'target groups' and thus an *indirect* means of generating political power (although his analysis does bring out the low 'exchange value' of protest as a power resource) (Lipsky 1970).

## Structuralists on protest

In structuralist urban sociology many social changes are seen as being responses of various kinds to pressure from protest groups, the most developed of which are called social movements. According to Castells 'the study of urban politics can be broken down into two analytical fields, fields which in reality are indissolubly linked; urban planning, in its various forms, and urban social movements' (Castells 1976b, p 149). What determines the development of protest groups and their success or failure are the contradictions (social problems/conflicts of interest) around which they arise, and the organization of the protest.

The simple existence of voluntary associations or evidence of protest behaviour tells us very little about the political system. We need to characterize protest behaviour as a type of *action* and to establish the meaning of membership in group organizations. The primary focus is thus on the *effects* of urban social movements, the material outcomes which result for those involved. 'The genesis of an organization does not form part of the analysis of social movements, for only its effects are important' (Castells 1976b pp 169–70).

Methodologically Castells identifies four stages in the study of urban social movements. First the analyst must identify the *stakes* involved, by which structuralists mean the objective (class) interests. These are then coded in terms of a sophisticated model of urban and social structures. (*The urban system* is viewed in terms of four basic categories, consumption 'which expresses the reproduction of labour power', production, exchange and management of urban system. *Urban practices* are also related to 'the system of economic, political and ideological places of *the social structure* and

to the different relations between these places which define the systems from an internal point of view' (Castells 1976b, p 160). Practices are related to *social organization* in terms of three dimensions: ecological forms, social stratification and the organizational system.) When the stakes have been identified and coded the analyst must identify the social groups intervening in relation to each stake and code them using the same concepts. Thirdly, the organizations involved are then characterized and their articulation to the system of actors determined. Finally the analyst must 'proceed to a concrete analysis of the situation which at the same time will amount to a demonstration of a law, in so far as the situation realizes the law by being made intelligible through the inter-relating of the real elements subjected to our theoretical coding' (Castells 1976b, p 171).

Empirical studies within this framework have provided fascinating analyses of the micro-politics of urban systems and produced fairly strong hypotheses directly related to the theoretical structure. It is suggested, for example, that protest movements based on economic contradictions will be more likely to secure fundamental changes of the system; that protest which is worker-dominated is more likely to be confrontational in tactics and goals, while protest movements with a social base cutting across class lines will tend to acquire middle-class leaderships and be integrative or co-optive in function; that protest organizations to be effective need to take what Castells calls 'a correct line' between fragmented failure to link contradictions (reformist ideology) and 'fusing them together in a single totalizing opposition (revolutionist utopia)' (Castells 1976b, p 171).

## Comparing pluralist and structuralist approaches

Both pluralist and structuralist analyses make assumptions about decision-makers' responses to protest rather than studying them directly, and yet both ascribe a key role to protest in securing changes in urban conditions and policies.

Parenti (1972) has convincingly argued that pluralists *assume* democratic public spirited values amongst decision-makers and that their apparent reliance on political mechanisms to secure responsiveness is illusory. Thus the possibility of indirect influence being exerted by élite groups on elected officials is rejected but a major role is given to indirect influence from the mass of ordinary electors. Combined with the image of a weak polity this leads to the ascription of a decisive role in securing changes of public policy to 'public opinion', the anticipated prospect of electoral defeat or damage and to popularity generally. Similarly protest movements are almost always ascribed influence when they attempt to influence policy, although some pluralists down-grade protest and stress instead the ability of individuals or groups to secure changes through the conventional channels of consultation such as the party system, the interest group process and the redress of grievances by elected representatives.

Structuralists have a similarly one-directional view of influences on social change, although they reach it from radically different assumptions and maintain it for very different reasons. Basically the literature assumes that decision-makers will make no concessions unless forced by protest movements. This is logical within the structuralist theoretical position when what is in question are fundamental anti-systemic changes posing direct threats to dominant values and interests. But as Pickvance shows, structuralists go beyond this to argue that decision-makers will not make even minor concessions unless forced to. Any social change on an issue where protest occurs, however small the change, is thus seen as being the effect of protest activity. This position reaches absurdity in cases such as Lojkine's analysis of the reasons for heavy investment in the Parisian public transport network around 1959, when the French state is alleged to have 'anticipated protest which was to emerge ten years later but which did not exist at the time' (Pickvance 1976, see also Lojkine 1972). It is difficult to imagine the most ambitious pluralist claiming this much for the rule of anticipated reactions.

The non-achievement of urban or political effects or of reforms in turn reflects the deficiencies of the protest or urban social movement according to structuralist analysis; for example, it could reflect the pursuit of an incorrect line, or the failure to develop around fundamental contradictions or to link contradictions effectively. At its extreme it offers an explanation of protest failure which is almost identical to pluralist accounts. Olives (1976, p 178) argues: 'Given the presence of organization-type of action and social base, it appears that it is the size of the stake which is decisive in determining whether urban or political effects are achieved or not.' Where the stakes are high mobilization is possible and within certain constraints, protest succeeds; where they are low it fails. Protest failure

thus reflects primarily the relative unimportance of the issues involved. As the pluralists have argued all along the political system faithfully responds to interest-based activity, it is just that the activity needed is a powerful, organized challenge to the system. Most reassuring of all, the analyst can be confident that important issues will generate political action, although not necessarily *appropriate* action.

The assumption of unidirectional influence reduces the study of protest to a very simple stimulus–response model in both the pluralist and structuralist views, but there can be no theoretical basis for making such simplistic judgements about causality, particularly as both groups of writers freely interpret the causal links they posit in terms of power.

I would argue in contrast that an effective study of protest movements should aim to illuminate a fundamental power relationship within the urban system, particularly one usually tending to keep an issue latent. It must be directed towards the verification of a counterfactual of the form — 'in the absence of $A$'s power over him, $B$'s actions would be $y$ rather than his observed behaviour, $x$'. It is in the nature of power that such verification will be possible only in conditions where the power relationship breaks down or is weakened sufficiently for $B$'s actions in the absence of a constraining power relationship to be assessed. But precisely because such a study illuminates the basic structure of the non-protest situation it can provide a valid basis for deductive generalization.

I turn now to a case study in which I shall try to demonstrate how analysis of urban protests can illuminate latent issues and structures of social power which might otherwise remain invisible.

**The context of protest: high-rise housing in Britain**
Over 1.4 million people now live in 450 000 high-rise flats in Britain, all of them built since 1953, mostly at the height of the 1960s public housing drive: 92% of these dwellings were built in large towns and cities, 80% of them in the conurbations and at least 40% in London alone. A statistical analysis of the influences on local authorities' construction policies suggests that they were built particularly in low-amenity inner-city areas exclusively for people in housing need. According to the Department of the Environment, 'probably a majority if not most residents in high-rise flats were rehoused there from slum dwellings' (Adams and Conway 1975). Of the four million people displaced by

clearance between 1953 and 1973 a very large proportion were thus rehoused in high-rise flats.

The expansion of high-rise housing can best be characterized as a boom: at the height of the boom in 1967 over two-thirds of all new public housing in Greater London was provided in high-rise flats. The boom was the result of intense pressure on local authorities by major contractors which centred particularly around the industrialized building campaign of the 1960s, and this pressure was throughout endorsed by national government (Conservative *and* Labour), and by the design professions, especially architects. The 'by-products' of the boom were the growth of industrialized housing, the repetitive use of standard contracts, package deals, negotiated contracts, serial contracts and a very rapid increase in contract sizes. The adoption of high-rise flats defined a public housing market in which the large firms scooped the pool by virtue of their technological sophistication and larger plant. An increasing degree of control over public housing passed to the major firms whose architects by the mid-1960s were designing over 20% of all council housing and nearly 70% of industrialized housing, mainly high-rise flats.

Industrial influence was so successful because, with the help of professional legitimation, high-rise housing could be represented as a *technological shortcut to social change*. Etzioni and Remp (1973) defined this as a 'solution' to a problem which permits economies to be made in resource allocation for managing the problem, or permits the problem to be tackled more directly. High-rise housing, so it was claimed, allowed a direct attack to be mounted on the abject housing conditions of poor inner-city residents. The problem was tackled *in situ* without diverting land resources from rural to urban use, and without any equalization of housing standards across urban areas, in particular without building any working-class housing in middle-class suburbs. Additionally high-rise high-density redevelopment of slum areas avoided the redesign of the extremely conservative planning system and the reorganization of local government which a redistribution of land for housing the inner-city population would have made necessary. It thus entailed the minimum necessary change of the *status quo*. The costs of high-rise dwellings in terms of housing forgone were substantial. High-rise construction costs were 50% more per dwelling and 80% more per square foot of dwelling space than two-storey houses. The extra costs of high-rise flats were met largely by generous State subsidies

and higher rents to council tenants. Of the total cost of high-rise construction of £1000m to £1500m approximately 40% was due to the increased cost of high building.

For the people rehoused from slum dwellings or the housing list, high-rise flats represented an obvious increase in housing amenity, but a massive *decline* in the standards of public housing provision. Sociological studies show that high-rise flats are quite unacceptable accommodation for families with children and are probably inadequate as dwellings for the elderly, yet these groups of households made up the vast majority of residents. For all types of household, coerced rehousing from 'slum' dwellings into high-rise flats typically meant the gain of basic amenities such as decent kitchens, bathrooms and WCs. But it frequently entailed the loss of a garden, a decline in space standards and the transition from a familiar neighbourhood to vast impersonal estates located in decaying neighbourhoods and equipped with inadequate facilities. The social impact of high-rise housing has been disastrous. Over 5% of all high-rise flats (i.e. 20 000 dwellings) have reached such a poor condition that they are officially described as 'difficult to let', even though most blocks are barely ten years old (Hillman 1976). One in five high-rise residents is dissatisfied with his accommodation, most high-rise estates are fairly intensely disliked and the building form appears to reduce dramatically the frequency of contacts between neighbours.

For the inner-city working class the advent of high-rise housing meant the acceptance of an extremely unpopular form of housing (Habrakan 1971). Yet the people concerned almost completely failed to influence policy, to protest or exercise voice options. Their reaction was overwhelmingly quiescence, silence, adaptation. To analyze why this was so we must focus on a particular local setting in which the logic of the relationship between the public housing apparatus and its nominal 'clients' can be explored. What follows is an empirical approximation to an experiment in which the basis of a power relationship is suddenly destroyed and a group of actors suddenly and unexpectedly gain an enhanced measure of autonomy in defining — or attempting to define — their own choices and futures. The case study is radically different from the normal history of rehousing into high-rise flats and it is precisely because it is so atypical that we can use it to analyze the general picture of non-protest and apparent lack of conflict.

## The case study: Newham

The empirical study of protest covered here concerns the London Borough of Newham, situated in London's working-class and industrial East End. A solidly Labour controlled authority, Newham in the mid-1960s embarked on a massive slum-clearance programme to rebuild the area's old and decaying housing stocks. In the situation of high demand on construction industry resources at this time, Newham Council's leadership decided that the only way to get their programme implemented was to involve a large national building contractor using an industrialized system of building. In 1963 the authority committed itself to a 1000-dwelling contract with a company called Taylor Woodrow–Anglian, a key subsidiary of the country's third largest building company. This involved building nine 24-storey high-rise blocks on large clearance areas. Between 1963 and 1968 Newham expanded this commitment to a total of 2200 dwellings, accounting for the bulk of its programme.

In May 1968 the first two blocks of the original nine-block contract were completed and construction of the others was well under way. But five weeks after the new tenants moved in, a small gas explosion on one of the upper floors of the second occupied block (named Ronan Point after Newham's housing vice-chairman) produced a disaster of national significance. The explosion triggered a partial collapse of the block on one side, killing five people and injuring others. Given the prevalence of methods of high-rise industrialized or system building in public housing at this time, the incident generated intense concern. The central government's Ministry of Housing set up a public inquiry into the collapse.

Within Newham itself, the Ronan Point collapse sparked the growth of an urban protest movement amongst residents of the clearance area who were supposed to move into the remaining blocks of the first Taylor Woodrow–Anglian contract. At different points in the neighbourhood, called Beckton, concerned residents independently of each other started collecting signatures for a petition against being rehoused in the system-built high-rise flats. They soon came together and, as one of them described it, 'from then on the whole thing sort of snowballed. People knocked on our doors asking to sign and we decided to form a fighting committee' (*Newham Recorder*, 23 May 1968). This was the first time clearance-area residents had possessed some form of collective organization in Newham. The Beckton protestors presented a petition signed by 700 of the area's 2000 residents to the Newham Council stating, 'under present conditions we will flatly refuse to leave our present slums to enter modern slums.' The Council's chief administrator replied only 'whether the blocks become slums or not will depend on the people who live in them.'

The Council's Labour leadership reacted to the inquiry into Ronan Point by refusing to discuss any of the issues raised by the partial collapse until after the Inquiry reported. This was ostensibly a response to possible legal difficulties in trying to get out of or change their contract commitments with Taylor Woodrow–Anglian, and the possibility of damages being incurred if unfounded criticisms of the system of building were to be made. But it meant that the Council leadership refused to discuss the Beckton protest. A local meeting called by the protestors (now formed into an action committee) was boycotted by all except one Labour ward councillor. The 200 people at the meeting made it obvious that they would not be rehoused into the system-built high-rise flats. 'At one point in the meeting nearly everyone was shouting that they were afraid to go into the tower blocks.'[1] The ward councillor was very evasive in his answers and had a rough ride. He explained the failure to stop work on the blocks by saying 'If the Inquiry finds everything in apple pie order we would lose a lot of money' (in compensation to the builders). After the meeting broke up in disgust the Beckton committee announced that as they were being 'cold-shouldered' by the local Labour party they were turning for help to the Conservatives, who announced their support of the residents' protest. This was an unwise move since the six Tory councillors had little influence and all the Labour councillors, including the discontented backbenchers, now felt free to ignore the group.

Meanwhile Taylor Woodrow-Anglian made rapid progress on the contract. Two months after Ronan Point, the firm completed the third and fourth of the nine blocks, into which the Council moved tenants with a blaze of reassuring publicity. Two weeks later an interim report was issued by the government Inquiry warning that system building methods had been found to contain a tendency to unacceptable weaknesses under certain conditions, and advising local authorities to disconnect gas supplies immediately to all blocks built using these methods. Newham Council still did nothing to halt work on the first nine-block contract, although they did pull out of follow-up orders for high-rise flats, switching to low-rise schemes with the same firm and building method. By the time the final Inquiry was published after six months the fifth and sixth blocks were virtually complete and the foundations of the last three blocks were already in place, making it difficult and expensive to change the building form.

The final Inquiry report destroyed the basis of Newham Council's 'wait-and-see' policy. Sweeping recommendations for new safety standards of construction were proposed plus massive strengthening operations on all system-built blocks erected under previous inadequate building regulations. The report precipitated a major crisis in Newham's housing programme. Over three hundred tenants had to be moved out of the still occupied system-built blocks and the Council's housing and policy committees argued for three months about whether to try to revise their contract with Taylor Woodrow–Anglian, possibly incurring a damages and compensation claim from the firm of 'as much as £1 million.' The Labour Council leadership pressed for completing the original contract, arguing:

We must bear in mind that we have committed the contractors to heavy expenditure, possibly as much as £500 000 and that our responsibility is a dual one. We are legally contracted for the works and cannot throw it out offhand. It is our duty to the homeless and the people on the housing list to go ahead with a scheme in the near future (Councillor W Watts quoted in the *Newham Recorder*, 12 December 1968).

The first argument captured the finance-conscious Conservative and ratepayers' councillors and the completely specious second argument whittled away the dissident Labour backbenchers. Although the decision over the last three blocks concerned the fate of at least 1000 Beckton residents the Newham leadership blankly refused to meet the protest committee and their case was not voiced by any councillors. Eventually (January 1969) Council negotiators narrowed their options to either completing the original contract or reducing the block height to 15 storeys, building additional low-rise flats and paying the firm agreed compensation of £50 000. The Policy Committee observed that under the second option the blocks 'would still be essentially high-rise' (London Borough of Newham 1968, p 279) and voted to complete the original contract, claiming that they could not ask the ratepayers to pay £50 000 for nothing. Reinforced by a three-line Labour whip this view secured a unanimous Council vote of endorsement. No Council member apparently pointed out that the reduction in height to 15 storeys would mean that at least 350 fewer people would have to live in high-rise flats. The retreat before paying out compensation is particularly ironic since this was a very small fraction of the extra costs which Newham incurred by failing to stop work on the blocks between May and November 1968. Strengthening contracts, which had to be given to TW–A since the blocks used their system, eventually cost £1 million (half of which was paid by the government), while rebuilding and strengthening Ronan Point cost a further £300 000.

Shortly after this decision the Council voted Taylor Woodrow–Anglian a further £5 million contract for 1000 system-built low-rise flats in order to try to get the Borough's wrecked housing programme back on target as quickly as possible. To this was soon added £1.3 million of work on strengthening contracts for the finished high-rise flat blocks.

Meanwhile the Council leadership had still refused to have any dealings with the Beckton residents' committee (apart from a brief meeting before the Inquiry report), despite extensive press publicity for their case. In February 1969 *when all the decisions about their housing had already been taken* the committee were told in a letter:

The Chairman of the Housing Committee is of the opinion that very little could be achieved at this stage by meeting your committee. But it is his firm intention at the appropriate time to call a meeting of all the residents' associations concerned to explain the position. It is hoped of course, that your association will be represented at the meeting.

The protest leader pungently described this as 'complete eyewash' since no other associations of clearance-area residents existed and no such meeting was ever called. He told the press: 'Approaching the problem in an orderly and gentlemanly manner is getting us nowhere. If you can't get anywhere peacefully and with commonsense, what is left?' What was left was a steadily escalating series of demonstrations and meetings which were doomed to failure since all the relevant decisions had already been taken. A Conservative councillor sympathetic to the residents asked the Council in March to consider guaranteeing clearance-area families freedom of choice about accommodation. The Housing Committee Chairman peremptorily declared that clearance-area residents would have to take the best dwellings available: 'In asking for a hard and fast assurance Councillor —— is being immature, unrealistic, impractical and obstructive.' At this point members of the protest committee in the public gallery disrupted the proceedings and were thrown out by the police. The Beckton residents' relations with the

Council leadership were henceforth non-existent and they were later publicly denounced by the (supposedly impartial) mayor when opening a new tower block. In May the issue of freedom of choice was raised again as a formal motion but this was voted out by the Council without even reaching the Housing Committee. After this meeting the Beckton committee picket got involved in an alleged assault on one of their ward councillors after he had explained his vote against the motion by saying that they were 'trying to hold the Council up to ransom'. With this extensively publicized incident the local Conservatives, embarrassed at their inability to control the increasing violence of the protest committee's actions and pronouncements, quietly washed their hands of the affair and the improbable alliance came to an end.

During the summer of 1969 the issue went off the boil until in October the Beckton protest committee were at last invited to 'discuss' their rehousing with the Council leadership, only to be told that whatever their views the bulk of the residents would be rehoused in the last three blocks, construction of which would finish in March 1970. The committee left the meeting in an angry mood, declaring 'the big crunch will come when the Council give us new addresses and we refuse to go. This organization was formed specifically to fight an attempt to move us into tower flats and nothing has happened to weaken our determination'. The protestors began planning their reaction to rehousing and mobilizing support to resist the move to the tower blocks, in vain as it turned out since nothing at all happened.

In the event the Council's forecast of rehousing dates was very optimistic and in February 1970 the Council organized two 'goodwill' public meetings, the main purpose of which was to tell the Beckton residents that their rehousing had been postponed for a further six months at least by delays in building the strengthened blocks. In the event these became the first and only occasions on which there was any opportunity for residents to tackle the Council leadership in public. At the first meeting 100 people listened to the Housing Chairman's opening speech in 'tense silence' while the protest committee picketed the meeting.[1] When the first questioner was told by a Housing Department official that virtually all the residents would be rehoused in the system-built high-rise flats, he replied amidst cheers, 'You can take that and stuff it!' The Chairman immediately threatened to close the meeting and was answered by a chorus of 'It's up to you!' Asked

if residents who refused one allocation in a tower block would be given another allocation, he replied 'above derisive cheers': 'The Council do not usually make a second offer if the first is considered sufficient and the refusal reason thought to be inadequate.' After a running fire of bitterly hostile questions one man shouted, 'You claim you're bettering us but you're not. You're nicking space off us — you are going to give us less than we started with. It's a bloody farce!' As the tone of the meeting became very angry and many people began to leave, the Housing Chairman quickly closed the proceedings and the Beckton committee claimed a moral victory for their boycott. The second meeting was even briefer and more embittered.

The protest committee followed up their apparent success by holding a meeting of women members who reiterated their intention to oppose rehousing by force if necessary, and by organizing a picket of the Council at which many councillors refused to accept copies of their statement of objection. But their support had already begun to decline. Many of the residents reacted to the Council's immovable stance at the public meetings with a feeling of despair, and as the rehousing date receded further into the future so the protest movement began to decay. People had already begun to leave the area. Families prepared to move to flats in other parts of the borough began to be rehoused. People ineligible for rehousing left in search of more permanent accommodation and by July those who remained were overwhelmingly anxious to be moved out of the area.

The climax of this process of decay came when the first families to move to houses received their allocations. Two of the protest leaders were offered and accepted houses and the committee fell apart, split by bitter personal animosities. The immediate consequence of the break-up was that the local press coverage of the issue ceased almost at once. In August a member of the defunct committee secured some publicity for the continuing plight of the area's residents who still had no information about when rehousing would begin. The Housing Department promised to hold a meeting to let the residents know firm dates, which was never apparently held. Instead the residents were told individually that they would be rehoused over the next seven months as the tower blocks became ready for occupation. By April 1971 nearly 1000 people had moved into the three blocks, leaving a few families holding out for houses or ground-floor flats in the now empty area, surrounded by vandalized homes and the wreckage of the former community.

For the Beckton residents this result represented an unmistakable, total defeat. They had failed to influence the Council's policy towards them in any significant way and their own organization had collapsed. Attempts by one of the protest committee to set up an organization independent of the Council in the new flats failed and a weak tenants' association with a constitution drawn up by the Council was established. Nor did the Beckton protest have much influence elsewhere in Newham. Late in 1971 the residents in another large clearance area mounted a demonstration to protest about the 'blitz conditions' in which they were forced to live and about their rehousing in flats instead of houses. This petered out very quickly with no discernible effect. 'Normal relations' between the Council and residents in clearance areas had thus been re-established.

## Non-protest on mass housing

The ripples from the Ronan Point collapse spread widely throughout Britain's mass housing apparatus after May 1968. The industrialized building market virtually collapsed within three years, the high-rise boom came to an end and the ideological underpinnings of public housing controlled by élite groups took a severe battering. But only in Newham was the destruction of the image of industrial–professional expertise so complete and the prospect of coerced rehousing so immediate that a protest movement developed.

Even with this basic momentum, the Beckton residents failed to produce a change in their housing. Their movement had no leverage on the power of the public housing apparatus and it could not gain any in the face of repressive responses. Even if the protesters had influenced decision-making they would not have done so because they forced the authority to take notice of them, but because actors in the authority were willing to be influenced. The fact that these actors *did not want to be influenced* is indicative of the strength of the structural forces tying the locality into the general development of public housing policy. The basic tendency of this development, towards the reproduction of an unequal *status quo* with the inner-city working class at the bottom of the pile, was far too strongly entrenched to be capable of alteration either by the protest movement or by the local authority.

Precisely because it was so atypical the Beckton protest illuminates the general picture of latent conflict on mass housing issues in the 1950s and 1960s. It provides objective evidence that in the absence of the normal coercive power relationships between the public housing apparatus and 'slum' residents, the latter would have chosen a different housing future, most basically one controlled by them. Thus non-protest on mass housing issues should be understood as the product of continued domination, politically as well as ideologically, the reflection of a powerless situation constantly reproduced by the routine exercise of power by the public housing apparatus.

## Note
[1] The account of this meeting is taken from the local newspaper.

## References

Adams B and Conway J 1975 The social effects of living off the ground *Department of the Environment Information Paper No 9*

Almond G and Verba S 1963 *The Civic Culture: Political Attitudes and Democracy in Five Nations* (Princeton, NJ: Princeton University Press)

Castells M 1976a Theory and ideology in urban sociology *Urban Sociology: Critical Essays* ed. C G Pickvance (London: Tavistock)

—— 1976b Theoretical propositions for an experimental study of urban social movements *Urban Sociology: Critical Essays* ed. C G Pickvance (London: Tavistock)

—— 1973 *Luttes Urbaines* (Paris: Maspero)

Dahl R A 1956 *A Preface to Democratic Theory* (Chicago: University of Chicago Press)

—— 1958 A critique of the ruling élite model *American Political Science Review* 52 463–9

—— 1961 *Who Governs? Democracy and Power in an American City* (New Haven: Yale University Press)

Etzioni A and Remp R 1973 *Technological Shortcuts to Social Change* (New York: Russell Sage Foundation)

Habrakan N J 1971 *Supports: An Alternative to Mass Housing* (London: Architectural Press)

Hillman J 1976 Faulty Towers *The Guardian* 14 January

Lipsky M 1970 *Protest in City Politics: Housing and the Power of the Poor* (Chicago: Rand McNally)

Lojkine J 1972 *La Politique Urbaine dans la Region Parisienne, 1945–72* (Paris: Mouton)

London Borough of Newham 1968 *Minutes* 58

Olives J 1976 The struggle against urban renewal in the Cité d'Aliarte (Paris) *Urban Sociology: Critical Essays* ed. C G Pickvance (London: Tavistock)

Parenti M 1972 Power and pluralism: a view from the bottom *Journal of Politics* 32 501–30

Pickvance C G 1976 Marxist approaches to the study of urban politics: divergences among some French studies *International Journal of Urban and Regional Research* 1 219–255

# 22 The Rise and Fall of Public Housing in the United States: a Case Study of Programmed Failure in Policy-making

*by E J Meehan*

Like so many other social programmes in the United States, public housing was a product of the 'new deal' era. The basic assumptions and procedures that guided its development and operation were hammered out in the 1930s. The approach that emerged in the Housing Act 1937 and was reproduced in its essentials in the Housing Act of 1949 dominated the public housing programme for the next four decades. Two basic modes of operation were used to supply housing services to low-income families. The original prototype, the 'conventional' public housing programme, was characterized by the use of a public agency to develop, own and operate housing facilities. By the end of 1974, more than 860 000 units of housing, 66% of the total public housing supply, had been developed by this method. Another 230 000 units, about 18% of the total, were produced by a variant known as 'turnkey' housing: these units were developed by private interests and then sold to a public agency, usually a local housing authority.

The Housing Act of 1965 authorized an alternative approach to supplying public housing for the poor, the leasing of privately owned facilities. By the end of 1974, leasing accounted for more than 13% of the total supply, some 169 000 units altogether (table 22.1) (p 208). The changes in operating philosophy introduced by the Housing and Urban Development Act of 1974 foreshadowed a major increase in the role played by leased housing in the overall public housing programme.

Source: Reprinted with permission from *A Decent Home and Environment: Housing Urban America* copyright 1977, Ballinger Publishing Company.

Indeed, if the 1974 act were to have been implemented systematically and rigorously and without major alteration it could have put an end to the traditional form of public housing in a relatively short period.

## Conventional public housing

The public housing programme incorporated into the Housing Act of 1937 was very much a child of the times. The first Roosevelt administration faced a staggering array of problems when it took office in 1933: the economy was stagnant, unemployment was extremely high and public confidence in government and private enterprize alike had been shaken by the stock market collapse and its aftermath. The housing industry was but one of many sectors of the economy in urgent need of assistance. In the cities, large concentrations of aged and dilapidated buildings lacking in amenities and badly needing repairs posed a hazard to the social health of the inhabitants, and, so it was believed at least, a danger to the social health of the wider community. Everywhere there was an acute shortage of decent low-income housing, and overcrowding had been common for years. In the circumstances, it is hardly surprising that the public housing programme was construed as a multipurpose activity, a way of simultaneously reducing the level of unemployment, assisting the beleaguered housing industry, eliminating slums and increasing the supply of cheap and decent housing available to the poor.

Public housing provided a major occasion for breaking the established mode of economic activity. The profit-maximizing production and distribution system then operating responded only to effective demand (desire *plus* capacity to pay); it clearly failed to provide for the housing needs of large segments of the population. That is, it had produced both housing that was too expensive for many, as well as the many for whom the housing was too expensive. How to deal with the situation? Any effort to maintain private ownership and to control the price of housing to the consumer would certainly founder unless production costs were controlled or rents subsidized. Cost control was anathema to owners and building unions alike, and direct cash subsidies, much favoured by real estate interests, were considered far too expensive over the long run, for they tended to over-stimulate demand for an inadequate supply of decent housing and provide unwarranted returns to the owners of poor or marginal quality housing. Public ownership offered a more efficient means of divorcing the cost of housing to the tenant from the cost of development. Its capital could be obtained at preferred rates, taxes avoided and any return to capital as profit retained in the public coffers.

In principle, public ownership of housing facilities enables the government to supply housing according to need rather than capacity to pay, absorbing the difference between rents charged and actual costs, but public ownership alone is not enough. Had the federal government been able to create enough decent housing to supply all of society's needs and then underwrite the cost of operation to a point where the tenant could readily afford the rent actually charged, programme administrators would only have needed to identify the needy and house them as expeditiously as possible. But if the housing had been of poor quality, if the supply had been significantly less than the need, or if operating costs had been inadequately subsidized and could not have been met from tenant rents without imposing serious hardship, the programme could not have operated properly. If the quality had been poor, the supply could not have been marketed or the cost to tenant would have been excessive; if the supply had been inadequate, access would have had to be restricted in some manner, probably by income.

## Development policies

The programme created by the 1937 Housing Act was built on a local–federal arrangement from which the state authorities were virtually excluded once enabling legislation was passed. Over time Congress at a national level tended to concentrate on four basic areas of housing policy, leaving most other matters for administrative decision-makers; the amount of housing authorized for addition to the public housing stock; the cost of developing the facilities; the fiscal arrangements that controlled programme operations; and the conditions of tenancy. Federal administrators exercised a great deal of control over the programme through budgetary oversight, allocation of resources for new developments and auditing and monitoring of everyday operations. Local governments could influence the programme in their areas through the formal cooperative agreement required by statute and less formally through appointment of the governing body that directed the programme. In practice, local governments had veto power over certain critical aspects of development, notably size, location, design and staffing of the facilities. The local housing authority (LHA), under the overall direction of the governing body, had a voice in site selection, choice of architect, design and so on, plus fundamental control over such aspects of daily operations as tenant selection, maintenance and repairs, legal actions against tenants, staffing, and so on. Until the 1970s, the LHAs were given wide latitude in operations by the federal administration; however, central control over details of policy increased substantially after 1969.

The direct costs of public housing to the local community were spelled out in a required federal–local agreement. The LHA was granted tax exemption in return for a payment in lieu of taxes amounting to 10% of gross rent less utility costs. Whether the payment covered the actual cost to the community — particularly during those periods when rents were high and occupancy virtually complete — is much debated and probably not answerable. However, the collective benefit obtained from public housing certainly outweighed any direct cost to the community beyond what was expended before the developments were built.

There were two curious gaps in the federal government's development policies. Firstly, little effort was made to control the quality of the housing produced; secondly, the cost of the site was not limited by any of the statutes. The 1937 Housing Act placed a ceiling on the overall cost of each dwelling unit and on the cost of each room, but these established limits could be adjusted upward by as much as 25% in 'high construction cost' areas, usually the larger cities where powerful construction unions operated. In 1949, limits on overall costs per unit were abolished, leaving cost per room as the sole criterion for controlling development cost. In 1970 Congress adopted a 'prototype' cost base for dwelling units of various sizes and types of construction. Again, the limits could be exceeded by 10% without waiver and another 10% with federal approval. Land costs remained uncontrolled and no qualitative construction controls were specified in the statutes.

The effect of weak control over development costs and housing quality is readily foreseen. Site costs were often unconscionable, particularly in the 1950s when the rate of development was high and projects were deliberately located in cleared slum areas — and forced to absorb the clearing costs. In the long run, the economic inefficiencies generated by land speculation were probably less important than the tendency for LHAs to pay premium prices for apartments so shoddily built that a choir of angels could not abide in them regularly without producing serious disrepair. Poor quality cannot be ascribed to federal miserliness. In St Louis, the housing authority paid for its projects at rates that equalled or exceeded the cost of luxury housing in the suburbs. It did not receive anything approaching *quid pro quo* from the housing industry.

The principle underlying the failure is not hard to find. A profit-maximizing economic system is also performance-minimizing. Unless performance criteria for the end product are specified fully (in which case there is likely to be a major increase in cost), the producer is bound by the rules of the game and perhaps by the 'facts' of economic life to maximize profits within the limits of the contract price. The most accessible source of additional profit is the quality of the product.

*Conditions of tenancy*

The more important congressional policies relating to tenancy in public housing dealt with eligibility for admission and continued occupancy, rent levels, definitions of tenant income and priorities assigned to different classes of applicants. The consistent central concern has been the amount of income that a prospective tenant was allowed to earn and the manner in which excludable income was identified and calculated. Under strong pressure from real estate interests, the prime goal of federal policy was to exclude from public housing anyone with enough income to obtain housing on the private market. Understandably, the real estate interests (i.e. private landlords and developers) sought to keep the income level of public housing tenants at a minimum, and since income limits were partly determined by local rent levels based on information supplied by the real estate industry, they were often quite influential. Over the long run, pressure on tenant income levels, taken in conjunction with other fiscal policies, contributed materially to the aggregation in public housing of a highly dependent population whose incomes were very low in relation to the rest of the community and changed much more slowly than general wage and price levels. Between the early 1950s and the mid-1970s the consumer price index for St Louis more than doubled, for example, but the median family income for public housing tenants rose from about $2400 per year to just over $3700 per year, far less than was needed even to keep pace with inflation.

The procedure used to determine income limits for those admitted to public housing was fairly complex. The LHA went to the local housing market (newspapers, agents, brokers, etc) to determine the current price of decent rental accommodations of various sizes. The income needed to 'afford' such housing was calculated by a five-to-one ratio. That is, if the going price for a three-bedroom apartment was $100 per month, it was assumed that a family with an income of $500 per month could afford it. The maximum allowable income for admission to public housing was 80% of that amount or $400 per month. Federal policy required a '20% gap' between the income sufficient to afford needed housing and the maximum allowable income for admission to the developments. If other factors remained constant, the tenant whose income increased by 25% over admission limits was required to leave public housing. He would then be earning enough to 'afford' the housing he needed on the private market. In principle, the policy was expected to keep everyone with enough income to purchase housing on the private market out of public housing. The lack of qualitative controls over estimates of available housing tended to produce significant over-estimates of the amount of 'decent' housing available and significant under-estimates of the going price. The relative value of the facilities afforded by public housing, even in areas where the stock was badly deteriorated, was attested by the length of housing authority waiting lists.

Once income limits required for admission were satisfied, the various priorities established by Congress or administrative action came into play. Preference was given to war veterans and persons displaced from their homes by public actions, such as slum clearance, by both the 1937 and 1949 acts. In 1954, a hostile Congress limited admission strictly to persons displaced by public action, but the policy proved untenable and was rescinded the following year. The impact of priority assignments probably varied from city to city but seems not to have been very great. Between 1966 and 1973, for example, fewer than 12% of all families entering public housing had been displaced by public action, and only 1.2% were uprooted by either urban renewal or housing development. In 1956 the elderly were given priority in admission and an increase in construction costs of $500 per room was allowed for housing designed specifically for elderly use. That priority was extended to the disabled shortly afterward.

The negative priorities involved in racial segregation were ignored by both Congress and the administrations. When the programme began, developments in many cities were racially segregated and blacks were excluded from white projects, or vice versa, as a matter of course. Formal segregation ended in 1954 when the US Supreme Court refused to overturn a California ruling which held that admission to public housing could not be refused on racial grounds. Within fifteen years, social and economic conditions in many locales combined to resegregate public housing, this time with respect to the rest of the community, by aggregating large numbers of minority group members in the developments.

The rent paid by public housing tenants was linked to family income and not to the amount of space occupied. A large family with little income might have paid less for a five-bedroom apartment than was charged a smaller family with a relatively larger income for an apartment with a single bedroom. Technically, the portion of tenant income paid for rent was not limited, though in practice the administrators tried to maintain rents at minimum levels. As housing authority operating costs soared after 1960, minimum and average rents climbed steadily, far more rapidly than tenant income. By 1969 such cost pressures had created very serious problems for most housing authorities. Some of the very poorest tenants were forced to pay as much as three-quarters of their income for rent, and payments equal to half of gross income were common. That condition, among others, led to the 1969 rent strike in St Louis and to significant disturbances elsewhere. The outcry produced one of the so-called Brooke amendments in 1969 which limited the amount that could be charged a tenant in public housing to 25% of adjusted income — less for persons of very low income. Why this was considered proper is uncertain: by European standards, 25% of income is a very high rent level. In 1974, the rental structure was simplified,

though the base was retained: rent control could vary from 25% of income for persons earning four-fifths of the median income in the area, to 5% of income for persons with very large families or very low incomes (less than half of the area's median income). The act required LHAs to fill at least 30% of their units with families from the very low-income group. However, the act also required each LHA to collect at least 20% of the total income of its tenants as rent.

*Fiscal policies*
The public housing programme created by Congress moved toward financial disaster as inexorably and predictably as any Greek tragedy. The quality and cost policies virtually ensured physical structures of minimal quality. Regulations imposed on tenant income guaranteed a very modest rent yield to the LHA. The self-destructive system was completed by adopting fiscal policies that were foreseeably unworkable and sticking to them for more than thirty years in the face of all evidence. To put the matter as starkly as possible, the federal government undertook to pay all capital (i.e. initial building) costs on public housing as they came due; it guaranteed the mortgage, leaving all other expenses as the responsibility of the local housing authority. The LHA's sole source of income was rent; no operating subsidy whatever was provided by Congress. Had the matter stopped there, the programme could not have survived very long. Economic collapse was rendered more certain by imposing four additional financial burdens on the LHA, to be met out of its rental income alone.

Firstly, utility costs (i.e. heating, lighting and water) were included in the rent charged the tenant. From the tenant's point of view utilities were a free good; rent was unaffected by the amount used. Predictably, utility costs became a major burden on local housing authorities even before prices began to climb in the energy-scarce climate of the 1970s. Secondly, 10% of gross income from rent, less utility costs, had to be turned over to local government each year as payment in lieu of taxes. Thirdly, the amount of reserves that any LHA could accumulate was limited by administrative action to 50% of one year's rent. That very effectively precluded the LHA from building up the funds needed for capital replacement or even for major maintenance such as roof repairs. As the apartments aged, major maintenance was deferred and handled piecemeal. Unfortunately, the amount of damage to a building or area increases directly and exponentially with the time delay between damage and repair. Maintenance deferral is an open invitation to vandalism. Fourthly, for any year in which an LHA 'showed a profit' (that is rental income exceeded expenses plus transfers to reserves), the surplus was used to pay capital costs.

From 1945 until 1953, the federal government actually paid less than 50% of the capital costs of public housing and in the peak years of 1948 and 1949 the LHAs paid nearly 85% of their own capital costs. Having salted the wounds inflicted by its actions, Congress added insult to injury in 1954 by requiring the LHAs to repay 55% of capital costs from rental income. The requirement had little real significance given the steady deterioration in the financial position of LHAs.

The housing programme could have succeeded only if costs, rents and tenant incomes had remained in relatively stable relationship for fairly long time periods. Since economic activity tends to very rapid shifts in costs and prices and the incomes of the poor do not keep pace with rising costs, the logic of the fiscal arrangements guaranteed a cost–income squeeze in any period of rapidly advancing wages and prices.

Even without inflation, the fiscal apparatus could not have succeeded. The LHA's income was a function of the price of housing on the local market and the income of its tenants: expenses depended on the size, quality, durability, design and so on of the developments, the kinds of tenants who occupied the premises and basic trends in the overall economy. There was no reason to suppose that the income needed to operate the developments would have been generated out of the interplay of this set of pressures.

In the private sector, the rental income needed for successful operation of multifamily apartments was calculated by a rule of thumb which stated that roughly one-sixth (17%) of development costs had to be generated each year in rent. Public housing authorities rarely succeeded in obtaining as much as 6% of their investment in annual rentals; in St Louis, the return averaged just over 4% for more than two decades.

## Changing the basic pattern: subsidies and privatization

The operating pattern for public housing established in 1937 continued with only minor changes until the 1960s. Both opponents and supporters of public housing tended towards what might best be called 'mindless

incrementalism' in their approach to policy-making. Instead of reasoned decisions to increase or decrease resources allocated to specific purposes on the basis of careful study of the effects of operation, mindless incrementalism is marked by increases or decreases in allocations that are unrelated to performance or may even be perverse in the light of performance.

In fairness to supporters of public housing, incrementalism was in some degree forced by the opposition; both nationally and locally, hostility to public housing remained widespread, vocal, powerful and implacable. It was difficult just to obtain a simple increase in the number of units authorized in a given fiscal year. Supporters of the programme may well have felt that asking for any policy innovations was too risky. The opposition in Congress and across the country was strong enough to delay enactment of a new housing law from 1945 until 1949, even though the proposed bill had strong Republican support. The public housing provisions of the 1949 Housing Act survived in the House of Representatives by only five votes, and though Congress authorized development of 165 000 units each year for five years in the 1949 act, only 250 000 units were actually funded over the entire decade of the 1950s. As President Eisenhower's administration (1952–60) grew more hostile to governmental intervention in the economic sphere, mere survival of the housing programme became increasingly less certain.

On the other hand, the special circumstances in which public housing operations began in the 1940s made the first generation of developments (authorized in 1937) conspicuously successful in the early years, masking the fact that they were actually living on borrowed time. World War II brought about a rapid shift in population to war-production urban centres and placed enormous pressure on the existing housing supply. Relatively full employment and higher earnings, coupled with special dispensation that allowed the use of public housing by war workers, created a bonanza for local housing authorities not entailed by federal policies. War workers kept the apartments full and incomes far in excess of the levels intended meant high rents and ample incomes for LHAs operating new developments that required little major maintenance or capital replacement. Even after the war ended occupancy remained high, some of the over-income workers remained, and most of the tenant body was employed; hence, rental income remained relatively high until the end of the 1940s. Difficulties began when over-income tenants were forced from the developments, occupancy began to fall and maintenance costs and capital replacement needs began to rise rapidly.

## The changing social environment

While public housing policy remained more or less frozen in its original mould, operating on principles borrowed from the traditional private sector, social change was proceeding rapidly in the housing developments as in the wider community. World War II largely accelerated certain trends established earlier, such as the movement of population from the central city to the suburbs. Persons in the higher-income brackets headed for the suburbs early in the century: federal housing and tax policies after 1945 encouraged blue-collar and white-collar workers to follow suit. New and old industries alike moved to the cheaper land on the city's outskirts as they expanded: a burgeoning trucking industry and expanded highway construction programme hastened the process by increasing accessibility. The erosion of the city's tax base was hastened by urban renewal, transportation construction and subsidies to housing and other construction. Despite the countless billions of dollars poured into central cities in an effort to preserve what was mistakenly identified as 'the city' as a whole, anticipated (private) investment did not materialize. The inner city became increasingly an isolated clump of older business facilities surrounded by a widening belt of deteriorated housing occupied mainly by the very poor, the black and the permanently unemployed. Neighbourhoods previously characterized by long-term residence and stable social behaviour crumbled and fragmented as the elderly died and the younger workers moved to the suburbs seeking homes they could afford, desirable schools and neighbourhoods and physical separation from the expanding inner-city ghetto. The end of the World War II employment boom hardened the differences between the inner city and the suburb. The recession of 1956–7 had a profound impact on most large cities, and for some the recession of 1961–2 was merely a continuation. Lack of employment, particularly for the young, the black and the disadvantaged, increased out-migration of the working-age population, leaving behind a body of residents increasingly dependent on public assistance: the permanently poor, the very young and very old, the disabled and the relatively helpless.

As in other things, public housing developments mirrored the course of events in the wider community. They ceased to be a stop-over for the working poor *en route* to a family-owned dwelling and became a haven for concentrated masses of dependent persons with little possibility of improving their lot through their own efforts. The indicators of their helplessness are classic: real wages that lagged persistently and significantly behind local and national levels; unemployment rates several times the national average; extreme transience in employment; marginal jobs; frequent and often sustained reliance on public assistance; and heavy concentrations of the very young and the very old, usually members of a minority group, often comprising two or three generations of public housing tenants. Tragically, helplessness was actually increasing at the very time when the dependent population was being urged most strongly to entertain rising expectations about the quality of life it lived and the life transformation of the population of public housing undermined the survival capacity of the LHAs in every part of the country. There was a steady decline in the number of employed workers and a steady increase in the number of families wholly or partially dependent on public assistance. Predictably, the amount of income available to each family tended to decline relative to wage and price rates in the wider community. The number of female heads of households increased with a concurrent expansion of the number of relatively undisciplined young persons living on the territory. In border cities such as St Louis, racial segregation based on income became the rule. Public housing became the prime repository for the very poor, the black, the elderly, the female head of household and her children, the unemployed and unemployable. Rents declined, expenses soared and meager reserves were soon expended: the financial position of the LHAs weakened rapidly and seriously.

Local housing authorities were caught in a broad flow of events they could not hope to master, bereft of resources, married perforce to inadequate and relatively inflexible federal policies, harassed by a clientele that desperately needed their services yet increasingly could not afford them. The cities were helpless, caught in the same whirlwind: the state governments for the most part, looked the other way. Declining productivity in the housing industry worsened the LHA situation by increasing building and maintenance costs without improving quality. The real costs of housing began to rise rapidly just as public housing development began expanding. Various contributory factors can be identified: profit-maximizing entrepreneurs used technological improvements to maximize profits, union power forced the inefficient use of expensive labour. It cost more and more simply to maintain the quality of a basic unit of housing; any effort to reduce costs apparently lead to an overall decrease in construction standards. Declining revenues forced maintenance to be deferred, which led to a deterioration in physical conditions, which stimulated vandalism, which further depressed the quality of the housing supply. The end result was too often a ghastly landscape of mutilated buildings, broken glass, empty apartments, abandoned automobiles, litter and garbage, a wasteland hostage to the criminal, vagrant, truant and street gang, a hazard to the passer-by and a nightmare to the resident.

## Subsidies

Money alone could not solve all the problems of the LHAs, but without significant additions to their income any efforts to improve design, tenant selection, management, maintenance, or other aspects of performance were futile. While the federal government allowed periodic increases in construction costs, modest rent increases and in rare cases allocated special funds for refurbishing some of the more conspicuous disasters, no regular operating subsidy of any kind was available until 1961, and there was no effective operating subsidy before 1972. Yet so long as the LHAs were wholly dependent on rental income, low tenant income meant that they could not derive enough revenue to operate the developments. Under the pressure of rising costs and deteriorating physical plant, the LHAs were literally forced to behave like the slum landlords they had become, increasing rents while services declined until the tenants finally balked and refused to pay. The only possible source of relief was a subsidy: the only realistic source of subsidy was the federal government. The poor were being squeezed dry.

Ironically, but perhaps typically, the first subsidy did little to ease the basic problem when it finally arrived. Instead, Congress subsidized the provision of housing for a whole new class of tenants by offering a bonus for housing the elderly. The Housing Act of 1956 allowed an extra $500 per room for the construction of housing for the elderly: the Housing Act of 1961 provided the LHAs with an additional $120 per year for each elderly family housed in the developments. In combination, the two subsidies ushered in the era of the elderly poor as favoured darlings of the public housing

programme; in response, their numbers increased spectacularly in the 1960s and 1970s. Since the only special features required for 'elderly' housing seems to have been a few feet of handrail in halls and bathrooms and a warning device to be pulled (if time permitted) should cardiac arrest occur, the increased construction allowance was a significant windfall for the builder. The operating subsidy provided the LHA with a parallel bonus, doubly sweetened by the highly desirable characteristics of the elderly as tenants: they usually have no young children; they are not prone to vandalism or violence; they pay their rent regularly; they cause little wear and tear on the premises; and they are almost universally regarded as worthy of assistance, for with some few exceptions, they *are* someone's mother or father and that, in the American scheme of things, guarantees virtue and deserving. And to put icing on the cake, the elderly desired only small apartments and lived readily in high-rise buildings: hence, they were ideally suited to the kind of housing the industry was tooled up to build in the 1960s. The orgy of construction for the elderly that followed was paralleled by a declining rate of construction in 'family' housing.

While the move to housing the elderly is readily explained, it is somewhat less easy to justify. Granted that the elderly poor required assistance, it is uncertain that their need was any greater than that of poverty-stricken families. In any case, no one bothered to inquire, nationally or locally. The result was a major reward for one class of prospective tenants (the elderly) and a significant reduction in the effort to service another class of tenants (dependent families). In St Louis, for example, the percentage of elderly families in public housing rose from perhaps 15% in 1955 to about 30% in 1970 and then more than 55% in 1975, while the total number of units available actually declined.

The next major change in the fiscal structure was made in the Housing Act of 1969: the act limited the rent that could be charged any tenant to 25% of adjusted income. The maintenance and operations subsidy needed to compensate the LHAs for lost revenue was not added until 1970, and significant and regular payments were not forthcoming until 1972. When the operating subsidy finally did arrive, it was inadequate!

Beginning in 1970, the federal government also tried to improve the physical condition of the developments by providing modernization funds that could be used for capital replacement and major repairs. Some of the larger LHAs also received grants that would help restore reserve balances badly depleted during the previous decade. Again, the resources available were nowhere near the level of funds actually needed.

*Privatization*

The second major change introduced into public housing during the 1960s was increased privatization of various aspects of operations, that is, the transfer of some functions to the private sector usually by authorizing contract relations between LHAs and private organizations for performance of needed services. The cause of the change was the obviously distressed condition of most large housing authorities. The justification was implicit in the received wisdom of the society, primarily the belief that private organization is *prima facie* more efficient than public. In 1974, legislation made privatization the central thrust of federal policy for the immediate future.

Three basic elements in the privatization of public housing first appeared in the Housing Act of 1965. Firstly, the act authorized 'turnkey' construction, the development of apartments by private entrepreneurs on their own sites for sale to LHAs. The developer contracted with the housing authority to supply facilities; provide a site, architectural and construction services and all other facilities needed: and deliver a finished product ready for use. Ostensibly a response to complaints about the quality of design and construction of conventional public housing turnkey development proved very popular; most new construction for LHAs after 1965 was turnkey. How closely the programme lived up to expectations is uncertain; development time was probably reduced, but in St Louis there was little evidence of significant improvements in design or construction, and while site costs were in some cases lower, construction cost remained at the allowed maximum. There is little reason to suppose that the LHAs could not have achieved the same results with even greater economy had they been given a freer hand in operations.

The second form of privatization authorized by the 1965 act was sale of public housing to tenants. Ordinarily, such sales were limited to detached, semi-detached, and terraced housing: high-rise apartments were specifically excluded from the home-ownership programme. Finally, public housing could be sold to non-profit organizations so long as it continued to operate as low-income housing. Neither procedure was widely used.

Thirdly, in 1967, the federal government authorized contracts between LHAs and private firms for management services in public housing, with or without the provision of routine maintenance. A standard fee was paid for each unit placed under management. Sometimes the LHA retained responsibility for maintenance: alternatively, the management firm also supplied routine maintenance. In all such cases, responsibility for capital replacement and major repairs remained with the LHA.

Both contract development and contract management of buildings by private firms have been widely used in public housing. Their popularity with LHA management is understandable. Private firms do have some genuine operational advantages over public organizations: they can locate and employ needed skills more quickly, purchase with greater facility, experiment more, and cut losses more quickly — there is less inertia effect in private operations, other things being equal.

But the use of contractual services was not without costs. The LHA lost a substantial part of its control over daily operations, and in fact the nature of the relation created a significant lag time between detecting pending trouble and forcing the contractor to produce a solution. The primary weapon available to the LHA was termination of a firm's contract, and that was not very useful against minor contract infractions such as late reports, modest performance delays, and so on. The manager who was an employee of a contracting firm was far less accessible to the LHA director than an employee of the LHA. The most serious fault with contractual management, however, was its long-range impact on the quality of governmental services, for if it is a fact that governments perform poorly and inefficiently in the managerial arena, purchasing such services on the private market will only guarantee that the inefficiency is prolonged indefinitely and the long-term cost of operations is increased. If governmental operations were for some reason beyond all hope of improvement, there would be no alternative: otherwise, such a policy of despair is unwarranted.

**Leased housing**

The long-term future of conventional public housing in the United States dimmed unmistakably in the 1970s. Construction of public projects, whether conventional or turnkey, came to a halt, and an ominous provision for closing out badly damaged developments was added to the housing legislation. By mid-1976 Pruitt and Igoe in St Louis were being demolished. Despite such notorious examples, it would be a serious mistake in judgement to label all conventional public housing, however developed, an utter failure. Nevertheless, that attitude is common among legislators, tenants, the general public, academics and even former supporters of the public housing programme. The distinction between failure of a specific effort (public housing as it was practised in the United States) and failure of a general strategy for supplying low-income housing has been ignored consistently. Some genuine failures, inadequate treatment in the media and widespread ignorance of the particulars of programme operations combined to reinforce the belief that abandoning conventional public housing was the course of wisdom. Official policy in the 1970s clearly accepted the same premises.

Leasing of privately owned facilities for use as public housing offered a quick and effective device for transferring public housing into the private sector; it has been supported by opponents of public ownership since the mid-1930s. Leasing was supported by the US Chamber of Commerce, the National Association of Real Estate Boards, and various Republicans in Congress known for their hostility to public housing. It began as a very small scale programme for using housing stock already in existence as a means of supplementing the more conventional developments; the initial quota was a modest 10 000 units per year for a four-year period. It was justified, in Congress and to the wider public, as a way of reducing costs and making greater use of the existing housing stock which would enable housing authorities to respond more quickly and efficiently to tenants' needs. By limiting the number of units that could be leased in any single building or estate to 10% of the total, the sponsors expected to disperse tenants more widely within the general population and thus avoid some of the stigma that had become attached to 'projects'. Finally, supporters of leasing argued that owners of substandard housing would be encouraged to rehabilitate their holdings in order to qualify for participation in the programme; thus, it would contribute to the improvement of neighbourhoods and an upgrading of the housing stock.

If the LHA could have controlled the quality, location and price of housing units leased for use by low-income families, the lease programme would have been a valuable adjunct to conventional housing, assuming that the purpose of an LHA was to supply as much of the need for housing as resources permitted. If the private housing market could have provided the LHA with a supply of adequate housing of acceptable quality, location, size, availability and cost, then leasing would have been extremely helpful. In general, leasing is likely to be more expensive than building or purchasing, particularly in the long run, because lease costs will include profits, local taxes and the cost of financing on the private money market. A number of factors can interfere in the leasing process. The supply of housing may be inadequate with respect to quality, size, location or cost, and even if housing is available owners may not be willing to lease their property to public agencies for use as low-income housing. Finally, the tenants themselves may object to certain aspects of the programme such as dispersal and separation from neighbours, or distance from home, neighbourhood, church, schools, and the like.

As it turned out, the private market did not supply the volume of housing needed by the LHAs (taking the nation as a whole) and that opened the door to a line of development far removed from what Congress had authorized and much more difficult to justify as an alternative to conventional development. What began as a program for leasing existing housing, for which there is ample economic justification, turned very quickly into a programme for private developers constructing new housing to be offered for lease, for which little if any real justification can be offered. This move required some fundamental changes in the public housing programme: the income level of the target population had moved upward, and the state government had entered the picture as a potential developer and funder.

### Lease of existing housing

The major form of leasing existing housing required prior approval by local government; it could not be undertaken by the LHA of its own volition. But since leased property was privately owned and paid full taxes, approval was fairly easy to obtain, other things being equal. Of course, some locales would have nothing to do with public housing in any form whether or not taxes were paid, but these were exceptions. Leasing did not require an LHA, though in most cases established housing authorities administered the programmes. Leasing authorizations were allocated by local branches of the federal department responsible. The LHA that received an allocation (upon request) advertised its readiness to lease and sought owners willing to supply apartments and prospective tenants wishing to rent them. Leases were negotiated for 12–36 months in the early days of the programme, but in 1966 the time was extended to 5 years and in 1970 to 15 years. Some of the 'existing housing' being leased after 1970 was quite new, or even purpose-built. The owner was paid a 'fair market rental' set by the federal government. Rent levels had to be consonant with general area rents, and the housing had to conform to local standards and building codes. Qualitative criteria were minimal: heating, lighting and cooking facilities had to be provided; the neighbourhood had to be free of 'characteristics seriously detrimental to family life', and reasonable access to schools, transportation, shopping, churches and so on was required. Usually, the LHA performed simple maintenance, leaving all extraordinary repairs and services to the owner.

Ordinarily, the LHA collected rent from the tenant and paid the owner the rent plus an additional subsidy obtained from the federal government. The amount of the subsidy was calculated using a very complex scheme known as the 'flexible formula,' intended to ensure that no more was paid for leased housing than would have been paid if the LHA had constructed the facility. If the leasing programme operated at a deficit, the LHA could also receive small subsidies normally provided for very large or elderly families.

At first the LHA decided who qualified to become tenants, and allocated households to particular dwellings, though sometimes building landlords took on this role. Later changes placed some definite responsibilities on the tenant family for the first time in the history of the public housing programme. In the past the prospective tenant had only to apply, establish eligibility and wait for an opening. Now, qualified tenants were given certificates of eligibility, which committed the LHA to housing assistance payments on the tenant's behalf. Responsibility for finding an apartment, however, lay with the tenant; the LHA could assist only in hardship cases.

The new regulations significantly reduced the role of the LHA in leasing. While the LHAs determined tenant eligibility and issued certificates, they had little operational responsibility; their principal functions were to conduct an initial inspection of the premises and to process the housing assistance payments. They also retained final control over eviction proceedings.

The 1974 Act further increased the owners' control over the leasing programme. He could now evict tenants with LHA approval. The authority of the LHA was much reduced, the tenants' responsibilities were increased, and the owner controlled the bulk of the operation. Indeed, the LHA was specifically enjoined from any action that would 'directly or indirectly reduce the family's opportunity to choose among the available units in the housing market.' Yet it seems reasonable to assume that the prime limit on the individual's 'freedom of choice' was likely to be the inability of the LHA to intercede on his behalf. While not quite so serious as refusing to allow a physician to choose his patient's medicine, much the same principle seems involved in both areas. The potential bargaining power of the LHA as collective purchaser and government agent was given away deliberately.

Finally, the search and lease arrangements were almost an open invitation to collusion between owner and tenant at the expense of the public treasury. For if the fair market rents were high enough so that the owner was adequately rewarded from the subsidy payment alone, as seems the case in practice, then the owner's interests would best be served by retaining the tenant regardless of the rent actually collected (by the owner, under the terms of the act). Such arrangements would be almost impossible to detect. Finally, carelessness or bias in the inspection of units would have the effect of converting the programme into a support system for marginal local slums with little possibility of forcing an improvement in quality or terminating the lease.

*Construction for lease*
The Housing Act of 1965 clearly intended that leasing should apply only to the existing supply of housing: it would serve as an adjunct to the conventional housing programme. That intention was transposed or transformed by the federal government into active support for construction of new housing for lease and for rehabilitation of existing housing. They argued that once new housing was built it became part of the 'existing' stock, and thus could be included in the programme. The result of this interpretation was a sharp increase in the amount of new construction for lease, particularly in states where public housing was difficult to site. Between 1969 and 1974, some 61 000 units of housing were added to the lease programme; of that number, more than 75% (46 123) were new construction, fewer than 0.5% were substantially rehabilitated, and 24% (14 500) were part of the existing housing

stock (table 22.1). This caused some anxiety among programme supporters, and an effort was made to maintain an even balance between new construction and lease of existing housing.

Until 1974, the regulations governing administration of new construction were the same as those applied to lease of existing stock, but the owner of the new apartment complex was in a much stronger position *vis-à-vis* both tenant and LHA than his counterpart who owned existing housing.

The owner retained total managerial control over new construction units and could contract with either an LHA or a private firm for managerial services. In effect, lease of new housing provided for the reintroduction of the traditional landlord system with only minor modifications. The owner paid for taxes, utilities and other services, was responsible for all maintenance, processed applications, selected tenants and collected rents. After 1974, he also determined tenant eligibility, verified it periodically, set the amount of family contribution and subsidy and terminated tenancy subject only to minor delay by the LHA.

The LHA's functions in the new construction for lease programme were minimal. It could serve as a co-sponsor of a development but could then exercise no managerial function over the property. When it did manage, it was at the owner's request. Federal regulations made joint sponsorship complex and difficult, and the central administration clearly favoured direct applications from developers and individual State agencies. Selection of developers, site and plan approvals and fiscal arrangements were all made entirely through local branches of the federal housing department.

The key to a strong construction-for-lease prog-ramme is the method of financing available and the opportunity for profits it provides. From the point of view of the developer, access to capital at preferred rates or the right to depreciate the investment quickly and thus generate tax losses is the key to success. Both of these techniques found a place in the construction-for-lease programme, often combined in the same development. To obtain preferred borrowing rates, the LHA's power to issue tax-free bonds was exploited and the creation of state government agencies with similar powers was encouraged.

Even more spectacular returns to the developer were produced by 'equity syndication'. The key to success here was the Internal Revenue Service's willingness to allow accelerated depreciation on multifamily properties intended for use by low-income families. In effect, the IRS ruling allowed the developer to depreciate the combined value of his 'up front' investment *and* his mortgage in a very short time, usually five years. Since this was generally more depreciation credit than one person could use unless his annual income was enormous, a portion of the depreciation credit could be sold or assigned to members of a limited liability company formed specially for that purpose. As an example, a developer who invested $500 000 of his own money and borrowed $4 500 000 more to build a $5 million project (a common ratio of investment to borrowing in the industry) could depreciate the property at a rate of $1 million per year for the first five years of operation: in effect, he buys $100 000 worth of depreciation per year for each of five years for only $50 000. He need only find individuals with large personal incomes and sell them the depreciation he has cumulated in the development. For the person with a large income, the depreciation is a good buy: the developer, obviously, can charge rather more than his own investment. At the end of five years, when the property value is reduced to nil, the project can be sold and the return treated as capital gains, taxed at a much lower rate than personal income.

**Table 22.1** Units of public housing authorized, 1966–74

| Type of housing | 1966 | 1970 | 1974 | % |
|---|---|---|---|---|
| Conventional housing | n.a. | 826 220 | 868 430 | 67 |
| Turnkey housing | n.a. | 150 460 | 230 530 | 19 |
| Leased housing | 9 460 | 114 370 | 169 040 | 14 |
| *of which*: | | | | |
| New housing | n.a. | 37 180 | 75 410 | |
| Rehabilitated | n.a. | 27 850 | 23 980 | |
| Existing stock | n.a. | 49 340 | 69 650 | |
| All public housing | 809 890 | 1 119 620 | 1 316 130 | 100 |

Note: These figures have been slightly rounded

### Reprise

The public housing programme in the United States was seriously damaged if not mortally wounded in the decade between 1965 and 1975. The justification for abandoning the effort that is heard most often, particularly from former supporters of the programme, is 'we tried it and it didn't work.' But the businessman or executive who tried a particular brand of machinery in various plants throughout the country without providing enough resources to maintain it properly, and knowing that the knowledge required to operate it properly was not available, then announced that the machinery was no good, sold it at great loss, and purchased far more expensive and less effective substitutes would, if exposed, be rightly denounced and discharged by the stockholders. Yet that is precisely the kind of condemnation without trial that public housing has suffered. In this case, both the stockholders (the tax-paying general public) and the consumers (the tenants) have a genuine grievance against the policymaker. At the national level, the programme has been characterized by lack of foresight, ignorance, failure to learn and dogmatic adherence to principle in despite of the observable consequences of action.

At the beginning of 1976, the public housing programme had effectively been divided into three parts and had lost much of its original impetus. The working poor, the original target population for public housing, had long since lost interest in the facilities. The dependent poor who replaced them in the mid-1950s were a poor risk and lacked clout; by mid-1976 they were increasingly herded into conventional housing developments that resembled nothing so much as downtown Indian reservations presided over by Indian agents and some few collaborators. The elderly poor, who became the primary target of the turnkey construction programme in the 1960s and 1970s, occupied the lion's share of the newer apartments. The third main arm of public housing, the leasing programme, transferred programme control to the private landlord in very large measure. Moreover, policy could be expected to squeeze the very low-income tenant as well as the working tenant out of public housing, other things equal. The Brooke amendment of 1969 had that effect in practice, and the requirement that LHAs collect at least 20% of tenant income in rent had the same impact. Programme regulations tend to press the landlord to seek higher-income tenants within the limits of his discretion since that policy will maximize income and minimize collection losses over time. The target population for the programme as it appeared in 1976 was one solid notch higher in income level than was the case in the 1960s; that difference, though slight, spells significant deterioration in the housing available to the very poor and an increase in its real cost.

If trends apparent by mid-1976 have continued, public housing will have been transposed into a direct rent subsidy programme within a private market framework. That is precisely the situation that public housing was originally intended to remedy. It involves loss of the economies attainable through collective purchasing and of the ability of the collectivity to deal with landlords on an equal footing. Further, the possibility of increasing the available knowledge of how best to provide decent housing for the poor with improved efficiency must also be discarded. In those circumstances, the facilities available for the poor are likely to decrease over the long run. Failure by the private sector to generate enough construction to take care of replacements and population expansion, which is likely given the vagaries of the industry and the different economic conditions found in different regions of the US will impose enormous hardship on those in society least able to withstand it.

The history of public housing in the United States is most discouraging for anyone seriously concerned with the willingness and capacity of democracies to deal with fundamental human problems. Clearly, the institutions responsible for the decisions that guided the development and operation of public housing failed miserably. Worse, they seem incapable of pursuing identifiable goals persistently or improving their performance with experience. They have not learned. And behind the failure of particular institutions lies the grim spectre of a society utterly unwilling to accept the social and economic burdens required for even a modest effort to redress the more grievous inequities in the established order. Here, perhaps more than anywhere else, is where public housing's failure was decided. Until genuine changes are made in society's fundamental priorities, efforts to alleviate the suffering of the unfortunate will continue to fail, or at best succeed imperfectly and be expensively disguised as solutions to other problems that society *is* willing to tackle — with the additional risk that such accomplishments will be taken as surrogate for the humane considerations that ought to guide collective action in these spheres. Of those changes, there has been no sign whatsoever.

# 23 Policy-making on Inter-urban Highways in Britain, 1945–79

## by M Painter

### The growth of the roads programme

In May 1946 the Labour government in Britain announced an ambitious ten-year programme of trunk road construction, including the building of some new roads on the German 'autobahn' model. These new roads were to form a 1000 mile network connecting the major industrial and commercial centres. However, this programme, like previous road programmes, was not carried out. To all intents and purposes central government expenditure on trunk-road construction and reconstruction was nil from 1947 until 1954.

The government took the view that roads were a major priority but that regrettably economic circumstances prevented a start. Successive ministers called for patience. In the Ministry of Transport a small band of roads men had been drawing plans and putting them away in drawers ever since the mid-1930s. Strong pressures were vigorously brought to bear on the government in the early 1950s for the release of funds for road building. Finer (1958) described the organization and tactics of the 'roads lobby' during this period, with the British Roads Federation (BRF) taking a leading role, and directing its attention in particular at Parliament. From 1952 on, debates on roads initiated by back benchers became increasingly frequent. They were bi-partisan and hardly a voice of dissent was heard amidst the clamour for better roads. In 1954 the breakthrough came and a small-scale start on new road construction was announced by the Conservative government.

Over the next twenty years this programme grew incrementally into something quite different in scale and political impact. The initial decision was the first step in a continuing programme involving a large long-term allocation of resources (table 23.1). A critical non-decision was involved in the way the programme was launched. There was no questioning of the basic argument that traffic growth was the justification for a roads programme. This is not surprising: as Finer (1958) put it,

What else *can* a government do but plan new roads? The only alternative would be to restrict private motoring, and that is, politically, quite impracticable (p 58).

In this context the need for a roads programme was self-evident, and the logic of arguments for a growing programme was unassailable.

The narrow definition of the issue, and the consequent logic of programme justification, was carried to further lengths as time passed. The Ministry took great pains to elaborate on and promulgate the basic simple argument that roads were needed to relieve traffic congestion, which would get worse as traffic grew. Evidence of traffic growth thus in itself became an argument for roads.

The use of words such as road 'deficiency' and 'need' were part of the language of programme justification. They introduced a spurious air of certainty and objectivity into the process of decision-making, with the intention of asserting a technical and 'rational' basis for programme growth. Procedures were subsequently developed to measure the benefits of time saving and reductions in operating costs accruing from new roads. These measures came to form the basis of a cost-benefit procedure. A Highways Economics Unit, set up in the Ministry in 1966, developed this procedure and it was introduced as standard practice for all large proposed road schemes in 1973, using a computer programme called COBA.

Source: Painter M 1980 Whitehall and roads: a case study of sectoral politics *Policy and Politics* **8** 163–86.

**Table 23.1**  Major highway (motorway) mileage in England,
1961–77

| Highway mileage | 1961 | 1966 | 1971 | 1976 | 1977 |
|---|---|---|---|---|---|
| Programmed | – | 521 | 568 | 547 | n.a. |
| Under construction | 120 | 118 | 282 | n.a. | 96 |
| Opened in year | – | 38 | 108 | 121 | 31 |
| In use | 122 | 384 | 676 | 1185 | 1216 |

An extensive critique of these techniques is not relevant to this case study, but it is important to see how they helped reinforce the basis for a narrow 'roads only' perspective. The evaluation methods all aimed to measure the benefits of roads *qua* roads, consisting of economic savings accruing to road users. Within this perspective the main decision criteria tended to be economic. Attention was focused on the measurable costs and savings. The narrowness of this approach came in for considerable criticism from 'environmentalists' who argued that cost/benefit analysis (CBA) could not cope with amenity or environmental 'intangibles' and that these factors were hence under-weighted in decisions. At first, this criticism was answered by claims that once techniques of measurement were refined, these factors could be incorporated in the CBA evaluation. Thus in 1970 the Ministry commented:

a generally accepted and practical method of evaluating amenity considerations would be an invaluable tool for rational decision-making.

Even though defenders of COBA came to agree that there were limits to its scope, and that environmental considerations and other 'intangibles' had to remain outside the CBA framework, these considerations inevitably remained an afterthought or an appendage to the economic evaluation. The logic of the programme and the procedures designed to carry it forward were based on calculations of 'value for money', with benefits to road users on one side of the equation and construction costs on the other side. The appearance of sophistication in fact hid a gross over-simplification of the complexities of the roads issue. They helped perpetuate a deception that these were not political choices, but economic necessities.

These techniques were only one factor in the policy process. In determining the overall level of resource commitment and the details of the programme, a variety of other considerations were important. But the promulgation of these techniques performed important political functions in helping to legitimate the programme's expansion. The Treasury encouraged the use of these evaluative procedures, and in Parliament, members of Estimates and Expenditure Committees expressed support and admiration for the Ministry's pursuit of economic rationality.

Another important step for road exponents was to force governments to make fairly firm commitments to long-term physical targets. In the early days, ministers were reluctant to announce long-term plans for fear of repeating the embarrassment of the 1946 roads plan. Thus early objectives were specific and short-term. As the programme extended over time, more general objectives, and the language of planning, became current. In 1957 it was announced that a special forward planning section was working on 'the trunk-road master plan'. In 1961 this became a commitment to 1000 miles of motorway and 1700 miles of modernized trunk roads. By 1971 the government envisaged 3500 miles of strategic trunk routes, including 2000 miles of motorway.

The existence of long-term objectives was vital to the road plan. To the outside world they were presented as 'brave new futures' and contributed to the building of a general level of public support. In Whitehall, they provided a basis from which to bargain for annual resources. Within the ministry, they promised a degree of predictability and stability in the organization of work, the pre-planning of major schemes and the mobilization of private-sector resources. They performed a major function as part of an internally generated process of building commitments in order to maintain programme growth.

One additional source of inertia and commitment became important as the programme grew. The ministry relied on private construction companies for building major trunk-road schemes. The availability of these resources was a necessary prerequisite for programme growth. The interdependency between the ministry and the industry led to a regularizing of relationships between the two sides, and construction industry interests in the broadest sense became a source of pressure for growth. The industry required a degree of predictability to plan investment and work, and the ministry was happy to cooperate at this general level. From 1963 on, the ministry adopted selective tendering, and subsequently took to circulating to its trusted list of contractors quarterly lists of works likely to begin in the next couple of years. In this way informal understandings were built up about the size, content and geographical spread of the future programme.

So far we have traced various aspects of the roads policy and programming processes which helped maintain the programme's momentum into the early 1970s. A further crucial aspect of establishing and maintaining such a programme was the construction of suitable and effective organizational tools. During the 1960s a large organizational empire concerned solely with roads was built up within the Ministry of Transport. During these years two important complementary changes occurred: first the integration of professional and administrative hierarchies, entailing an upgrading of the status and influence of professional engineers, and second a virtual 'hiving-off' of a self-contained roads wing in the ministry. The first was a pre-condition for the second, as it permitted the creation of a self-contained integrated hierarchy for road-programme administration and technical work. In 1965 the roads 'wing', which was at that time under the joint command of a Deputy Secretary from the administrative side and the Director of Highway Engineering from the professional side, gained as its head a Director General of Highways, a professional civil engineer. This emergence of a self-sufficient roads wing facilitated the development of an *esprit de corps* and a strong sense of 'mission', reinforced for many by a civil engineering professional ethos.

In the field, the roads wing of the ministry set up special agencies to supervise the trunk-road programme across the country. Much of the technical personnel for design work came from local government or from private consultant firms. In the early years of the programme these resources were rather poorly coordinated, and it was not easy to exercise uniform controls from headquarters. Hence in 1967/8 six Road Construction Units were set up to cover the whole country, in which local government engineering staff on a seconded basis and central government technical and administrative resources were brought together under the direct control of the ministry's roads wing. Hence, both at headquarters and in the field, road programme work was contained in one special-purpose organizational structure, single mindedly concerned with building roads.

## The roads policy sector

The system of making policy set out above can be viewed as a 'policy sector'. The basic defining characteristic of a policy sector is the subject matter, or 'class

of decision problems'. This provides the focus for organization, for programmes, for political interactions and for decision-making. It also helps determine who benefits and who bears the costs. Policy sectors do not have precise boundaries, but there is an important distinction between that which is 'relevant' to the policy sector subject matter and that which is 'irrelevant'. This is often the same as identifying friends and enemies. Defining relevance, or attempting to restrict relevance, is a crucial part of problem formulation and programming, and the less self-contained a problem is said to be the more unstable and conflict-ridden the policy sector is likely to become. Administrative routines, technical criteria and programme logics will be challenged. In fact, the process of problem definition and redefinition becomes a crucial subject matter of conflict. 'Outsiders' with different perspectives are likely to succeed with these challenges if they can demonstrate that programme outcomes are not meeting needs. The major consequences of a policy sector being established can be traced out in terms of its overlap and coordination with other policy sectors — in the case of the roads policy sector this involved wider transport issues.

Inter-urban roads are one part of a transport system, but in terms of transport policy they have been a distinct part. The major obstacle to an integrated transport policy has been the government's reluctance to restrain motor-vehicle traffic growth. 'Transport policy' within this constraint has been made up of separate policies for each of the transport modes. A major study on transport needs undertaken by the 'Hall Group' (Ministry of Transport 1963) argued that the demand for roads could be only very slightly affected by railway policy. Even if rail managed to gain a significant growth in goods traffic, the volume involved would not be sufficient to significantly dampen the growth in road traffic, because of the small initial share of rail in overall traffic (Ministry of Transport 1963). Moreover road and rail met different 'needs'. Road carried most of the short-haul goods traffic, and this accounted for a very large proportion of road traffic. In 1974 the Department of the Environment claimed that 'the demand for investment in the one mode of transport is largely independent of the demand for the other' (House of Commons 1974). Governments adopted a view that both should be developed to meet the demands being placed on them. Overall efficiency would be maximized by offering good services by both, even in areas where there was some competition. Transport policy was predominantly conceived of in such a way as to include a self-contained roads policy to meet demand and reduce congestion.

In the 1960s, roads programmes began to be justified in terms of other policies which could plausibly be associated with them. Regional policy considerations were used to promote schemes in depressed areas, and in the early 1970s claims began to be made for the importance of inter-urban roads in 'improving the environment', specifically through relieving congestion in country towns (particularly 'historic towns') by building new through routes and by-passes. Subsequently this objective was given new weight as protest grew against the growth of heavy lorry traffic in towns, and the primary trunk-road network came to be referred to in terms of 'special lorry routes'.

These modifications to the underlying objectives of the roads programme were in part surface changes for public consumption to take account of changing fashion. But they also significantly helped maintain programme momentum in the face of growing difficulties. While economic criteria and the logic of demand and congestion still underpinned the programme, final decisions on commitments were not necessarily based on routine applications of these criteria. Political opposition to particular schemes caused delays, and technical difficulties sometimes added to the problems of bringing to fruition the 'most needed' schemes. Practical and political considerations such as an even spread of schemes across the country or the general difficulty and expense of building motorways in the congested southeast led to important 'compromises' on these strictly economic criteria. Indeed the more 'legitimate' criteria that could be taken on board in bringing schemes forward, the more chance there was of programme expansion. With special objectives, urgency could be counselled to overcome delays. So multiple criteria were not unwelcome, and 'overlap', so long as it added to these criteria, was not a threat.

In 1970 the old Ministry of Transport was merged at central government level into the much larger Department of the Environment (DoE), supposedly designed to improve the coordination of policies with major environmental aspects. This entailed an implicit challenge to the roads mens' autonomy and the highways wing showed a marked resilience against the integrationist ideas behind the creation of the DoE. A proposal to integrate the planning side of highways with regional policy and planning directorates was successfully warded off with the argument that highway planning was already 'done', and the roads men needed to be left alone to get on with the job. The highways wing was at first brigaded alongside planning in the DoE but in September 1972 it was reunited with the rest of transport. In 1976, just prior to the dismemberment of the DoE, transport and planning were again placed together, but throughout, the highways directorate remained intact and enjoyed sufficient independent status to be considered at ministerial level as a separate entity.

## Implementation and increasing vulnerability

If the roads men, even in the DoE, were not severely threatened by integrationist ideas or demands at the centre, the artificiality and ultimate unrealism of self-containment did emerge during the process of implementation. In the localities where the impact of road building was felt roads became amenity issues. This broadened the issue and made a 'roads only' perspective untenable. The intrusion of amenity issues finally was felt at the centre as these issues came to be championed by an increasingly vociferous 'environmental movement'. But it was the cumulative impact of local protest that set the challenge going. Controls and procedures necessary to bring schemes to fruition broke down and led to the internal dissolution of the programme. An indication of this, indeed its crucial manifestation, was the increasing length of delays experienced in completing individual schemes, or what the ministry called 'slippage'. In 1957 it was estimated that schemes took 'four years or more' to come to fruition; in 1969 the figure mentioned was seven years; by 1973 it was said to be taking 'up to ten years'; and in 1978 'fifteen or more years'. The generation of programme momentum was critically affected.

There were three major steps in the growth of opposition to road schemes, each widening the scope of the issue and hence creating greater difficulties. At first, local objections came largely from individual property owners, often farmers, who defined the issue in monetary or 'livelihood' terms and sought compensation. This often took the form of negotiated detailed route changes or design concessions (such as bridges across new roads to link two parts of one farm property). Generalized opposition from agricultural interests tended to be defused by the process of 'successful' negotiation of compensatory measures for separate property owners. Procedural concessions, such as a 1959 provision for advance payment of compensation, assisted in the successful management of this form of objection.

During the 1960s a more threatening form of opposition grew, led by local and national amenity societies, but generally the scope of conflict was restricted to areas which posed no fundamental challenge to the ministry's programme. The ministry dealt with the 'environmental issue' almost solely in terms of aesthetics and design. A Landscape Advisory Committee was set up in 1956 to comment on the design and impact of every trunk-road scheme. In 1970 the Council for the Protection of Rural England questioned the effectiveness of this body and claimed that it could be 'used as a scapegoat for any proposal unfavourably received by the public' (*The Times*, 19 September 1970). The ministry certainly cited the committee in its attempts to ward off criticism, and its narrow terms of reference ensured that its advice was restricted to measures for fitting proposed roads into the landscape, on the assumption that they were going ahead.

The third stage in the growth of opposition occurred in the 1970s with an increasing militancy among local objectors. This was due to a number of factors, primarily a growth in frustration over the apparent one-sidedness of the procedures for objection coupled with a growing 'fundamentalist' strain in the demands being made. The latter was associated with the 'environmental movement' which mounted a concerted attack on the motor car and motorways. National organizations such as Friends of the Earth entered the conflict, and in 1974 a National Motorways Action Committee was set up. Local protesters began to attack the roads programme as a whole, challenging the need for a road rather than arguing about its route. Containment through 'participation' and the presentation of a more conciliatory and politically sensitive face to the localities were of no help when the roads men still stuck to the government's network objectives and refused to compromise on the 'need' for a scheme. The scene was set for confrontation at local public inquiries. They were transformed from 'quiet, calm and orderly affairs' where individual grievances about a scheme were aired, into highly charged political arenas where objectors challenged the whole basis of the roads programme. This also entailed challenging the legitimacy of the inquiry procedure itself.

This legitimacy had been called into question because of the clear advantages the ministry enjoyed. Their's was the proposal 'on the table' and the normal procedures restricted argument to its shape and form,

not its need; inspectors, working within ministry guidelines, frequently took a restrictive view on the scope of valid arguments and the rights of 'non-statutory' objectors; the ministry men enjoyed overwhelming resources of time, expertise and knowledge while often restricting access to information for objectors trying to frame a case; and so on. The experience of such inquiries often left objectors with a sense of outraged impotence. During 1975 and 1976 a number of major inquiries were brought to a halt by procedural disputes caused by claims from objectors that they should be allowed to challenge ministry policy and the 'need' for a scheme, and these disputes occasionally precipitated acts of physical disruption by angry protesters who turned some inquiries into fully fledged political demonstrations (Levin 1979). In response, the government in 1976 initiated a review of inquiry procedures. For the road programme as a whole these events not only further slowed down the implementation process, they added significantly to the pressures causing the political decline of the programme.

## The Department of the Environment's budgetary decisions and the decline of the road programme

During the 1970s, from 1973 on, successive ministers depicted the slowing down of the rate of growth of the programme, and its ultimate real decline, as a process of reappraising transport priorities, and this claim was reiterated in the 1976 Transport Review. In 1970/1 motorways and trunk roads accounted for 7.64% of the DoE's expenditure. This fell to 4.56% in 1977/8. The sources of this decline, however, are not to be found in a process of central reappraisal so much as in the quite distinct pressures being exerted in different policy sectors to which the centre made piecemeal responses.

In November 1973 the minister Mr Peyton announced a switch 'mainly from urban-road to rail', and the PESC White Paper (1973) referred to a slower rate of growth for trunk-road spending with a 'major switch to support for public transport' (HMSO 1973). These announcements were not easily translated into immediate cuts because of programme inertia. But by this stage the programme was in effect slowing down of its own accord. Thus in 1973 and in successive years it was very vulnerable to cuts in a growing Whitehall mood of limiting expenditure growth. Within the DoE the inability of the roads men to spend their money made them easy prey for competing programme advocates. Budgetary decisions in the DoE were made in such a way as to enhance this vulnerability, as the Secretary of State occupied an arbitral role in fixing shares. Housing and public transport were both sectors in which strong pressures were being exerted for more funds. In the case of public transport, in 1973 the Minister was presented with a report from the Railway Board which sought greater financial assistance due to continuing losses. The government chose not to embark on sweeping rail closures, and promised support. At the same time, a decision to keep down rail fares as an anti-inflationary measure led to a rapid rise in fare subsidies. In 1970/1 the subsidy to rail passenger transport was £114m; in 1973/4 it was £195m and in the following year it was £390m (all at constant 1975 prices). Bus operation subsidies increased even more steeply. In other words the apparent 'reappraisal' and change in priorities was an outcome of independent pressures for funds from public transport, and the coincidental vulnerability of the road programme. What appeared to be a deliberate central reappraisal was in fact an outcome of sectoral politics.

Nevertheless, the result of these sectoral pressures and changes was in the end an overall shift in thinking and policy in the transport field. The Consultative Document of 1976 and the White Paper of 1977 in effect summarized the current state of official thinking arising from the changes in sectoral transport politics in the 1970s. As an outcome, the underlying logic and the objectives of the road programme were rejected. The network objectives were abandoned in favour of a 'more selective approach', and this greatly weakened the roads mens' arguments at local inquiries that the 'need' for a particular scheme was an unquestionable policy 'given'. The notion that road building would have to continue to grow to meet traffic growth — the underlying programme logic — was rejected, and a fixed ceiling of expenditure (about £300m at 1977 prices) was projected into the early 1980s. Attention would be focused on 'specific and local problems of the trunk road network' (White Paper 1977).

These policy changes were accompanied by organizational ones. The post of Director General of Highways was abolished in 1977 and the roads wing was integrated with local transport divisions under a single deputy secretary. A review of the RCUs was set in motion with the object of dismantling them in the 1980s. Within the Department of Transport (established in 1976 when the DoE was split) a special Policy Review Directorate began to look more closely at aspects of policy coordination in the transport field than had been the case previously. As a result of a major inquiry into trunk road appraisal methods, which was critical of some of the techniques, a less mechanistic and more pluralistic appraisal method was promised 'to bring out more clearly how decisions rest on judgements about the relative importance of considerations of different kinds' (White Paper 1977). Finally, following the review of inquiry procedures, some important concessions to the demands of objectors were made.

## Conclusion

The history of trunk-road policy-making in Whitehall can best be understood as a case of sectoral politics. Policy was formed and challenged in the context of a distinct trunk-roads policy sector. The problems of maintaining programme momentum were tackled in the 1960s through technical devices aimed at increasing the capacity of policy-makers in the sector to control its immediate environment, and they had the political effect of increasing the degree of self-containment and isolation of the roads men. There was an element of tragedy in their efforts: the more they applied their sectoral logic to solving problems of implementation, the more they set themselves up for ultimate destruction. They were vulnerable to disruption, and their efforts encouraged protesters to adopt increasingly disruptive tactics. In spite of the seeming inexorability of the programme the problem of 'slippage' was its major undoing.

## References

Department of the Environment 1971 *Roads 1970-71* (London: HMSO) p 11

Finer S E 1958 Transport interests and the roads lobby *Political Quarterly* 29 47–58

Levin P H 1979 Highway inquiries: a study in governmental responsiveness *Public Administration* 57

Ministry of Transport 1963 *Roads 1962/63* (London: HMSO) p 6

PESC White Paper 1973 Cmnd 5519, *Public Expenditure to 1977-8* (London: HMSO) p 60

*Transport Policy, A Consultation Document*, 1976, 2 vols (London: HMSO)

White Paper, 1977 *Policy for Roads in England 1978*, Cmnd 7132 (London: HMSO)

# Name Index

# Subject Index

Page numbers in italics refer to tables or diagrams.